Edited by
Farid Chemat and Jochen Strube

Green Extraction of Natural Products

Related Titles

Tiwari, B.K., Brunton, N.P., Brennan, C.S. (eds.)

Handbook of Plant Food Phytochemicals

Sources, Stability and Extraction

2013
Print ISBN: 978-1-444-33810-2 (Also available in a variety of electronic formats)

Bart, H., Pilz, S. (eds.)

Industrial Scale Natural Products Extraction

2011
Print ISBN: 978-3-527-32504-7 (Also available in a variety of electronic formats)

Dayan, N., Kromidas, L. (eds.)

Formulating, Packaging, and Marketing of Natural Cosmetic Products

2011
Print ISBN: 978-0-470-48408-1 (Also available in a variety of electronic formats)

Dewick, P.M.

Medicinal Natural Products

A Biosynthetic Approach

Third Edition

2009
Print ISBN: 978-0-470-74168-9 (Also available in a variety of electronic formats)

Dunford, N. (ed.)

Food and Industrial Bioproducts and Bioprocessing

2012
Print ISBN: 978-0-813-82105-4 (Also available in a variety of electronic formats)

Attokaran, M.

Natural Food Flavors and Colorants

2011
Print ISBN: 978-0-813-82110-8 (Also available in a variety of electronic formats)

Civjan, N. (ed.)

Natural Products in Chemical Biology

2012
Print ISBN: 978-1-118-10117-9 (Also available in a variety of electronic formats)

Edited by Farid Chemat and Jochen Strube

Green Extraction of Natural Products

Theory and Practice

WILEY-VCH
Verlag GmbH & Co. KGaA

The Editors

Prof. Dr. Farid Chemat
Universitairè d'Avignon et des Pays du Vaucluse
INRA, UMR 408, Green Extraction Center
33 rue Louis Pasteur
84000 Avignon
France

Prof. Dr. Jochen Strube
TU Claustal
Institute for Separation & Process Technology
Leibnizstraße 15
38678 Claustal-Zellerfeld
Germany

Library of Congress Card No.: applied for

British Library Cataloguing-in-Publication Data
A catalogue record for this book is available from the British Library.

Bibliographic information published by the Deutsche Nationalbibliothek
The Deutsche Nationalbibliothek lists this publication in the Deutsche Nationalbibliografie; detailed bibliographic data are available on the Internet at http://dnb.d-nb.de.

Print ISBN: 978-3-527-33653-1
ePDF ISBN: 978-3-527-67679-8
ePub ISBN: 978-3-527-67681-1
Mobi ISBN: 978-3-527-67680-4
oBook ISBN: 978-3-527-67682-8

Cover Design Adam-Design, Weinheim, Germany
Typesetting Laserwords Private Limited, Chennai, India
Printing and Binding Markono Print Media Pte Ltd., Singapore

Printed on acid-free paper

Contents

Preface

Green extraction of natural products is a new concept that meets the challenges of the twenty-first century, protecting both environment and consumers and, at the same time, enhancing competitiveness of industries by becoming more ecologic, economic, and innovative and thereby sustainable. It is based on the discovery and designs of extraction processes which reduce energy consumption, allow use of alternative solvents and renewable natural products, and ensure a safe and high quality extract and final product. Within the green extraction approach, the concept of a "green extract" is introduced. This is an extract obtained by following processes that have the lowest possible impact on the environment (less energy and solvent consumption, etc.), and whose eventual recycling is planned for from the beginning (coproducts, biodegradability, etc.). This green extract should be the result of a whole chain of values in both senses of the term: economic and responsible, starting from production and harvesting of the plant, the transformation processes – not only solid–liquid extraction but also separation and purification – together with formulation and marketing.

This book is an attempt to advance practical objectives of "green extraction of natural products." The book has been made possible due to the collaboration between "Dechema ProcessNet – Germany" and "France Eco Extraction" associations but also because of the critical mass of international research and industrial teams who have contributed to establish a series of methodological and technological tools in the field of extraction of natural products to prevent and reduce petroleum solvents, fossil energy, and chemical wastes and hazards in extraction as a process including preparation of starting materials, drying, grinding, solid–liquid extraction, liquid–liquid extraction, separation, purification, formulation, until final packaging. Part of the contents are based on the significant amounts of materials accumulated from a Dechema training course on phytochemical process development and production held several times over the past years at the TU Clausthal with lecturers from industry and academia, some of who have contributed to this book.

This book attempts to summarize current knowledge on green extraction of natural products in terms of innovative processes, methods, alternative solvents, and product safety. It provides necessary theoretical background and details about green extraction with regard to techniques, mechanisms, protocols, industrial applications, safety precautions, and environmental impacts. This book is targeted at industry professionals as well as academicians engaged in separation and extraction engineering or natural product chemistry research, and graduate-level students. Each chapter would be complementary to other chapters and based on presentations by the reputed international researchers and professionals, addressing the latest efforts in the field.

We are convinced that this book will make a useful contribution toward the collection of accumulated knowledge in one place, and is the starting point for future collaborations in this new area of "green extraction of natural products" between research, industry, and education, covering a wide range of relevant applications: perfume, cosmetic, pharmaceutical, food ingredients, nutraceuticals, biofuel, and fine chemicals industries.

January 2015

Farid Chemat
Avignon University, INRA, France
Jochen Strube
Clausthal University of Technology, Germany

List of Contributors

Karim Allaf
University of La Rochelle
CNRS
Transfer Phenomena and Instantaneity in Agro-Industry and Building
Laboratory of Engineering Science for Environment (LaSIE FRE 3474)
17042 La Rochelle, Cedex 01
France

Tamara Allaf
ABCAR-DIC Process
BP12053
17010 La Rochelle, Cedex 01
France

Antoine Bily
ORTESA
LabCom Naturex-Avignon University
84000 Avignon, Cedex 9
France

and

Naturex
250 rue Pierre Bayle
BP 81218
84911 Avignon, Cedex 9
France

Simon Both
Clausthal University of Technology
Institute for Separation and Process Technology
Leibnizstraße 15
38678 Clausthal-Zellerfeld
Germany

Farid Chemat
Université d'Avignon et des Pays de Vaucluse
INRA, UMR 408
GREEN Extraction Team
84000 Avignon
France

and

ORTESA
LabCom Naturex-Avignon University
84000 Avignon, Cedex 9
France

Giancarlo Cravatto
Università di Torino
Dipartimento di Scienza e Tecnologia del Farmaco
10125 Torino
Italy

Reinhard Ditz
Clausthal University of Technology
Institute for Separation and Process Technology
Leibnizstraße 15
38678 Clausthal-Zellerfeld
Germany

Anne-Sylvie Fabiano-Tixier
Avignon University
INRA
UMR408
GREEN Team Extraction
Rue Louis Pasteur 33
84000 Avignon, Cedex 1
France

and

ORTESA
LabCom Naturex-Avignon University
84000 Avignon, Cedex 1
France

Hansjoerg Hagels
Boehringer Ingelheim Pharma GmbH & Co. KG
LPS Germany
Department of SCM and Engineering
Phyto Center
Binger Straße 173/
HPZ 6425-EG-00
55218 Ingelheim am Rhein
Germany

Urban Jenelten
Firmenich S.A.
Route des Jeunes
1211 Genève 8
Switzerland

Iraj Koudous
Clausthal University of Technology
Institute for Separation and Process Technology
Leibnizstraße 15
38678 Clausthal-Zellerfeld
Germany

Werner Kunz
University of Regensburg
Institute of Physical and Theoretical Chemistry
Universitätsstraße 31
93040 Regensburg
Germany

Alexei A. Lapkin
University of Cambridge
Department of Chemical Engineering and Biotechnology
New Museum Site
Cambridge CB2 3RA
UK

Nikolai Lebovka
Université de Technologie de Compiègne
Groupe Technologies Agro-Industriels
EA 4297
Département Génie des Procédés
BP 20529
60205 Compiègne, Cedex
France

and

NAS of Ukraine
Institute of Biocolloidal Chemistry named after F. D. Ovcharenko
42, Boulevard Vernadskogo
03142 Kyiv
Ukraine

Carlos A. Ledesma-Escobar
University of Córdoba
Department of Analytical
Chemistry
Annex Marie Curie Building
Campus of Rabanales
14004 Córdoba
Spain

and

University of Córdoba
Agroalimentary Excellence
Campus ceiA3
University of Córdoba
Reina Sofía University Hospital
Maimónides Institute of
Biomedical Research (IMIBIC)
14071 Córdoba
Spain

and

Technological Institute of
Veracruz
Chemical and Biochemical
Engineering Department
Av. Miguel Angel de Quevedo
2779
Veracruz Ver. 91897
México

María D. Luque de Castro
University of Córdoba
Department of Analytical
Chemistry
Annex Marie Curie Building
Campus of Rabanales
14004 Córdoba
Spain

and

University of Córdoba
Agroalimentary Excellence
Campus ceiA3
Reina Sofía University Hospital
Maimónides Institute of
Biomedical Research (IMIBIC)
14071 Córdoba
Spain

Mustafa Zafer Özel
University of York
Green Chemistry Centre of
Excellence
Department of Chemistry
York YO10 5DD
UK

Sandrine Périno
Université d'Avignon et des Pays
de Vaucluse
INRA, UMR 408
Green Extraction Team
84000 Avignon
France

Loïc Petigny
Université d'Avignon et des Pays
de Vaucluse
INRA, UMR 408
Green Extraction Team
84000 Avignon
France

and

BASF
Beauty Care Solutions France
69007 Lyon
France

Jean T. Pierson
Naturex
250 rue Pierre Bayle
BP 81218
84911 Avignon, Cedex 9
France

Natacha Rombaut
Avignon University
INRA
UMR408
GREEN Team Extraction
Rue Louis Pasteur 33
84000 Avignon, Cedex 1
France

and

ORTESA
LabCom Naturex-Avignon
University
84000 Avignon, Cedex 9
France

Jochen Strube
Firmenich Clausthal University
of Technology
Institute for Separation and
Process Technology
Leibnizstraße 15
38678 Clausthal-Zellerfeld
Germany

Martin Tegtmeier
Schaper & Brümmer GmbH &
Co. KG
Bahnhofstraße 35
38259 Salzgitter
Germany

Eugene Vorobiev
Université de Technologie de
Compiègne
Groupe Technologies
Agro-Industriels
EA 4297
Département Génie des Procédés
BP 20529
60205 Compiègne, Cedex
France

Joël Wajsman
BASF
Beauty Care Solutions France
69007 Lyon
France

1
Green Extraction: From Concepts to Research, Education, and Economical Opportunities

Farid Chemat, Natacha Rombaut, Anne-Sylvie Fabiano-Tixier, Jean T. Pierson, and Antoine Bily

1.1
Introduction

One of the principal aims of sustainable development of green processing is in the teaching of green chemistry in colleges, high schools, and academic laboratories of both developed and developing nations. The researchers from academia and professionals from industry have chosen not to ignore the potential consequences of green chemistry and processes and have realized that they have responsibilities in education, research, and acceptance for industrial implementation of green chemistry (analysis, extraction, synthesis, separation, etc.) [1]. They recognize that their research will affect the future of the planet with the creation of new products and processes that improve the quality of life and reduce environmental hazards [2–4]. The implementation of green chemistry technologies minimizes the use of materials that are hazardous to human health and environment [5], decreases energy and water usage, and maximizes efficiency (Figure 1.1).

Extraction of natural products has been used probably since the discovery of fire. Egyptians and Phoenicians, Jews and Arabs, Indians and Chinese, Greeks and Romans, and even the Mayans and Aztecs, all possessed innovative extraction processes (maceration, alembic distillation, etc.) used even for perfume, medicine, or food. However, during the 1990s, it was not easy to find literature concerning the dispersed efforts for greening the extraction practices. It was necessary to wait for the tremendous development of green chemistry made by the Environmental Protection Agency (EPA) and led by Paul Anastas [3], who published a series of fundamental books from 1994 trying to create a general conscience on the need for green chemistry.

Recent trends in extraction techniques have largely focused on finding solutions that minimize the use of solvent and energy, such as supercritical fluid extraction, ultrasound extraction, subcritical water extraction, controlled pressure drop process, pulse electric field, and microwave extraction. The tremendous efforts made in greening extraction processes can be evaluated through the consideration of books devoted to these aspects as can be seen in Figure 1.2. Theoretical and

Green Extraction of Natural Products: Theory and Practice, First Edition.
Edited by Farid Chemat and Jochen Strube.

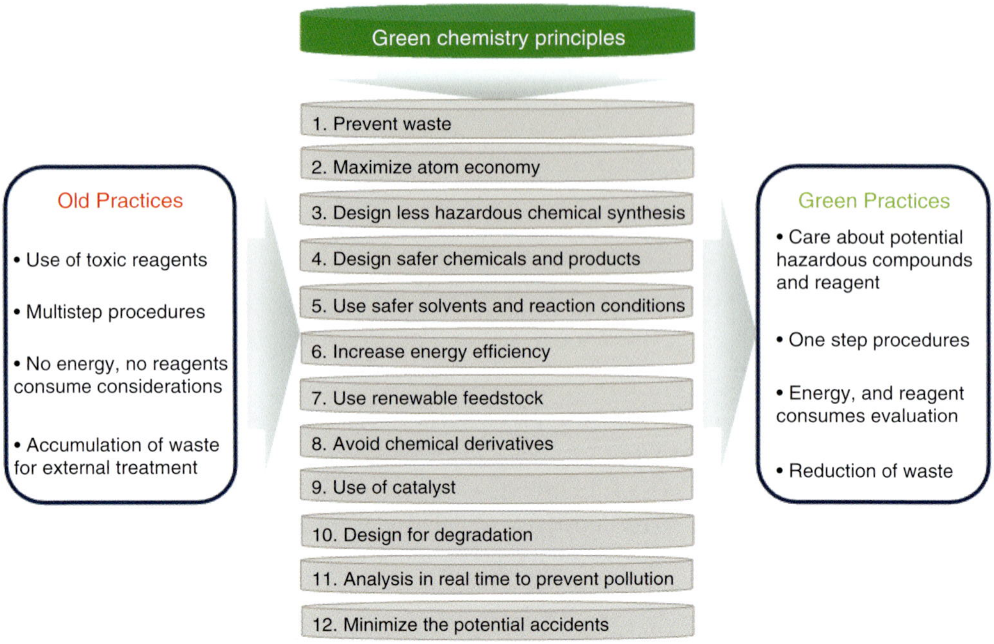

Figure 1.1 Impact of green chemistry in changing industrial and academic practices.

practical efforts are absolutely necessary to convince the members of the chemical societies about the need for such a revolution in our mentality and practice.

To meet the requirements of the market and to abide by the regulations in force, any extraction process must meet a number of quality criteria, contrary to some popular misconceptions; the "natural" state of an extract is no guarantee of its harmlessness to humans and the environment. In such a changing context, we must now switch over from a simple interest in data analysis to interest in models to a strong consideration of the environmental side effects of our practices as a consequence of the high demands of the extraction processes. This evolution or revolution in the extraction of natural products is summarized in Figure 1.3. Green extraction of natural products could be a new concept to meet the challenges of the twenty-first century, to protect both the environment and consumers, and at the same time enhance competition among industries to evolve more ecologic, economic, and innovative methods. Within the green extraction approach, the concept of the green extract is that of an extract obtained in such way as to have the lowest possible impact on the environment (lower energy and solvent consumption, etc.) and whose eventual recycling would have been planned for (coproducts, biodegradability, etc.). The green extract should be the result of a whole chain of values in both senses of the term, that is, economic and responsible, starting from the production and harvesting of the plant, the transformation process of extraction, and separation together with formulation marketing.

Extracting Bioactive
Compounds for Food Products
M. A. Meireles
ISBN: 978-3-527-32504-7

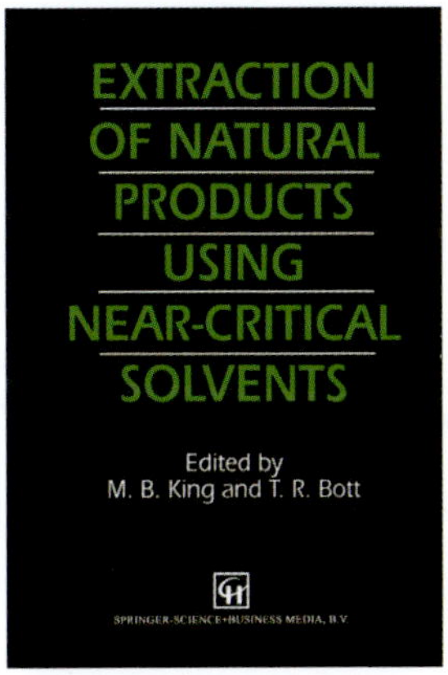

Extraction of Natural
Products Using Near-critical
Solvents
M.B. King
T.R. Bott
ISBN: 978940104974

Natural Product Extraction
M. Rostagno
J. Prado
ISBN: 978-1-84973-606-0

Industrial Scale Natural
Products Extraction
H. J. Bart
S. Pilz
ISBN: 9781420062373

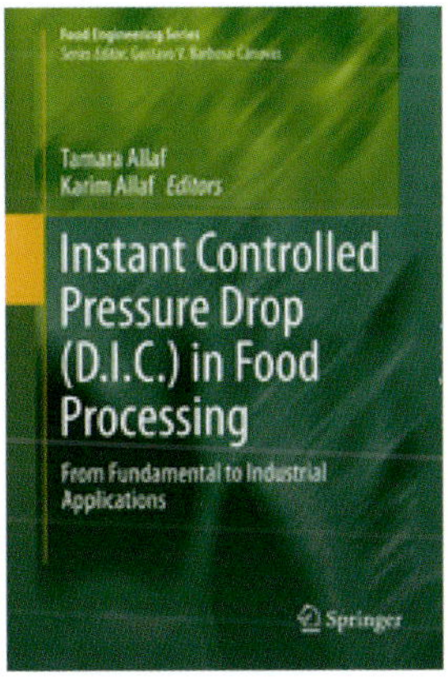

Instant Controlled
Pressure Drop (D.I.C.)
In Food Processing
T. Allaf
K. Allaf
ISBN: 9781461486688

Microwave Assisted
Extraction for
Bioactive Compounds
F. Chemat
G. Cravotto
ISBN: 9781461448297

Figure 1.2 Books devoted to green extraction of natural products.

Various industrial sectors, such as pharmaceutical, cosmetic, and food industries have increasing needs in natural products. They are obtained from plant or animal resources through extraction processes. In order to meet demand and fulfill regulations, the extraction processes are challenged to increase extraction efficiency (yield and selectivity toward compounds of interest), reduce or eliminate petrochemical solvents, together with moderate energy consumption. Within

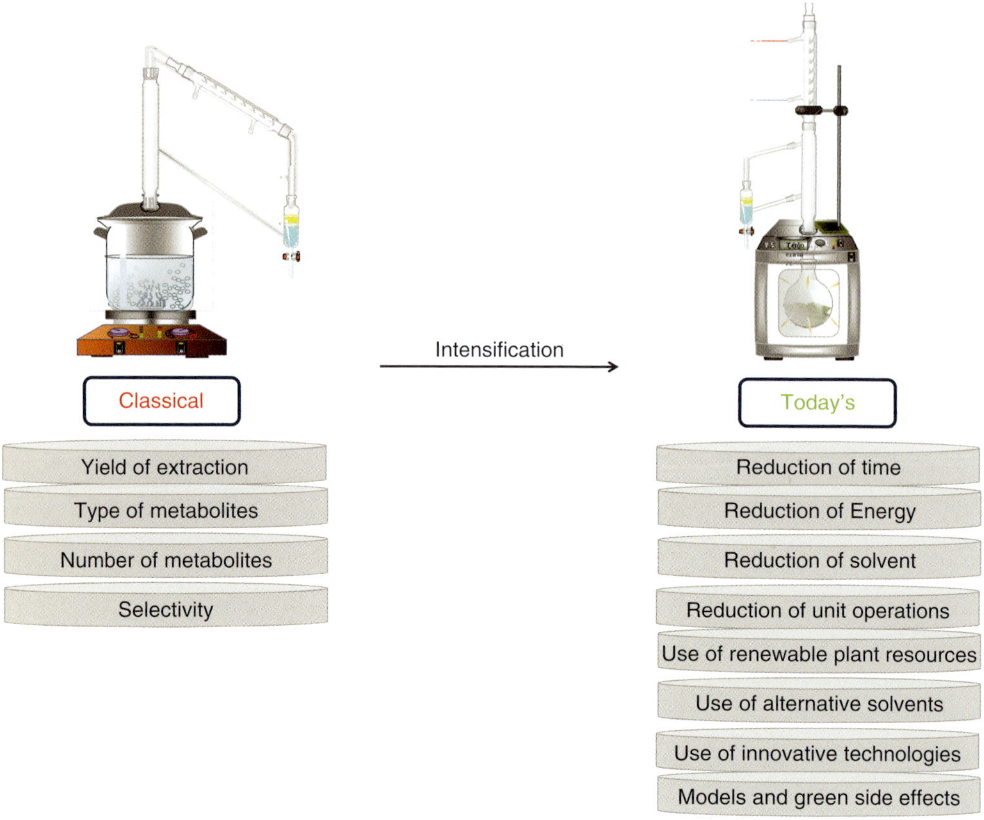

Figure 1.3 Extraction of natural products: evolution or revolution.

these constraints, green extraction has been recently introduced on the basis of green chemistry and green engineering further referring to modern "sustainable processes." In relation to green extraction of natural products, its definition has been given by Chemat *et al.* [2]: "*Green Extraction is based on the discovery and design of extraction processes which will reduce energy consumption, allows use of alternative solvents and renewable natural products, and ensure a safe and high quality extract/product.*" The listing of the six principles of green extraction of natural products should be viewed for industry and scientists as a guideline to establish an innovative and green label, charter, and standard, and as a reflection to innovate not only in process but in all aspects of solid–liquid extraction. The principles have been identified and described not as rules but more as innovative examples to follow, discovered by scientists and successfully applied by industry.

- *Principle 1:* Innovation by selection and use of varieties of renewable plant resources
- *Principle 2:* Use of alternative solvents, principally water or bio-based solvents

- *Principle 3:* Reduction in energy consumption by energy recovery, using innovative technologies
- *Principle 4:* Manufacture of coproducts instead of waste to include the bio- and agro-refining industries
- *Principle 5:* Reduction in unit operations, favoring safe, robust, and controlled processes
- *Principle 6:* Aiming for a non-denatured and biodegradable extract without contaminants.

This chapter aims at specifying the notion of natural products and illustrating how natural products can be used as ingredients in different industrial sectors. For this, we limit ourselves to compounds obtained from the plant kingdom. The biomolecule composition of a vegetable cell is usually divided into two major groups: primary and secondary metabolites. The former refers to key compounds in plant metabolism, and the latter to compounds which are involved in specific functions in the plant. Croteau *et al.* [6] introduced the term *plant natural product* to designate secondary metabolites, due to their biological activity and role in plant ecology. The expression "*natural product*" is today broadly used to designate extracts obtained from plants containing specific compounds, possessing a technological, functional, or nutritional application [7, 8]. Using the example of the orange tree as a veritable biorefinery, a few valuable metabolites are evidenced. The chemical structure and diversity of the different classes of metabolites in the plant kingdom are then reviewed. Some applications of natural products as ingredients are ultimately presented.

This chapter also presents as ultimate examples, the successful application of green extraction of natural products in academia as a vector of green teaching and research and also in industry, presenting a continuous challenge for innovation and competitiveness.

1.2
Orange Fruit is not Limited to Produce Only Juice?

The pattern from metabolites to ingredients can be illustrated through the example of the orange tree. The major industrial use of oranges is juice production. However, the known diversity of the phytochemical composition of the fruit and the orange tree allows considering the tree as a biorefinery of natural products. The following section describes the by-product valorization of orange juice production along with some examples of the phytochemicals identified in the orange tree and fruit.

With a total production of 68 million tons in 2012 [9], orange is one of the major crops in the processing industry. Out of the total production, 95% is used to manufacture orange juice. Considering that 3 kg of oranges are necessary to produce 1 l of orange juice, there is a great potential for valorization of the by-products generated [10]. Further, not only the by-products from the fruits but also the whole orange tree can be used for nutritional, pharmacological, or cosmetic purposes. In this section, the pathways for valorization of orange products are described. Different types of by-products can be generated from orange juice production: those from the orange tree and those from the orange itself. By-products generated by orange juice production are pulps, peels, and seeds [10, 11].

One of the main by-products obtained from the orange fruit is the pulp (42.5% w/w of orange waste) [9, 12]. Owing to their high fiber content, pulps are mainly used for livestock feeding. Moreover, inclusion of the pulp in cattle feed would support growth and lactation in a better way than would starch-rich supplements [13].

Accounting for 50% of the mass of the fruit, orange peels contribute to a high proportion of the by-products [11]. Peels can be processed by steam distillation [14] or by cold pressing [15] to extract essential oil. Essential oil is used in foods and beverages as a flavoring agent, in cosmetics, and in perfumes [16]. Peels also contain some valuable chemicals, for example, pectin [17], which used as a texturing agent and food stabilizer. Pectin can be incorporated in foods such as jam or jellies. Peels contain structural components such as cellulose [18] and hemicellulose. After biorefining, these molecules can be transformed into biofuels and biochemicals. The juice extracted from the peels is evaporated to give citrus molasses [19], marketed for livestock feeding [13], or is used in the fermentation process of alcohol. Further use of this by-product requires more refining as the phytochemicals in the juice contribute to a strong bitterness and dark color. As a refined product, orange peel syrup can be used as natural sweetener. Orange seeds represent a small part of the residue from orange juice production (from 0.5 to 5%). The main interest of the seeds is in their oil content (up to 40%). Extraction of orange seed oil has been performed by organic solvents [20, 21]. However, because the high oil content of the seeds, mechanical pressing could perhaps be used to recover oil.

Flowers, leaves, and the lignocellulosic biomass have found various applications. Flavoring substances such as essential oils, namely, Neroli and Petitgrain, are extracted from the flowers and leaves (less than 1% content [22]), respectively. Oils can be used in the cosmetics and food industries [23, 24]. It is noteworthy that the water distilled from flowers and leaves after the extraction of essential oils (orange blossom water and eau de brouts) can also be valorized in the same way as essential oils because of their aromatic characteristics. The lignocellulosic biomass issued from roots, twigs, and leaves can also be valorized by the production of biochemicals and/or biofuels [25, 26]. These inedible parts of the tree also contain some alkaloids, which may be used in the pharmaceutical industry.

The numerous uses of orange by-products are due to the phytochemical composition of the fruit and the tree. The following section examines the structure and localization of these natural products. The section focuses on the identification of the chemical structure of the phytochemicals of interest contained in the orange tree. Their localizations and structures are illustrated in Figures 1.4 and 1.5. The variety and structural complexity of these phytochemicals enables their

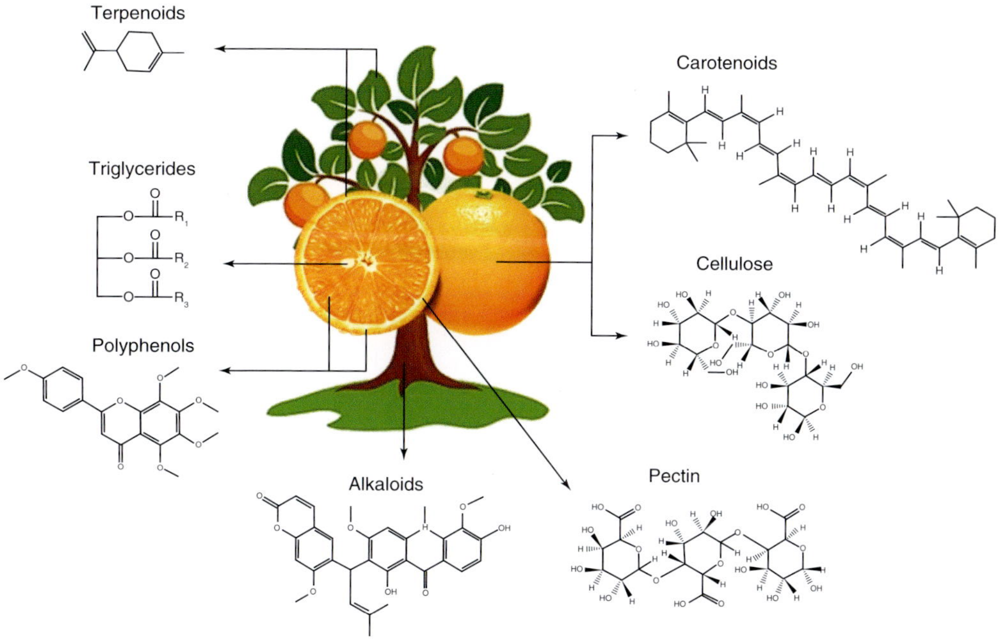

Figure 1.4 Localization and structure of phytochemicals contained in the orange tree and fruit.

classification according to their biochemical function into primary and secondary metabolites.

Two types of polymers issued from primary metabolites are constitutive of the by-products detailed in the previous section: cellulose and pectin. Cellulose, which is polymer of cellobiose, is a structural cell component [27] and, therefore, is present in the whole tree. Pectin would be located more specifically in the peel of the fruit. The lipidic fraction is mainly found in the seeds of oranges. Although seeds have a high oil content, the low amount of seeds in the orange fruit leads to an overall low lipidic fraction taking the whole tree into consideration.

Three common classes of secondary metabolites are identified in the composition of the orange tree: terpenoids, alkaloids, and polyphenols. Terpenoids are constitutive of essential oils and those found in oranges (Neroli and Petitgrain oils) have different terpenic profiles. Neroli oil is composed of linalool, limonene, beta-pinene, *trans*-beta-ocimene, linalyl acetate, and terpineol, while Petitgrain oil is mostly composed of linalyl acetate and linalool [23, 24, 28]. Orange peels also contain a terpenic fraction, which is exclusively composed of limonene. The high concentration of limonoids were identified at first in the fruits of the the Rutaceae family (e.g., *Citrus*), however those compounds were further identified in other plants [29, 30]. Carotenoids are another terpenoid class identified in the orange tree: β-cryptoxanthin, β-carotene, lycopene, and isomers of violaxanthin were detected, more specifically, in the fruit [31]. Alkaloids are located in the

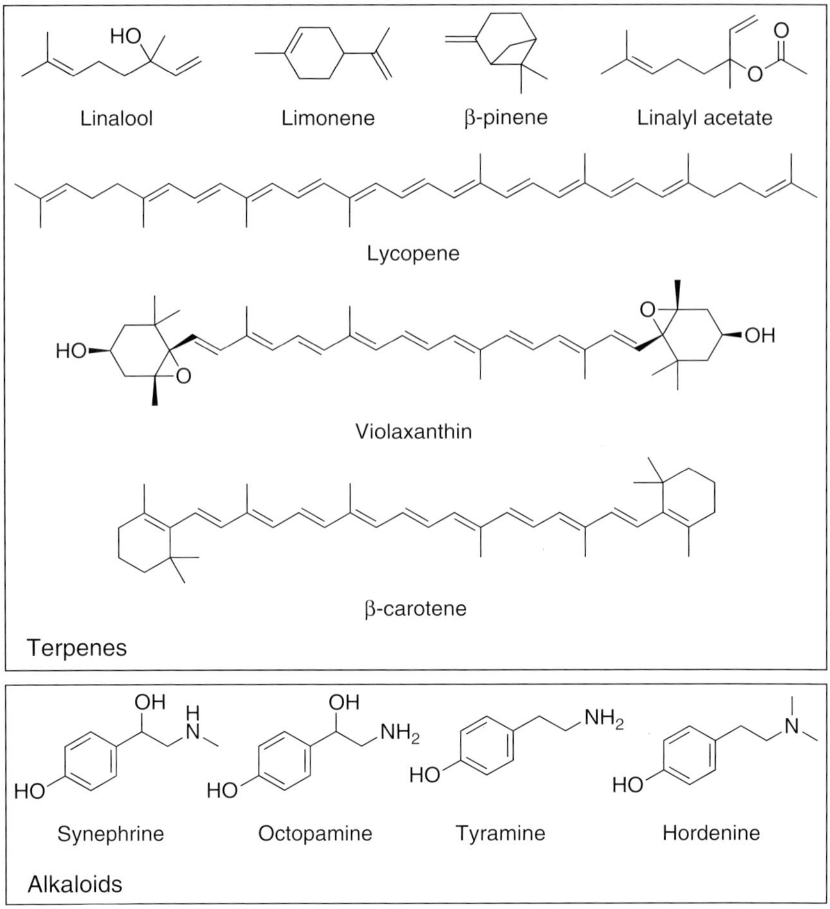

Figure 1.5 Structure of secondary metabolites (terpenes and alkaloids) contained in the orange tree.

leaves and the roots of the orange tree [32]. Owing to their biological activity, these compounds are of interest in pharmaceutical applications. Synephrine, which is an adrenergic and used for weight management, has been identified in the orange tree [33]. It is mainly isolated from citrus species and commonly found in juices. Although synephrine can be synthesized, the increasing demand for natural extract makes this alkaloid an important extraction target. Other alkaloids commonly reported from *Citrus aurantium* are octopamin, tyramine, *N*-methyltyramine, and hordenine [34]. Polyphenols are the third class of secondary metabolites found in the orange tree. Phenolic compounds identified in the orange fruit (located more specifically in peel and juice) are essentially flavanone glycosides such as nariginin, hesperidin, narirutin, and neohesperidin [35]. These polyphenols are specific of the *Citrus* family, and are of increasing interest because of their antioxidant potential. In a less specific occurrence, lignin, a structural component mostly found in the membranes of vegetable cells, is found in the whole orange tree.

This introductive example of the orange tree, through the valorization of by-products generated from orange juice production, allows us to distinguish different categories of molecules considered as natural products. The following section summarizes the categories of metabolites that can be identified in vegetables, aiming at a classification of natural products.

1.3 Chemistry of Natural Products

The term *natural product* designates any substance or chemical produced by nature. Considering vegetables, this wide definition can be restricted to any chemical vegetable component or phytochemical. In biochemistry, phytochemicals are commonly classified into two groups according to their function: primary and secondary metabolites. Primary metabolites include any compound essential to basic plant metabolism and thus survival. On the other hand, secondary metabolites are involved in ecological functions such as attractants for pollinators, in the protection against herbivorial and microbial infections, and in allelopathy (which refers to a positive or negative effect, e.g., the growth of one plant into another induced by phytochemicals). In this section, the well-known primary metabolites are briefly reviewed, focusing on secondary metabolites.

1.3.1 Primary Metabolites

1.3.1.1 Glucides

Glucides are metabolites mainly used for the storage and transport of energy and are located in every organ of a plant [27]. In cells, glucides are stored as starch which is a polymer of glucose. Glucides are also metabolic precursors of secondary metabolites and are constitutive of cell membranes. For instance cellulose, a macromolecule composed of a repetition of glucose moieties, is one of the main components of vegetable cell membranes. It is one of the most abundant compounds on the Earth.

Owing to their numerous functions in basic plant metabolism, glucides constitute a wide class of metabolites. The classification, chemistry, and structure of glucides are described in many biochemistry reference books [36, 37], therefore, their diversity is not reviewed in this section, but examples are given of glucides having industrial applications. Glucides are divided into two classes: oses and osides, which are polymers of oses or a combination of oses with a non-glucidic molecule, respectively [38].

Oses classically processed in industry are fructose, glucose, and oses derivatives such as polyols. One of the most-extracted oside is saccharose (e.g., isolated from sugar beet and sugar cane) [38]. The main polysaccharides which can be extracted from vegetables are fructosannes such as insulin, starch, cellulose, fibers, mucilages, and pectins [38]. These macromolecules can be used as such for food or nonfood application, or can also be chemically transformed to produce chemicals or ingredients (e.g., starch is the starting point of modified starches, dextrines, glucose syrups, glucose, and sorbitol). Fibers are a group mainly composed of cellulose, hemicellulose, pectins, and lignins; because of their indigestibility, these compounds qualified as "dietary fibers" [38]. Mucilages are highly hydrophilic colloids and can be used for their texturing properties (e.g., in pharmaceutical applications). Pectins are mostly located in fruits (apples, citrus fruits) and are structural macromolecules that constitute cell walls. They can be used for their jellifying properties [38].

1.3.1.2 Lipids

Similar to glucides, lipids have an energy storage function [27], and are the precursors of numerous metabolites. Lipids are a group of heterogeneous compounds (fatty acids, glycerolipids, cerides, etc.). In plants, lipids are stored in oleosomes as oils, which are mainly composed of triacylglycerols (esters of glycerol and fatty acids) [27]. Some lipid structures are illustrated in Figure 1.6.

Fatty acids are a large group composed of linear saturated and unsaturated hydrocarbonated chains with a carboxylic acid function. Commonly found saturated fatty acids range from butyric (C4) to arachidic fatty acids (C20) [39]. Longer chains exist and are synthesized through the same metabolic pathway by elongation, but are less common in the plant kingdom. Unsaturated fatty acids are classically produced by the addition of double bonds to the corresponding saturated fatty acids [8]. The number of unsaturations and their position along the carbon chain are used to classify unsaturated fatty acids. Palmitoleic (C16:1), oleic (C18:1), linoleic (C18:2), linolenic (C18:3), and arachidonic (C20:4) fatty acids are among the commonly found unsaturated fatty acids [27].

Through the combination of fatty acids with primary metabolites present in plants (e.g., glycerol), complex compounds are formed (mono-, di-, and triglycerides), such as glycolipids, glyceroglycolipids, phospholipids and phosphosphingolipids (based on sphingosine structure) [40], or aromatic polyketides such as anthraquinones [41] and naphthoquinones [42].

1.3.1.3 Amino Acids and Proteins

Amino acids are of great importance in metabolism, as they are the building blocks of enzymes, peptides, and structural proteins. Amino acids are also precursors of numerous secondary metabolites (alkaloids, betalains, some polyphenols, etc.) [38]. Proteins play a major role in human and animal nutrition. However, the applications as the final product or extract composed specifically of amino acids, proteins, and enzymes are quite limited, and therefore are not developed in this section.

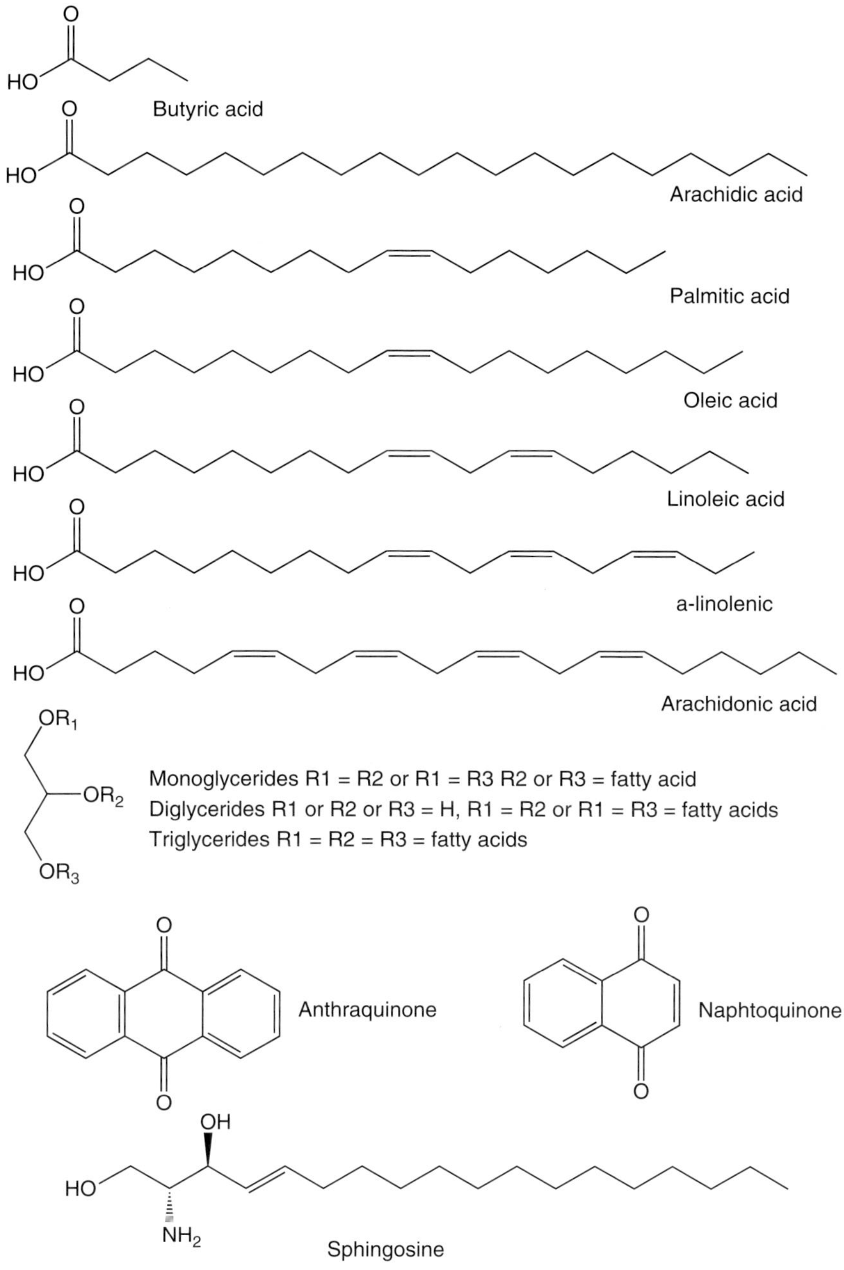

Figure 1.6 Chemical structure of some lipids.

1.3.2 Secondary Metabolites

Secondary metabolites are involved in the development of specific characteristics of the plant such as color and flavor (e.g., flowers), therefore, playing a role in the ecological functions of a vegetable [6]. Three types of phytochemicals are classified as secondary metabolites: terpenoids, alkaloids, and phenylpropanoids with allied phenolic compounds [6]. Since secondary metabolites are involved in specific functions, their number and diversity will depend on the plant being considered.

1.3.2.1 Terpenoids

Terpenoids cover a great range of diverse compounds: up to 25 000 terpenoids have been identified [6]. Because of their flavoring properties, terpenoids are mainly known as major components of essential oils [43]. Isoprenoids are the building blocks of the terpenoid family and are synthesized through either the mevalonate pathway or the deoxyxylulose pathway [44]. Terpenoids are classified according to the number of isoprene units. The main classes are listed in Table 1.1, with some structures represented in Figure 1.7.

The tetraterpene carotenoid family contains the carotenoid and the xanthophyll family, which are major pigmentation components and which have also been found to possess nutraceutical properties. Modifications of these terpenoid classes lead to specific subclasses of compounds found to be bioactive principals of certain plant varieties. Some of the terpenoids of interest are listed in the following. Gingkolids such as Gingkolide B are highly modified diterpenes from the exclusive Gingko biloba species [45]. Steviosides are steviolglucosides (tricyclic diterpene), which are of high interest in the sweetener industry [46]. Limonoids like limonin, are degraded triterpenoids called tetranortriterpenoids due to the removal of four carbons from a side chain [30]. Quassinoids such as bruceantin, are another example of triterpenoids, which have lost 10 carbons and are not to be confused with diterpene structures [47]. Steroids are tetracyclic-modified triterpenoids, which represent a class of bioactive compounds. Their typical structure is based on the cholesterol structure and the main sterols in plants are phytosterolssuch as stigmasterol, sitosterol, fucosterol, ergosterol, and campesterol [48].

Table 1.1 Classification and examples of terpenoids according to the number of isoprene units.

Category	Example of terpenoid groups
Monoterpene (C10)	Linalool (acyclic), limonene (cyclic), pinene (bicyclic)
Sesquiterpene (C15)	Farnesol (acyclic), bisabolene (cyclic), α-eudesmol (polycyclic)
Diterpenes (C20)	Phytol (acyclic), retinol (cyclic), taxadiene (polycyclic)
Sesterpene (C25)	Geranylfarnesol
Triterpene (C30)	Sapogenin (polycyclic)
Tetraterpene (C40)	β-carotene (red pigment), xanthophyll (yellow pigment)

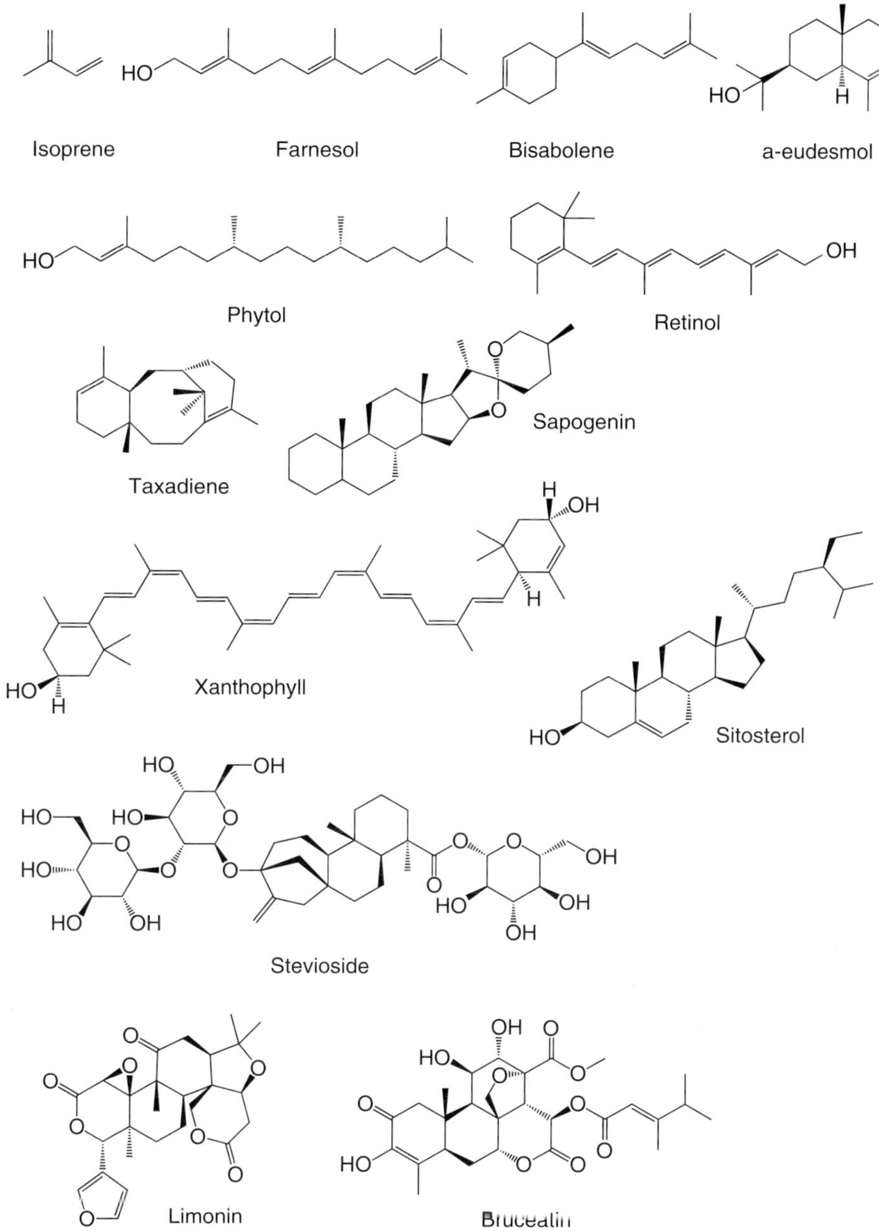

Figure 1.7 Examples and chemical structure of some terpenes.

1.3.2.2 Alkaloids

In general, alkaloids are defined as nitrogen-containing compounds. They possess strong pharmacological effects which results in a close monitoring of edible products that may contain such compounds [49]. Alkaloids are biosynthesized starting from various amino acids and it is easier to classify them regarding to their precursor, rather than based on their structure [8]. Examples of alkaloid chemical structures are indicated in Figure 1.8.

Alkaloids originating from L-glutamate transformed into L-ornithine contain the pyrrolidine ring system as well as the tropane ring system, which is found in cocaine [39], atropine [41], or scopolamine [49]. The pyrrolizidine ring system is another structure derived from the L-arginine pathway. The second group of alkaloids derives from the L-lysine amino acid which forms the piperidine core structure. One of the best-known alkaloids based on this structure is the piperine found in black pepper [50]. But L-lysine is also precursor to the quinolizidine structure [51], which is not found in common fruits and vegetables but can potentially exist in more specific ethnic vegetables at some trace levels. The same goes for alkaloids derived from the indolizidine structure [52]. Pyridine alkaloids, as the vitamin B family, are more widespread in the human diet (e.g., niacin) [53], contain such a core structure and originate, from the amino acid aspartic acid. Alkaloids obtained from tryptophane can be built with the indole core structure (tryptamine [54], harmine [55], serotonin [56]), and more rarely, as pyrroloindole alkaloids or derived into ionic compounds like betalains [57], for example, betanin. The biosynthetic pathway of betalains, however, suggests that its synthesis starting from a tyrosine and not from tryptophan.

Anthranilic acid, as intermediates to tryptophane, can also lead to alkaloids found in nature (quinazoline, quinoline, and acridine) [58] but none of importance in fruits and vegetables. The same goes for the imidazole-structured alkaloids, which are derived from histidine [59]. The last alkaloid family we report are the purine-derived alkaloids, which are extremely well known for their occurrence and wide consumption [60]. They are mostly represented by caffeine, theobromine, and theophylline. All other alkaloids are generally combinations of the pure alkaloids described here with combination into other families such as terpenoid alkaloids (taxol) [2] or steroidal terpenoids (tomatidine) [61].

1.3.2.3 Polyphenols

Polyphenols are well known as nutrients from nature, because of their ubiquitous occurrence; and their pharmacological significance makes them major research subjects in the nutraceutical sciences [62–66]. As they have been reviewed thoroughly by different authors, we briefly recall the main classes of polyphenols.

Phenolic acids constitute a wide class of polyphenols. The simplest phenolic acids contain a carboxylic acid function with one or more hydroxyl substitutions on the benzene ring (Figure 1.9). The most common phenolic acids are gallic acid, used as is or as an ester (galloyl), chlorogenic acids such as 4-chlorogenic acid, and feruloylquinic acids such as 4 feruloyl-D-quinic acid, which originate from *trans*-cinnamic acid [67]. Other commonly found phenolic acids are differently

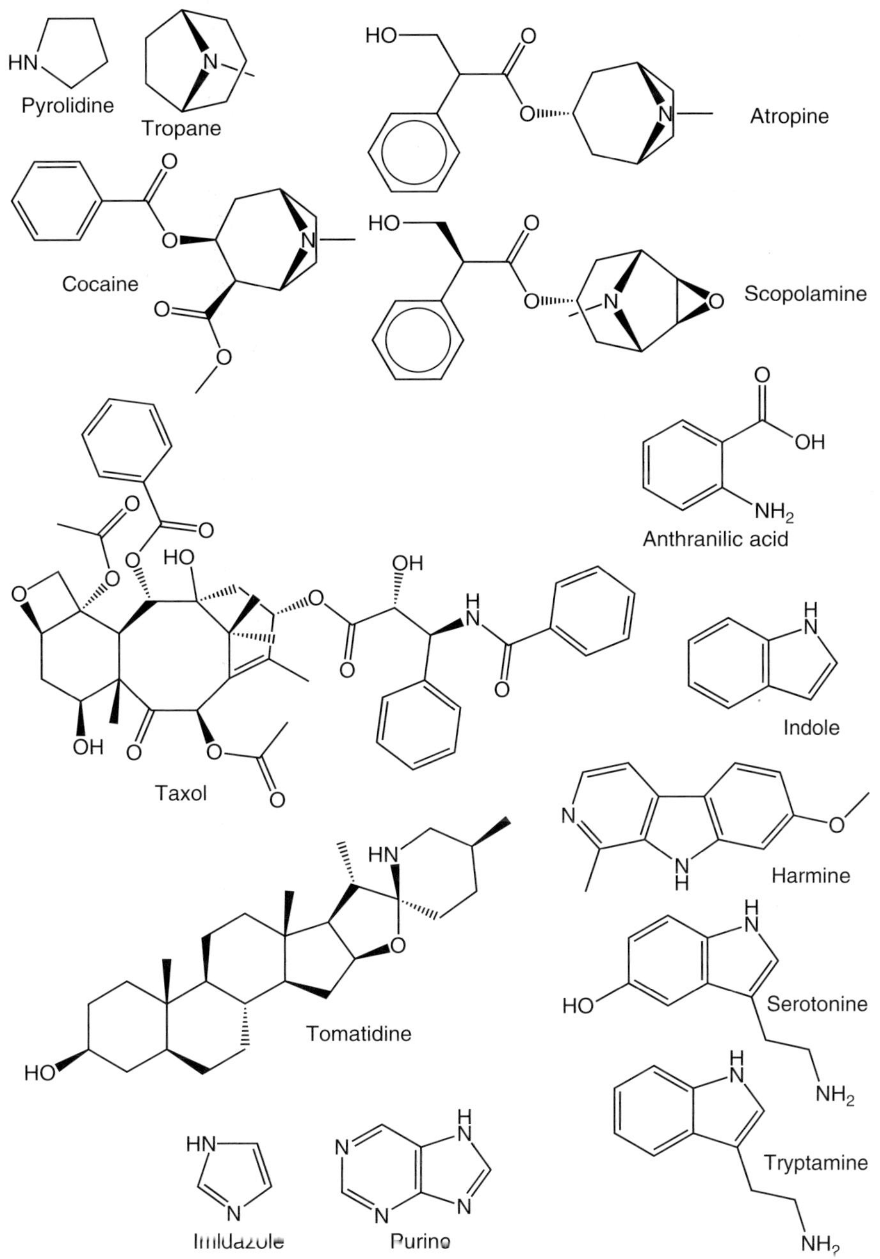

Figure 1.8 Examples of alkaloid structures and molecules.

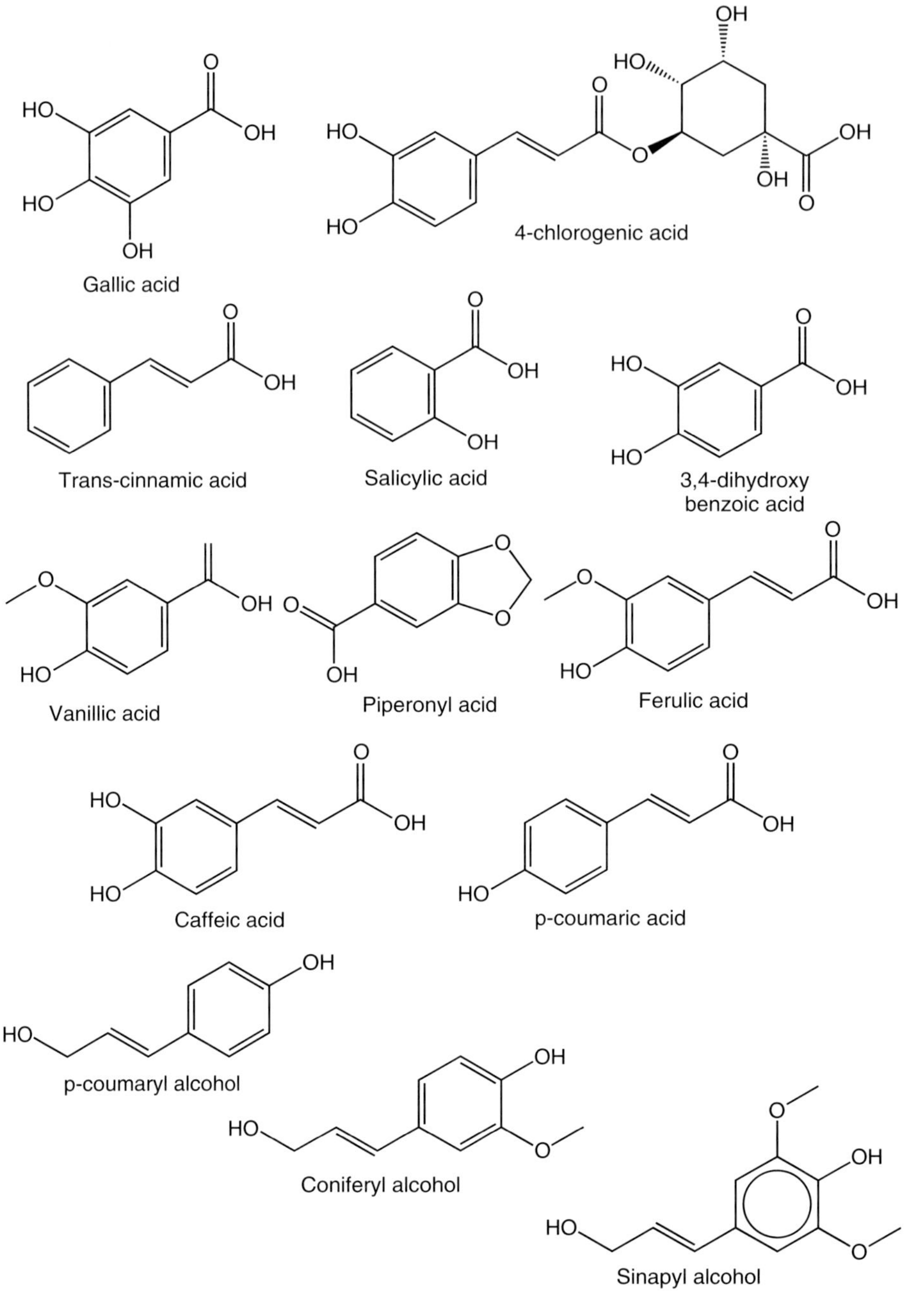

Figure 1.9 Examples of polyphenol structures (phenolic acids).

substituted ones, including hydroxybenzoic acids such as salicylic acid [68], hydroxycinnamic acids such as 3-coumaric acid [69], protocatechuic acids such as 3,4-dihydroxybenzoic acid, vanillic and piperonylic acids [70], and cinnamic derivatives such as, ferulic, caffeic, and p-coumaric acids [62]. Owing to their early occurrence in the main pathways for the biosynthesis of phytochemicals, they are quite ubiquitous and well known either as stand-alone compounds or as reagents in the food industry and in nutraceutical supplements [71].

More complex structures based on phenolic acids possess longer chain substitution, extending the range of phenolic acids (Figure 1.10). These compounds are hydrolyzable tannins, which are gallic acid esters with a glucose central structure. Tannic acid [72] is a good example of highly substituted hydrolyzable tannin, yielding 10 gallic acid moieties per glucose moiety. Their chemical composition and geometry make hydrolyzable tannins class an extremely diverse family.

Apart from the hydrolyzable tannins, the largest family of polyphenols is the flavonoids [73–75]. They are characterized as containing a benzopyrone moiety as central core structure. The family is subdivided into the following subfamilies: coumarins, flavones, isoflavones, neoflavones, and all subsequent derivatizations of those core structures. All of them have been isolated from nutritionally relevant sources. Flavones contain the following groups of compounds: flavonol, flavanol, flavanone, and flavanonol [76–78]. Flavonol dimers or trimers also lead to the proanthocyanidin group [79]. And finally, the very specific group of the anthocyanins (as the aglycone core structure of the anthocyanidins) is an ionic version of the flavones with the chemical specificity of the flavyllium core structure [62, 71]. Other compounds that are part of this greater family are the xanthones [80, 81] and the stilbenes [82], which are present in specific genus and families of edible plants.

Apart from the mentioned polyphenol classes, structural polymerized polyphenols are constitutive of cells. Lignin is an important part of the secondary cell wall for plants and accounts for 30% of all nonfossil carbon fuel on earth [83, 84]. It is found at a higher concentration in wood and wooden structures, but as structural cell component is also located in leaves, flowers, and fruits. Lignin is a macromolecule composed of three monolignol molecules, namely, *p*-coumaryl alcohol, coniferyl alcohol, and sinapyl alcohol, all of which are methoxylated to different degrees [83].

1.4 From Metabolites to Ingredients

Utilization of plants either for medicinal, industrial, or food purposes is related to their biomolecular composition. While primary metabolites are ubiquitous to the plant kingdom, by having functions essential to plant metabolism and constituting the basis of human and animal nutrition, secondary metabolites occur at much lower concentrations levels and, because of their biological function, can be specific of a plant specie. Secondary metabolites, however, benefit from

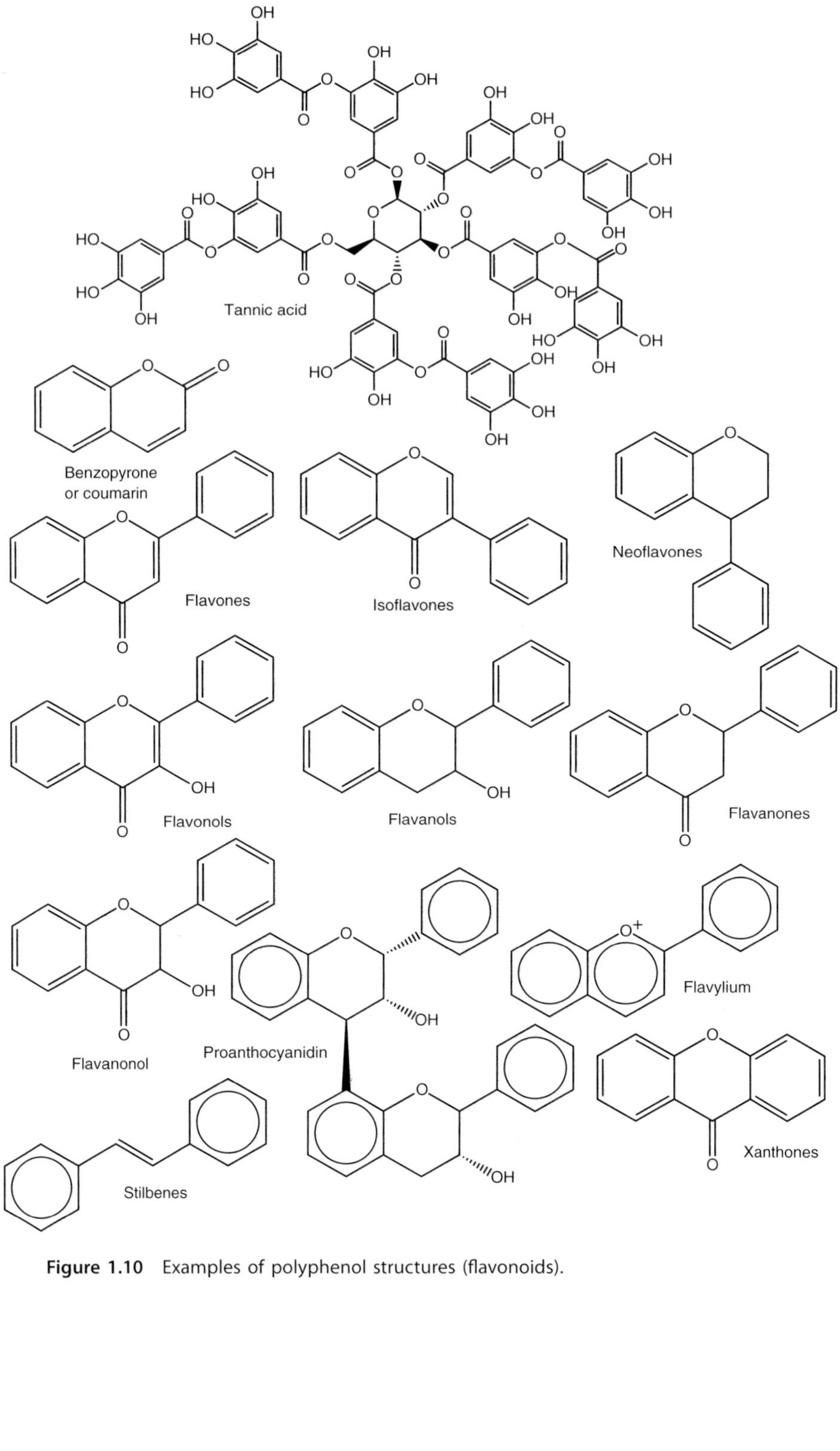

Figure 1.10 Examples of polyphenol structures (flavonoids).

a strong interest as some compounds affect biological systems [85]. This characteristic enables to qualify such compounds as bioactive [85]. Bernhoft [86] proposed a definition of bioactive compounds: secondary plant metabolites eliciting pharmaceutical or toxicological effects on humans and animals. Owing to their reactivity and chemistry, bioactives are used in numerous applications. The following section details the use of some secondary metabolites as ingredients in the pharmaceutical, nutritional, and food sectors. The uses of bioactive compounds can be classified according to their technological role: coloring agents, functional food and nutraceuticals, preserving agents, flavors, fragrances, and edible oils [7]. Some bioactive applications examples are presented in Table 1.2.

The natural coloring agent category refers to dyes, pigments, or any substance that produces color [7]. Natural colors found in nature range over yellow, orange, red, blue, and green. The bioactive compounds responsible for the production of such colors differ according to their chemical structure, physical properties (e.g., solubility), and localization in the plant (flowers, leaves, berries, etc.) [7]. All compounds have in common a chromophore (molecule-specific structure) that absorbs light in the wavelength range of the visible region known as *pigment* [100]. Some natural colorants such as chlorophylls contain metallic ions (e.g., Mg, Fe, Zn), responsible for the overall color of the complex [101]. Considering the color diversity, different major groups of chemicals were identified as responsible for coloring: carotenoids (terpens), chlorophylls, anthocyanins (polyphenols), betalains (alkaloids), and curcuminoids [100]. These groups of molecules are commercially available natural pigments and are used as alternative to synthetic pigments. However, natural colorants are less stable than the synthetic ones as regards to physical (light, temperature, and oxygen), chemical (alkalis, acids, oxidizing, and reducing agents), and biological (enzymes, microorganisms) factors [100, 101].

Flavors and fragrances may mostly be attributed to essential oils contained in aromatic plants. In industry, essential oils are formulated to produce flavor and fragrances for a wide range of end uses such as in soaps, cosmetics, perfumes, baked foods, ice creams, aerosols, syrups, and pharmaceutical preparations [102]. Essential oils are characterized by a strong odor, and appear as volatile and clear liquids. As essential oils usually have a lower density than water, they are commonly extracted by hydro or steam distillation [16]. The concentration of essential oils in plants varies between 0.01 and 15% (w/w) [7], depending on the type of plant and the part of the plant considered. Different plant organs can be used to extract essential oils: seeds (anise [103]), barks (cinnamon [104]), woods (rosewood [105]), rhizomes (ginger [106]), leaves (rosemary [107]), flowers (lavender [108]), peels (orange [87]), and roots (valerian [109]). From a chemical point of view, essential oils are a complex mixture of compounds. Terpenes are the most common compounds found in essential oils and more specifically the monoterpenoid subclass (90% of identified terpenoids [7]). Aromatic and aliphatic compounds of the shikimate, polyketide, and alkaloid groups are also constitutive of essential oils (for example, aldehydes, alcohols, phenols, methoxy derivatives, and methylene dioxy compounds [7, 110]).

Table 1.2 Applications in the food, nutrition, and cosmetics industries of some secondary metabolites.

Metabolites	Vegetable source	Extract	Examples of applications	References
Terpenoids				
Monoterpenoids				
Limonene, linalool	Genus *Citrus*	Liquid, volatile extracts	Major components of essential oils issued from the different vegetable sources, applications in cosmetics (perfumes, deodorizers, body care products) and food (flavoring) industries	[87]
Geraniol	*Melissa officinalis* L.			[88]
Citronellol	*Pelargonium graveolens*			[89]
Myrcene	Hops (*Humulus lupulus* L.)			[90]
Menthol	Cornmint, peppermint			[91]
Tetraterpenoids				
Carotenoids				
Bixin Norbixin	*Bixa orellana* L.	Annatto: bixin-rich oil extracts, norbixin-rich extract water soluble	Yellow-red pigments, coloring agent E106b (butter, margarine, cheese, cereals, fish products)	[92]
Ketocarotenoids Capsanthin Capsorubin Zeaxanthin Violaxanthin Antheraxanthin	Red peppers (*Capsicum annuum* L.)	Paprika: oleoresin, powder	Orange-red pigments, coloring and flavoring agent E160c • Food industry (soups, sausages, cheese, snacks, salad dressing) • Nutraceutical and pharmaceutical industries	[93]

Crocin Picrocrocin Safranal	*Crocus sativus*	Saffron	Orange-red pigments, coloring and flavoring agent (beverages, bakery, and confectionary products, curry products, soups)	[94]
β-Carotene	Carrots (*Daucus carota* L.) Palm fruit	β-Carotene-rich oil-soluble extract	Food pigment, nutritional/pharmaceutical use (e.g., pro-vitamin A)	[95]
Polyphenols				
Flavonoids				
Anthocyanins	Grape extracts (*Vitis vinifera* L.)	Powder	From blue to red food pigments E163 (soft drinks, sugar confectionery, dairy products, fruit preservatives (jams, canned fruit)	[96]
Isoflavones	Soybeans	Powder	Use as antioxidants and nutritional applications	[97]
Lignans	Flaxseeds (*Linum usitatissinum* L.)	Powder	Nutritional applications	[98]
Alkaloids				
Caffeine	Coffee beans	Powder	Food and beverage ingredient, pharmaceutical applications	[99]

While all food products and food ingredient play an essential role in nutrition and sensory characteristics, some food component may provide an extra functionality, conferring to the overall product health benefits or desirable physiological effects beyond basic nutrition [111]. These functional foods are obtained by enrichment in a given bioactive, which is used as an ingredient able to promote or provide a beneficial effect on health such as acting on cardiovascular diseases, cholesterol reduction, degenerative diseases, among numerous potential health applications [7]. In functional foods, nutraceuticals may be distinguished by a concentration of the bioactive compounds considered as a health-promoting compound. These compounds may also be concentrated, serving as nutritional supplements, known as nutraceuticals [112]. In plants, the bioactives found in functional foods are mainly used for their different properties: polyunsaturated fatty acids and phytosterols may also be encountered. While all of these bioactive compounds are used for their specific health benefits (fibers, organic acids, polyunsaturated fatty acids [113], phytosterols [114], and organosulfides [115]), some may also be used additionally for a technological property: antioxidants (phenolics, tocopherols [116, 117]) and colorants (anthocyanins, betalains [118, 119]).

Edible fats and oils are respectively solid and liquid materials composed predominantly of glyceryl esters of fatty acids (90–95%), traces of non-glycerol-based lipids are also identified (phytosterols, tocopherols, etc.) [120, 121]. Fats and oils have been much used in food industry. Some of the major applications include nutrition (more specifically regarding to essential fatty acids), cooking, spread products (ingredients in bakery products and butters, respectively), shortening, and biodiesel feedstock [122, 123]. The nonfood applications of fats and oils include ingredients for formulation of soaps, detergents, paints, varnishes, plastics, and lubricants, among others [124].

This section reviewed how natural products contained in plants can be used as ingredients. Far from being exhaustive, it can be identified that a wide range of metabolites can be used in various industrial sectors: mainly in food as complements or as additives (aromas, flavors, colorants) but also in many other industries such as the pharmaceutical and agricultural industries, or in medicinal extracts or feedstuff. The introductive example of the orange tree biorefinery underlines the huge potential of valorization of process by-products. By generalizing biorefining, the agricultural industry would gain much benefit by commercializing their unused metabolites left in the waste biomass.

1.5
Green Extraction from Research to Teaching

Education plays an important role in materializing the concepts and principles of green chemistry and engineering [125]. Teaching green chemistry serves several key functions. First, it provides fundamental knowledge related to new chemical products and processes, data that are necessary to develop cleaner technologies [126]. Moreover, these new products and processes developed in an academic setting can, in some cases, have direct applications to industry. Finally, academia serves as the primary means to educate students about the need to design green chemistry technologies, and provides them with the tools to do so.

For example, in Avignon University, to illustrate an application of green chemistry in teaching laboratories, we used a new green procedure employing ultrasound energy and microwave energy as the energy source to teach fundamental extraction concepts.

As an example, we have developed a new green procedure, using microwave energy as energy source, to teach the fundamental concepts of extraction of essential oils (Figure 1.11). Essential oils are volatile secondary metabolites that plants produce for their own needs other than nutrition. They are widely used

Figure 1.11 Practical work on green and conventional extraction techniques for Master students at Avignon University.

in foods, cosmetics, and pharmaceuticals [127]. In general they are complex mixtures of organic compounds that give characteristic odor and flavor to the plants. The objective of this teaching was to offer students the opportunity to compare the potential of this green technique for extraction of essential oil with a traditional hydrodistillation method (used in all the teaching laboratories all over the world) and to appreciate the benefits of using greener chemical methods. Green microwave extraction procedure is regarded as a greener method of heating because extraction times tend to be shorter than those required using the conventional procedure.

Microwave energy is well known to have a significant effect on the rate of various processes in the chemical and food industry [128]. Microwave extraction is an area of research which has an impact in several fields of modern chemistry. All the reported applications have shown that microwave-assisted extraction is an alternative to conventional techniques for such matrices. The advantages of using microwave energy, which is a noncontact heat source, for the extraction of essential oils from plant materials, include more effective heating, faster energy transfer, reduced thermal gradients, selective heating, reduced equipment size, faster response to process heating control, faster start-up, increased production, and elimination of process steps. Extraction processes performed under the action of microwave radiation are believed to be affected in part by polarization and volumetric and selective heating.

Microwave energy is a key enabling technology in achieving the objective of sustainable and green chemistry for research, teaching, and commercial applications. It has been shown that solvent-free conditions are especially suited to microwave-assisted organic synthesis, as reactions can be run safely under atmospheric pressure in the presence of significant amounts of products. When coupled to microwave radiation, solvent-free techniques have proved to be of special efficiency as green and economic procedures. Major improvements and simplifications over conventional methods originate from their rapidity, their enhancement in yields, and product purities. Microwave hydrodiffusion and gravity (MHG) of essential oil has been compared with the conventional hydrodistillation method, for the extraction of essential oil from fresh grapefruit peel. The method offers important advantages over traditional alternatives, namely, shorter extraction times, better yields, lower costs, greener features (as there is no residue generation and no water or solvent is used), and greener production (reduced CO_2 is rejected in the atmosphere). MHG is a laboratory exercise which easily teaches fundamental sustainable chemical lessons and successfully incorporates green chemistry into the teaching laboratory in both developed and developing nations.

At the same time, the idea has also been to develop a portable microwave-assisted extraction (PMAE) [129, 130] that can be used directly in the crop or the forest; this procedure is also appropriate for the teaching laboratory because it does not require any special microwave equipment.

In recent years, portable sample preparation apparatus have been developed, which can help with in-field extraction, sample preparation, analysis, and data evaluation. A portable instrument is defined as "easily" movable, convenient for carrying, and capable of being transferred or adapted in altered circumstances [131]. Miniaturization (downscaling) is an important factor in modern society reflected strongly in science and technology [132] including medicine, chemistry, environment, food safety [133], and so on. Miniaturization of a procedure can be achieved simply by reducing the dimensions of the systems used in earlier approaches or by developing completely new setups or techniques. When compared to conventional systems, miniature systems can perform similar methods with remarkably reduced consumption of plant matrix and solvents, size and power requirement, system cost, faster analysis time, and massively parallel analysis capability. It is an attractive method for application requiring on-field rapid assay [134, 135]. Given its reduced size, this microwave technique requires a small amount of sample in a small glassware system which is designed as a miniature alembic to be placed inside a microwave oven. In this way, the PMAE apparatus can provide information about the actual state of the fresh samples and their essential oil content. This procedure is also appropriate for the teaching laboratory, does not require any special microwave equipment. The experiment developed, in addition to providing fast information on real samples, allows the students to learn extraction, chromatographic, and spectroscopic analysis skills. It is a visual example of rapid, sustainable, and green extraction of natural products. It is suitable of achieving success if commercially introduced. and involves green chemical processing with microwave energy (Figure 1.12).

Extraction of natural products has a very long history and has been associated with commerce since ancient times. Archeological findings indicated that extraction processes were performed in Mesopotamia in 3500 BC [136]. During the same period, trading activities and exchange of goods were intensifying. The first solvents that were used were with water using hydrodistillation in Syria [136] or fermented beverage such as beer and wine in Ancient Egypt [137].

Nowadays, the extraction of natural products takes an important place in manufacturing processes of ingredients derived from natural products. Growing numbers of finished products formula contain botanical ingredients for their functional properties (coloring, antioxidants, flavor, and biological activity). The largest market that uses botanical extract are the pharmaceutical segment and food, followed by personal care. Botanical ingredients are a growing business with very dynamic markets such personal care with approximately 10% of estimated growth (especially cosmetics), foods (driven by the nutraceutical sector with approximately 5% growth in 2012), and the pharmaceutical area (2%).

The growth of the botanical extract industry is linked to several drivers including

- request for naturally derived components by the consumers;
- a need for standardization and stability of the ingredients when compared to raw botanicals.

Figure 1.12 Introduction to green extraction for primary schools at Avignon University.

A current trend in the different markets of food, personal care, and nutrition and health is the replacement of synthetic ingredients by naturally derived ingredients. This trend has been driven by consumer request for more natural products. The industry of natural products has been successful in providing natural solutions as replacement to food dyes, food antioxidants, texturizers, aromas, active compounds, and so on. It is interesting to note that the natural labeling does not provide clear indication or evidence about the environmental impact of the manufacturing process applied to the botanical raw material, nor does it indicate the type of process and level of processing behind a botanical ingredient. Furthermore, it does not allow the consumer to distinguish between hand-made unstandardized products and industrially made standardized products.

Green extraction is a research thematic that encompasses a comprehensive strategy based on the discovery and the design of extraction processes in order to reduce energy consumption and promote the use of alternative solvents and renewable raw material. Green extraction has been investigated mainly at laboratory scale by several research teams mostly in Europe. Green extraction represents a good opportunity to rationalize and formalize eco-friendly developmental and industrial practices across the industries.

Several industrials have shown interest in the principles of green extraction. Some of these principles are already in line with industrial practices and compatible with the primary interests of the botanical extraction industry. Some of these principles areas follows:

- *The use of alternative solvents.* The spectrum of solvents that are currently used in the industry varies according to the nature of the botanical raw material and targeted compounds, availability of the solvent, and its cost. The regulatory framework also defines what solvent can or cannot be used for a specific application. Almost all the solvents currently used are derived from crude oil, with the exception of water and ethanol produced by fermentation. Lipophilic solvents such as hexane are among the primary targets in the quest for alternative solvents. The use of alternative solvents also requires the involvement of solvent manufacturers in term of solvent cost reduction and registration. Some solvents already used do not need replacement solutions. For example, alcohol from fermentation is used as an extraction solvent and it meets the definition of a green solvent. Water is also the obvious choice of green solvent for polar compounds. Water is already largely used for manufacturing processes in many applications in food, personal care, or pharmaceutical ingredients. Although ecofriendly, water extraction processes require optimization on energy uses and recovery of intermediate polarity compounds (flavonoids, phenolic compounds, alkaloids) that are often the targeted compounds. Water is also a solvent that promotes hydrolysis and enzymatic activity during the manufacturing process.
- *Use of renewable raw material for botanical extracts.* Agricultural by-products are already used as starting raw material of botanical extractions (e.g., grape seeds and grape marc and ginseng fibers) by several industries. The use of agricultural by-products provides additional revenue to the primary producers (growers) and globally improves the whole agricultural value chain. The current trends in by-product valorization are to increase the value added by the process derived from the by-product and to move from low-value valorization (organic fertilizers, pellets for feedstock) to high-value specialty ingredients (functional ingredients, health ingredients).
- *Energy consumption by energy recovery and reduction of unit operations.* The natural products industry is a very competitive environment and to survive, the botanical extract manufacturers are already using optimized processes. The R&D, industrial, and cost control departments are constantly looking for more effective and cost-efficient ways to extract botanicals. Most of the industry members are aware that they need to reduce the carbon footprint of their processes. Therefore, the green extraction principles of reduction in energy consumption by energy recovery and reduction of unit operations and favoring safe robust and controlled processes are already existing objectives for industries especially on large volume items and/or commodity products. The arrival of new extraction technologies at production scale (pressurized liquid extraction, microwave-assisted extraction, etc.) will certainly benefit those industries in the quest for more efficient extraction process on a long-term basis.

Even though the concept of green extraction meets the demand of the final consumer in terms of a greener product, education needs to be imparted in order to explain what the benefits are for the final consumer. Consumers are not always familiarized with what is an extract or that solvents are used in manufacturing botanical ingredients. This will require popularization by the scientific community and industry members and avoid the shortcuts of "green washing."

How does green extraction applies to the botanical extraction industry and what remains to be done? We can take the antioxidant rosemary extraction processes as an example. Rosemary is a symbol of the success of a safe and efficient antioxidant botanical solution that has replaced partially synthetic preservatives in various food matrices (especially meat and beverages). Initially, rosemary extracts were declared as flavoring ingredients, in 1997, when Naturex, Raps, and Robertet decided to clarify the situation and invited all the rosemary extract producers to join forces in order to register rosemary as a food antioxidant within the European legislation. In 2010, rosemary extracts were classified as food additives by the European Commission and assigned the number E392 (Commission Directives 2010/67/EU and 2010/69/EU repealed in 2013 by EU regulation 231/2012 and 1333/2008). "Antioxidant: extracts of rosemary" are to be produced with one of the four extraction processes described in the regulation, by means of solvent extraction (ethanol, acetone, or ethanol followed by hexane) or supercritical carbon dioxide extraction, paying respect to purity criteria. The volume of rosemary extracted by Naturex are significant, it currently represents about 50% of the rosemary collected in Morocco.

1.5.1 Principle: Innovation by Selection of Varieties and Use of Renewable Plant Resources

The starting point of the industrial extraction process is the selection of the feedstock biomass that will have the highest concentration in targeted compounds or at the least the more cost-effective per kilogram of pure compounds. Naturex has been using rosemary leaves from Morocco and other locations for 20 years, and has maintained a long history of quantification of carnosic acid/rosmarinic acid content material harvested in different locations and at different time of years. Some specific varieties growing in the Atlas Mountains are especially high in carnosic acid content. Carnosic acid concentrations decreased at high temperatures and low precipitation rates [138] and acid levels decreased in parallel with dropping relative water content in the plant [139]. Rosemary is a perennial plant that supports very well the trimming of the aerial part. This wild harvesting is sustainable and renewable and is contributing significantly to Moroccan economy.

1.5.2 Principle: Use of Alternative Solvents and Agro Solvent

Several extraction processes are derived from rosemary leaves. Polar compounds such as rosmarinic acid are extracted with aqueous solvents and also methanol.

Lipophilic compounds such as diterpenes phenolic acids are extracted with acetone or ethanol solvent. Several versions of lipophilic extracts (containing carnosic acid) are obtained after a multistep process, starting with ethanolic extraction and followed by several steps of purification using various pH-differential precipitations, and solubilization [140]. The organic versions use ethanol derived from fermentation.

1.5.3
Principle: Production of Coproducts Instead of Waste to Include Biorefinery

The rosemary leaves used in most of our rosemary processes are hydrodistillated first and the obtained essential oil is further refined for flagrance and flavor applications. We note that this operation is performed in order to increase value chain but also for technological reasons. Most customers request deodorized and discolored versions of rosemary extracts in order to reduce organoleptic and color impact on their recipes and the current purity criteria of the antioxidant E392 includes a ratio between antioxidant compounds and volatiles compounds. Rosemary leaves contains several others valuable fractions: ursolic acid, rosmarinic acid, carnosic acid and derivatives, flavonoids, and lignocellulosic fibers. More work is needed, because although several fractions are already valorized (carnosic, ursolic, and rosmarinic acids), some fractions are not (fibers, waxes, etc.). Rosemary is a good model for applied research for a complete and comprehensive biorefinery process.

1.5.4
Principle: Prioritizing a Non-denatured and Biodegradable Extract without Contaminant

Once the product has been purified and dried, the native purified extract is analyzed according to a specification sheet that ensures that it meets the acceptance criteria. The conformity to the specifications is also matched with the target process yield (both mass yield and active yield). Naturex has been developing an extensive portfolio of testing methods for both rosmarinic and carnosic acids; these methods have been recognized as market standards and are used worldwide. Internal specification such as the ratio between carnosol and carnosic acid allows to ensure that carnosic acid has not been degraded in carnosol during the process.

1.6
Conclusions and Perspective

Green extraction is a good idea and comprehensive strategy that promotes energy consumption reduction and process efficiency, and minimizes environmental impact of manufacturing processes of botanical extracts. Green extraction will benefit from industrial involvement and exposure to industrial constraints. Industry members will benefit, as well, from green extraction with new technologies and new research tools to develop new products and to optimize existing products. The research work remaining is important and will require synergies and open communications between academic and private sectors in order to transform these principles into an industrial use and bring eco-extracts to a large number of consumers.

References

1. Armenta, S., Garrigues, S., and De La Guardia, M. (2008) Green analytical chemistry. *TrAC, Trends Anal. Chem.*, **27**, 497–511.
2. Chemat, F., Vian, M.A., and Cravotto, G. (2012) Green extraction of natural products: concept and principles. *Int. J. Mol. Sci.*, **13**, 8615–8627.
3. Anastas, P.T. and Warner, J.C. (1998) *Green Chemistry: Theory and Practice*, Oxford University Press, New York, 135 p.
4. Li, X. and Franke, A.A. (2009) Fast HPLC-ECD analysis of ascorbic acid, dehydroascorbic acid and uric acid. *J. Chromatogr. B*, **877**, 853–856.
5. Warner, M.G., Succaw, G.L., and Hutchison, J.E. (2001) Solventless syntheses of mesotetraphenylporphyrin: new experiments for a greener organic chemistry laboratory curriculum. *Green Chem.*, **3**, 267–270.
6. Croteau, R., Kutchan, T., and Lewis, N. (2000) in *Biochemistry and Molecular Biology of Plants* (eds B. Buchanan, W. Gruissen, and R. Jones), American Society of Plant Physiologists, Rockville, MD, pp. 1250–1319.
7. Cavalcanti, R.N., Forster-Carneiro, T., Gomes, M.T.M.S., Rostagnon, M.A., Prado, J.M., and Meireles, M.A.A. (2013) in *Natural Product Extraction: Principles and Applications*, Green Chemistry, Vol. 21 (eds M.A. Rostagno, J.M. Prado, and G.A. Kraus), Royal Society of Chemistry, Dorchester, pp. 1–46.
8. Dewick, P.M. (2002) *Medicinal Natural Products: A Biosynthetic Approach*, 2nd edn, John Wiley & Sons, Ltd, 520 p.
9. FAOSTATS *http://faostat.fao.org/site/567/DesktopDefault.aspx?PageID=567#ancor* (accessed 14 March 2014).
10. Rezzadori, K., Benedetti, S., and Amante, E.R. (2012) Proposals for the residues recovery: orange waste as raw material for new products. *Food Bioprod. Process.*, **90**, 606–614.
11. Kesterson, J.W. and Braddock, R.J. (1976) *By-Products and Specialty Products of Florida Citrus*, Bulletin 784, Agricultural Experiment Stations, University of Florida.
12. United States Agricultural Research Service (1956) *Chemistry and Technology of Citrus, Citrus Products and By-Products*, U.S. Department of Agriculture.
13. Bampidis, V. and Robinson, P. (2006) Citrus by-products as ruminant feeds: a review. *Anim. Feed Sci. Technol.*, **128**, 175–217.
14. Ferhat, M.A., Meklati, B.Y., Smadja, J., and Chemat, F. (2006) An improved microwave Clevenger apparatus for distillation of essential oils from orange peel. *J. Chromatogr. A*, **1112**, 121–126.
15. Minh Tu, N.T., Thanh, L.X., Une, A., Ukeda, H., and Sawamura, M. (2002) Volatile constituents of Vietnamese pummelo, orange, tangerine and lime peel oils. *Flavour Fragance J.*, **17**, 169–174.
16. Chemat, F. and Sawamura, M. (2010) in *Citrus Essential Oils* (ed M. Sawamura), John Wiley & Sons, Inc., Hoboken, NJ, pp. 9–36.
17. Masmoudi, M., Besbes, S., Chaabouni, M., Robert, C., Paquot, M., Blecker, C., and Attia, H. (2008) Optimization of pectin extraction from lemon by-product with acidified date juice

using response surface methodology. *Carbohydr. Polym.*, **74**, 185–192.
18. Bicu, I. and Mustata, F. (2011) Cellulose extraction from orange peel using sulfite digestion reagents. *Bioresour. Technol.*, **102**, 10013–10019.
19. Grohmann, K., Manthey, J.A., Cameron, R.G., and Buslig, B.S. (1999) Purification of Citrus peel juice and molasses. *J. Agric. Food Chem.*, **47**, 4859–4867.
20. Okoye, C., Ibeto, C., and Ihedioha, J. (2011) Preliminary studies on the characterization of orange seed and pawpaw seed oils. *Am. J. Food Technol.*, **6**, 422–426.
21. Aranha, C.P.M. and Jorge, N. (2011) Physico-chemical characterization of seed oils extracted from oranges (*Citrus sinensis*). *Food Sci. Technol. Res.*, **19**, 409–415.
22. Smadja, J. (2009) in *Essential Oils and Aromas Green Extraction and Applications* (ed F. Chemat), Har Krishan Bhalla & Sons, Dehradun, pp. 122–146.
23. Dugo, G., Bonaccorsi, I., Sciarrone, D., Costa, R., Dugo, P., Monde, L., Santi, L., and Fakhry, H.A. (2011) Characterization of oils from the fruits, leaves and flowers of the bitter orange tree. *J. Essent. Oil Res.*, **23**, 45–59.
24. Bonaccorsi, I., Sciarrone, D., Schipilliti, L., Trozzi, A., Fakhry, H.A., and Dugo, G. (2011) Composition of egyptian Neroli oil. *Nat. Prod. Commun.*, **6**, 1009–1014.
25. Demura, T. and Ye, Z.-H. (2010) Regulation of plant biomass production. *Curr. Opin. Plant Biol.*, **13**, 298–303.
26. Kumar, R., Singh, S., and Singh, O.V. (2008) Bioconversion of lignocellulosic biomass: biochemical and molecular perspectives. *J. Ind. Microbiol. Biotechnol.*, **35**, 377–391.
27. Guignard, J.L. (1996) *Biochimie végétale*, Dunod, Paris, 274 p.
28. Bousbia, N., Vian, M., Ferhat, M., Meklati, B., and Chemat, F. (2009) A new process for extraction of essential oil from Citrus peels: microwave hydrodiffusion and gravity. *J. Food Eng.*, **90**, 409–413.
29. Manners, G.D. (2007) Citrus limonoids: analysis, bioactivity, and biomedical prospects. *J. Agric. Food Chem.*, **55**, 8285–8294.
30. Roy, A. and Saraf, S. (2006) Limonoids: overview of significant bioactive triterpenes distributed in plants kingdom. *Biol. Pharm. Bull.*, **29**, 191–201.
31. Fanciullino, A.-L., Dhuique-Mayer, C., Luro, F., Casanova, J., Morillon, R., and Ollitrault, P. (2006) Carotenoid diversity in cultivated citrus is highly influenced by genetic factors. *J. Agric. Food Chem.*, **54**, 4397–4406.
32. Pellati, F. and Benvenuti, S. (2007) Fast high-performance liquid chromatography analysis of phenethylamine alkaloids in Citrus natural products on a pentafluorophenylpropyl stationary phase. *J. Chromatogr. A*, **1165**, 58–66.
33. Putzbach, K., Rimmer, C., Sharpless, K., Wise, S., and Sander, L. (2007) Determination of bitter orange alkaloids in dietary supplement Standard Reference Materials by liquid chromatography with atmospheric-pressure ionization mass spectrometry. *Anal. Bioanal. Chem.*, **389**, 197–205.
34. Pellati, F. and Benvenuti, S. (2007) Chromatographic and electrophoretic methods for the analysis of phenetylamine alkaloids in *Citrus aurantium*. *J. Chromatogr. A*, **1161**, 71–88.
35. Khan, M.K., Zill, E.H., and Dangles, O. (2014) A comprehensive review on flavanones, the major citrus polyphenols. *J. Food Compos. Anal.*, **33**, 85–104.
36. Harper, A.H., Murray, E.G.D., Mayes, P.A., and Rodwell, V.W. (2002) *Biochimie de Harper*, 25th edn, De Boeck, Presses de l'Université de Laval, 933 p..
37. Voet, D., Voet, J.G., Rousseau, G., and Domenjoud, L. (2005) *Biochimie*, De Boeck Supérieur, 1600 p.
38. Bruneton, J. (1987) *Eléments de phytochimie et de pharmacognosie*, Technique et Documentation Lavoisier, 585 p.
39. Hounsome, N., Hounsome, B., Tomos, D., and Edwards-Jones, G. (2008) Plant metabolites and nutritional quality of vegetables. *J. Food Sci.*, **73**, R48–R65.
40. Sugawara, T. and Miyazawa, T. (1999) Separation and determination of glycolipids from edible plant sources by

high-performance liquid chromatography and evaporative light-scattering detection. *Lipids*, **34**, 1231–1237.
41. He, X.-G. (2000) On-line identification of phytochemical constituents in botanical extracts by combined high-performance liquid chromatographic–diode array detection–mass spectrometric techniques. *J. Chromatogr. A*, **880**, 203–232.
42. Dzoyem, J., Kechia, F., and Kuete, V. (2011) Phytotoxic, antifungal activities and acute toxicity studies of the crude extract and compounds from *Diospyros canaliculata*. *Nat. Prod. Res.*, **25**, 741–749.
43. Meyer-Warnod, B. (1984) Natural essential oils: extraction processes and application to some major oils. *Perfum. Flavor.*, **9**, 93–104.
44. Dewick, P.M. (2002) The biosynthesis of C5–C25 terpenoid compounds. *Nat. Prod. Rep.*, **19**, 181–222.
45. Acton, Q.A. (2012) *Diterpenes – Advances in Research and Application: 2012 Edition*, Scholarly Editions, Atlanta, 109 p.
46. Wölwer-Rieck, U. (2012) The leaves of *Stevia rebaudiana* (Bertoni), their constituents and the analyses thereof: a review. *J. Agric. Food Chem.*, **60**, 886–895.
47. Houël, E., Bertani, S., Bourdy, G., Deharo, E., Jullian, V., Valentin, A., Chevalley, S., and Stien, D. (2009) Quassinoid constituents of *Quassia amara* L. leaf herbal tea. Impact on its antimalarial activity and cytotoxicity. *J. Ethnopharmacol.*, **126**, 114–118.
48. Ghanbari, R., Anwar, F., Alkharfy, K.M., Gilani, A.-H., and Saari, N. (2012) Valuable nutrients and functional bioactives in different parts of olive (*Olea europaea* L.) –A review. *Int. J. Mol. Sci.*, **13**, 3291–3340.
49. Cordell, G.A. (2013) Fifty years of alkaloid biosynthesis in Phytochemistry. *Phytochemistry*, **91**, 29–51.
50. Srinivasan, K. (2007) Black Pepper and its pungent principle-piperine: a review of diverse physiological effects. *Crit. Rev. Food Sci. Nutr.*, **47**, 735–748.
51. Bunsupa, S., Yamazaki, M., and Saito, K. (2012) Quinolizidine alkaloid biosynthesis: recent advances and future prospects. *Front. Plant Sci.*, **3**, article 239, 1–7.
52. Michael, J.P. (2008) Indolizidine and quinolizidine alkaloids. *Nat. Prod. Rep.*, **25**, 139–165.
53. Dembitsky, V., Poovarodom, S., Leontowicz, H., Leontowicz, M., Vearasilp, S., Trakhtenberg, S., and Gorinstein, S. (2011) The multiple nutrition properties of some exotic fruits: biological activity and active metabolites. *Food Res. Int.*, **44**, 1671–1701.
54. Servillo, L., Giovane, A., Balestrieri, M.L., Casale, R., Cautela, D., and Castaldo, D. (2013) Citrus genus plants contain N-methylated tryptamine derivatives and their 5-Hydroxylated forms. *J. Agric. Food Chem.*, **61**, 5156–5162.
55. Dhawan, K., Dhawan, S., and Sharma, A. (2004) Passiflora: a review update. *J. Ethnopharmacol.*, **94**, 1–23.
56. Badria, F.A. (2002) Melatonin, serotonin, and tryptamine in some egyptian food and medicinal plants. *J. Med. Food*, **5**, 153–157.
57. Stintzing, F. and Carle, R. (2007) Betalains – emerging prospects for food scientists. *Trends Food Sci. Technol.*, **18**, 514–525.
58. D'Yakonov, A.L. and Telezhenetskaya, M.V. (1997) Quinazoline alkaloids in nature. *Chem. Nat. Compd.*, **33**, 221–267.
59. Jin, Z. (2011) Muscarine, imidazole, oxazole, and thiazole alkaloids. *Nat. Prod. Rep.*, **28**, 1143–1191.
60. Ashihara, H., Sano, H., and Crozier, A. (2008) Caffeine and related purine alkaloids: biosynthesis, catabolism, function and genetic engineering. *Phytochemistry*, **69**, 841–856.
61. Andersson, C. (1999) *Glycoalkaloids in Tomatoes, Eggplants, Pepper and Two Solanum Species Growing Wild in the Nordic Countries*, Nordisk Ministerråd and Nordisk Råd.

62. Scalbert, A. and Williamson, G. (2000) Dietary intake and bioavailability of polyphenols. *J. Nutr.*, **130**, 2073S–2085S.
63. Gorinstein, S., Zemser, M., Haruenkit, R., Chuthakorn, R., Grauer, F., Martin-Belloso, O., and Trakhtenberg, S. (1999) Comparative content of total polyphenols and dietary fiber in tropical fruits and persimmon. *J. Nutr. Biochem.*, **10**, 367–371.
64. Duthie, G., Duthie, S., and Kyle, J. (2000) Plant polyphenols in cancer and heart disease: implications as nutritional antioxidants. *Nutr. Res. Rev.*, **13**, 79–106.
65. Saleh, Z., Stanley, R., and Nigam, M. (2006) Extraction of polyphenolics from apple juice by foam fractionation. *Int. J. Food Eng.*, **2**, article 2, 1–15.
66. Llorach, R., Martínez-Sánchez, A., Tomás-Barberán, F., Gil, M., and Ferreres, F. (2008) Characterisation of polyphenols and antioxidant properties of five lettuce varieties and escarole. *Food Chem.*, **108**, 1028–1038.
67. Soong, Y. and Barlow, P. (2006) Quantification of gallic acid and ellagic acid from longan (*Dimocarpus longan* Lour.) seed and mango (*Mangifera indica* L.) kernel and their effects on antioxidant activity. *Food Chem.*, **97**, 524–530.
68. Szajdek, A. and Borowska, E. (2008) Bioactive compounds and health-promoting properties of berry fruits: a review. *Plant Foods Hum. Nutr.*, **63**, 147–156.
69. Rice-Evans, C., Miller, J., and Paganga, G. (1997) Antioxidant properties of phenolic compounds. *Trends Plant Sci.*, **2**, 152–159.
70. Haminiuk, C., Maciel, G., Plata Oviedo, M., and Peralta, R. (2012) Phenolic compounds in fruits – an overview. *Int. J. Food Sci. Technol.*, **47**, 2023–2044.
71. Crozier, A., Jaganath, I., and Clifford, M. (2009) Dietary phenolics: chemistry, bioavailability and effects on health. *Nat. Prod. Rep.*, **26**, 1001–1043.
72. Santos-Buelga, C. and Scalbert, A. (2000) Proanthocyanidins and tannin-like compounds – nature, occurrence, dietary intake and effects on nutrition and health. *J. Sci. Food Agric.*, **80**, 1094–1117.
73. Kuhnau, J. (1976) The flavonoids. A class of semi-essential food components: their role in human nutrition. *World Rev. Nutr. Diet.*, **24**, 117–191.
74. Cook, N. and Samman, S. (1996) Flavonoids – Chemistry, metabolism, cardioprotective effects, and dietary sources. *J. Nutr. Biochem.*, **7**, 66–76.
75. Ross, J. and Kasum, C. (2002) Dietary flavonoids: bioavailability, metabolic effects, and safety. *Annu. Rev. Nutr.*, **22**, 19–34.
76. Schieber, A., Berardini, N., and Carle, R. (2003) Identification of flavonol and xanthone glycosides from mango (*Mangifera indica* L. Cv. "Tommy Atkins") peels by high-performance liquid chromatography-electrospray ionization mass spectrometry. *J. Agric. Food Chem.*, **51**, 5006–5011.
77. Hoffmann-Ribani, R., Huber, L., and Rodriguez-Amaya, D. (2009) Flavonols in fresh and processed Brazilian fruits. *J. Food Compos. Anal.*, **22**, 263–268.
78. de Almeida, A., Miranda, M., Simoni, I., Wigg, M., Lagrota, M., and Costa, S. (1998) Flavonol monoglycosides isolated from the antiviral fractions of *Persea americana* (Lauraceae) leaf infusion. *Phytother. Res.*, **12**, 562–567.
79. Dixon, R. and Xie, D. (2005) Proanthocyanidins – a final frontier in flavonoid research? *New Phytol.*, **165**, 9–28.
80. Suksamrarn, S., Suwannapoch, N., Ratananukul, P., Aroonlerk, N., and Suksamrarn, A. (2002) Xanthones from the green fruit hulls of *Garcinia mangostana*. *J. Nat. Prod.*, **65**, 761–763.
81. Balunas, M.J., Su, B., Brueggemeier, R.W., and Kinghorn, A.D. (2008) Xanthones from the botanical dietary supplement mangosteen (*Garcinia mangostana*) with aromatase inhibitory activity. *J. Nat. Prod.*, **71**, 1161–1166.
82. Boonlaksiri, C., Oonanant, W., Kongsaeree, P., Kittakoop, P., Tanticharoen, M., and Thebtaranonth, Y. (2000) An antimalarial stilbene from *Artocarpus integer*. *Phytochemistry*, **54**, 415–417.

83. Boerjan, W., Ralph, J., and Baucher, M. (2003) Lignin biosynthesis. *Annu. Rev. Plant Biol.*, **54**, 519–546.
84. Pfaltzgraff, L.A., De Bruyn, M., Cooper, E.C., Budarin, V., and Clark, J.H. (2013) Food waste biomass: a resource for high-value chemicals. *Green Chem.*, **15**, 307–314.
85. Azmir, J., Zaidul, I.S.M., Rahman, M.M., Sharif, K.M., Mohamed, A., Sahena, F., Jahurul, M.H.A., Ghafoor, K., Norulaini, N.A.N., and Omar, A.K.M. (2013) Techniques for extraction of bioactive compounds from plant materials: a review. *J. Food Eng.*, **117**, 426–436.
86. Bernhoft, A. (2010) A brief review on bioactive compounds in plants. Proceedings from a Symposium held at The Norwegian Academy of Science and Letters, Oslo, Norway.
87. Allaf, T., Tomao, V., Besombes, C., and Chemat, F. (2013) Thermal and mechanical intensification of essential oil extraction from orange peel via instant autovaporization. *Chem. Eng. Process.: Process Intensif.*, **72**, 24–30.
88. Saeb, K. and Gholamrezaee, S. (2012) Variation of essential oil composition of *Melissa officinalis* L. leaves during different stages of plant growth. *Asian Pac. J. Trop. Biomed.*, **2**, S547–S549.
89. Boukhatem, M.N., Kameli, A., and Saidi, F. (2013) Essential oil of Algerian rose-scented geranium (*Pelargonium graveolens*): chemical composition and antimicrobial activity against food spoilage pathogens. *Food Control*, **34**, 208–213.
90. Bernotiene, G., Niviskiene, O., Butkiene, R., and Mockute, D. (2004) Chemical composition of essential oils of hops (*Humulus lupulus* L.) growing wild in Aukštaitija. *Chemija*, **15**, 31–36.
91. Charles, D.J., Joly, R.J., and Simon, J.E. (1990) Effects of osmotic stress on the essential oil content and composition of peppermint. *Phytochemistry*, **29**, 2837–2840.
92. Scotter, M. (2009) The chemistry and analysis of annatto food colouring: a review. *Food Addit. Contam., Part A*, **26**, 1123–1145.
93. Giuffrida, D., Dugo, P., Torre, G., Bignardi, C., Cavazza, A., Corradini, C., and Dugo, G. (2013) Characterization of 12 Capsicum varieties by evaluation of their carotenoid profile and pungency determination. *Food Chem.*, **140**, 794–802.
94. Cossignani, L., Urbani, E., Simonetti, M., Maurizi, A., Chiesi, C., and Blasi, F. (2014) Characterisation of secondary metabolites in saffron from central Italy (Casia, Umbria). *Food Chem.*, **143**, 446–451.
95. Ahamad, M.N., Saleemullah, M., Shah, H.U., Khalil, I.A., and Saljoqi, A.U.R. (2007) Determination of beta carotene content in fresh vegetables using high performance liquid chromatography. *Sarhad J. Agric.*, **23**, 767–770.
96. Bridle, P. and Timberlake, C.F. (1997) Anthocyanins as natural food colours – selected aspects. *Food Chem.*, **58**, 103–109.
97. Setchell, K.D. and Cassidy, A. (1999) Dietary isoflavones: biological effects and relevance to human health. *J. Nutr.*, **129**, 758S–767S.
98. Saleem, M., Kim, H.J., Ali, M.S., and Lee, Y.S. (2005) An update on bioactive plant lignans. *Nat. Prod. Rep.*, **22**, 696–716.
99. Riddell, L.J., Sayompark, D., Penny, O., and Keast Russell, S.J. (2012) in *Caffeine: Chemistry, Analysis, Function and Effects* (ed V.R. Preedy), The Royal Society of Chemistry, Cambridge, pp. 22–38.
100. Delgado-Vargas, F., Jiménez, A.R., and Paredes-López, O. (2000) Natural pigments: carotenoids, anthocyanins, and betalains — characteristics, biosynthesis, processing, and stability. *Crit. Rev. Food Sci. Nutr.*, **40**, 173–289.
101. Wilska-Jeszka, J. (2007) in *Chemical and Functional Properties of Food Components* (ed Z.E. Sikorski), CRC Press, Boca Raton, FL, pp. 245–274.
102. Brud, W.S. (2009) in *Handbook of Essential Oils* (eds K.H.C. Baser and G. Buchbauer), CRC Press, Boca Raton, FL, pp. 843–853.
103. Rodrigues, V.M., Rosa, P.T.V., Marques, M.O.M., Petenate, A.J., and Meireles,

M.A.A. (2003) Supercritical extraction of essential oil from aniseed (*Pimpinella anisum* L.) using CO_2: solubility, kinetics, and composition data. *J. Agric. Food Chem.*, **51**, 1518–1523.

104. Li, Y.-Q., Kong, D.-X., and Wu, H. (2013) Analysis and evaluation of essential oil components of cinnamon barks using GC–MS and FTIR spectroscopy. *Ind. Crops Prod.*, **41**, 269–278.

105. Almeida, M.R., Fidelis, C.H.V., Barata, L.E.S., and Poppi, R.J. (2013) Classification of Amazonian rosewood essential oil by Raman spectroscopy and PLS-DA with reliability estimation. *Talanta*, **117**, 305–311.

106. Roy, B.C., Goto, M., and Hirose, T. (1996) Extraction of ginger oil with supercritical carbon dioxide: experiments and modeling. *Ind. Eng. Chem. Res.*, **35**, 607–612.

107. Allaf, T., Tomao, V., Ruiz, K., Bachari, K., ElMaataoui, M., and Chemat, F. (2013) Deodorization by instant controlled pressure drop autovaporization of rosemary leaves prior to solvent extraction of antioxidants. *LWT-Food Sci. Technol.*, **51**, 111–119.

108. Chemat, F., Lucchesi, M.E., Smadja, J., Favretto, L., Colnaghi, G., and Visinoni, F. (2006) Microwave accelerated steam distillation of essential oil from lavender: a rapid, clean and environmentally friendly approach. *Anal. Chim. Acta*, **555**, 157–160.

109. Safaralie, A., Fatemi, S., and Salimi, A. (2010) Experimental design on supercritical extraction of essential oil from valerian roots and study of optimal conditions. *Food Bioprod. Process.*, **88**, 312–318.

110. Charles, S. (2009) in *Handbook of Essential Oils* (eds K.H.C. Baser and G. Buchbauer), CRC Press, Boca Raton, FL, pp. 121–150.

111. Hasler, C.M., Bloch, A.S., Thomson, C.A., Enrione, E., and Manning, C. (2004) ADA reports, position of the American Dietetic Association: functional foods. *J. Am. Diet. Assoc.*, **104**, 814–826.

112. Herrero, M., Cifuentes, A., and Ibañez, E. (2006) Sub- and supercritical fluid extraction of functional ingredients from different natural sources: plants, food-by-products, algae and microalgae: a review. *Food Chem.*, **98**, 136–148.

113. Kolanowski, W. and Laufenberg, G. (2006) Enrichment of food products with polyunsaturated fatty acids by fish oil addition. *Eur. Food Res. Technol.*, **222**, 472–477.

114. Jones, P.J. and Abumweis, S.S. (2009) Phytosterols as functional food ingredients: linkages to cardiovascular disease and cancer. *Curr. Opin. Clin. Nutr. Metab. Care*, **12**, 147–151.

115. Santhosha, S.G., Jamuna, P., and Prabhavathi, S.N. (2013) Bioactive components of garlic and their physiological role in health maintenance: a review. *Food Biosci.*, **3**, 59–74.

116. Anna-Maija, L., Vieno, P., and Afaf, K.-E. (2002) in *Functional Foods* (eds G. Mazza, M. Le Maguer, and J. Shi), CRC Press, Boca Raton, FL, pp. 1–38.

117. Shahidi, F. (2005) Nutraceuticals and functional foods in health promotion and disease prevention, in *Proceedings of WOCMAP III*, Vol. 6: Traditional Medicine and Nutraceuticals, Acta Horticulturae 680, International Society for Horticultural Science, Belgium.

118. Stintzing, F.C. and Carle, R. (2004) Functional properties of anthocyanins and betalains in plants, food, and in human nutrition. *Trends Food Sci. Technol.*, **15**, 19–38.

119. Dillard, C. (2000) Phytochemicals: nutraceuticals and human health. *J. Sci. Food Agric.*, **80**, 1744–1756.

120. Xenakis, A., Papadimitriou, V., and Sotiroudis, T. (2010) Colloidal structures in natural oils. *Curr. Opin. Colloid Interface Sci.*, **15**, 55–60.

121. Ollé, M. (2002) Analyse des corps gras, Techniques de l'ingénieur, dossier P3325, p. 15.

122. Gui, M.M., Lee, K.T., and Bhatia, S. (2008) Feasibility of edible oil vs. non-edible oil vs. waste edible oil as biodiesel feedstock. *Energy*, **33**, 1646–1653.

123. Gunstone, F.D. (2002) *Vegetable Oils in Food Technology. Composition, Properties and Uses*, Blackwell Publishing Ltd, Oxford, p. 352.

124. O'Brien, R.D. (2003) *Fats and Oils: Formulating and Processing for Applications*, 2nd edn, CRC Press, Boca Raton, FL, pp. 234–291.

125. Peñas, F.J., Barona, A., Elias, A., and Olazar, M. (2006) Implementation of industrial health and safety in chemical engineering teaching laboratories. *J. Chem. Health Saf.*, **13**, 19–23.

126. Armenta, S., Garrigues, S., and De La Guardia, M. (2008) On-line vapour-phase generation combined with Fourier transform infrared spectrometry. *TrAC, Trends Anal. Chem.*, **27**, 15–23.

127. Araus, K., Uquiche, E., and Del Valle, J.M. (2009) Matrix effects in supercritical CO_2 extraction of essentials oils from plant material. *J. Food Eng.*, **92**, 438–447.

128. Wiesbrock, F. and Schubert, U.S. (2006) Microwaves in chemistry: the success story goes on. *Chim. Oggi – Chem. Today*, **24**, 30–34.

129. Chemat, F., Perino-Issartier, S., Petitcolas, E., and Fernandez, X. (2012) "in situ" extraction of essential oils by use of Dean-Stark glassware and a Vigreux column inside a microwave oven: a procedure for teaching green analytical chemistry. *Anal. Bioanal. Chem.*, **404**, 679–682.

130. Perino, S., Petitcolas, E., De La Guardia, M., and Chemat, F. (2013) Portable microwave assisted extraction. An original concept for green analytical chemistry. *J. Chromatogr. A*, **1315**, 200–203.

131. McMahon, G. (2007) *Analytical Instrumentation: A Guide to Laboratory, Portable and Miniaturized Instruments*, John Wiley & Sons, Ltd, Chichester, 318 p.

132. Contreras, J.A., Murray, J.A., Tolley, S.E., Oliphant, J.L., Tolley, H.D., Lammert, S.A., Lee, E.D., Later, D.W., and Lee, M.L. (2008) Hand-portable Gas Chromatograph-Toroidal Ion Trap Mass Spectrometer (GC-TMS) for detection of hazardous compounds. *J. Am. Soc. Mass Spectrom.*, **19**, 1425–1434.

133. Kim, S.J., Gobi, K.V., Harada, R., Shankaran, D.R., and Miura, N. (2006) Miniaturized portable surface plasmon resonance immunosensor applicable for on-site detection of low-molecular-weight analytes. *Sens. Actuators B*, **115**, 349–356.

134. Ryvolová, M., Macka, M., and Preisler, J. (2010) Portable capillary-based (non-chip) capillary electrophoresis. *TrAC, Trends Anal. Chem.*, **29**, 339–353.

135. Sui, X., Liu, T., Ma, C., Yang, L., Zu, Y., Zhang, L., and Wang, H. (2012) Microwave irradiation to pretreat rosemary (*Rosmarinus officinalis* L.) for maintaining antioxidant content during storage and to extract essential oil simultaneously. *Food Chem.*, **131**, 1399–1405.

136. Levey, M. (1955) Evidences of ancient distillation, sublimation and extraction in Mesopotamia. *Centaurus*, **4**, 23–33.

137. Bart, H.-J. (2011) Extraction of natural products from plants – An introduction, in *Industrial Scale Natural Products Extraction* (eds H.-J. Bart and S. Pilz), Wiley-VCH Verlag GmbH, Weinheim.

138. Luis, J.C. and Johnson, C.B. (2005) Seasonal variations of rosmarinic and carnosic acids in rosemary extracts. Analysis of their in vitro antiradical activity. *Span. J. Agric. Res.*, **3**, 106–112.

139. Munné-Bosch, S. and Alegre, L. (2000) Changes in carotenoids, tocopherols and diterpenes during drought and recovery, and the biological significance of chlorophyll loss in Rosmarinus officinalis plants. *Planta*, **210**, 925–931.

140. Naturex (1999) High purity carnosic acid from rosemary and sage extracts by pH-controlled precipitation. US Patent US 5859293 A.

2
Process Engineering and Product Design for Green Extraction

Simon Both, Reinhard Ditz, Martin Tegtmeier, Urban Jenelten, and Jochen Strube

2.1
Market and Market Development

The history of treatment of diseases with drugs shows that in nearly all cultures plant extracts provide the basis of medicinal therapy. In spite of enormous progress in the development of chemical-synthetic active pharmaceutical ingredients (APIs), herbal drugs are of high importance even today. In the past years, the demand for plant-based dietary supplements, flavors, agrochemicals, and cosmetics, which are always based on plant extracts, has grown considerably [1]. This is mainly due to the strong "back to nature" trend in industrialized countries as well as the linked understanding that plant-based products are more healthy, natural, and environmentally friendly [2, 3]. A corresponding orientation can be observed in companies operating successfully in this market. Focusing on the German market is mostly affordable only for small-scale companies, whereas larger enterprises at least require a European, but usually a worldwide market access. Because of this, pricing limits result for prevailing non-European markets, restricting the development of cost-intensive new know-how and related technologies.

In phytotherapy as well as in other plant-based product branches not only the complex mixtures but also purified extracts are gaining increasing importance. Many applied extraction processes have been developed empirically and are thus not optimally designed concerning operating conditions and choice of apparatus. Hence, there is a large optimization potential, for which proven and tested methods of design of experiments (DoEs) and physicochemical modeling should be applicable. Owing to the complexity of herbal feedstocks and their extracts, this transfer has not yet been completed. In the present chapter, the current issues in plant extraction are discussed and solutions are also presented, concerning the regulatory framework for, for example, phytopharmaceuticals.

Green Extraction of Natural Products: Theory and Practice, First Edition.
Edited by Farid Chemat and Jochen Strube.

2.2 Regulatory Framework

The main body of legislation for production of plant extracts is directed to the application of the extract and its secondary products. Thereby, a distinction between pharmaceuticals, food products, and other applications is made.

The German Pharmaceuticals Act and the Pharmaceuticals and Active Agent Manufacturing Ordinance (Arzneimittel- und Wirkstoffherstellungsverordnung, AMWHV) based thereon provide the regulatory frame for the former section. European standards such as the EU-GMP-guidelines (European Union) and the European Pharmacopoeia further amend the rules and standards and substantiate the principal remarks concerning the Good Agricultural and Collection Practice (GACP) for allocation of plant material and the good manufacturing practice (GMP) for the production processes. Before a drug is allowed to be marketed, an approval by the appropriate (national) authority is required. In Germany, this is the Bundesinstitut für Arzneimittel und Medizinprodukte (BfArM), for central European approval the European Medicines Agency (EMA), and in the United States the Food and Drug Administration (FDA). As a prerequisite, effectivity, harmlessness, and quality of the drug have to be proven, which is done in corresponding surveys and studies. Documentation is carried out in the Common Technical Document (CTD), which has been adopted by the International Conference on Harmonisation (ICH). In chapter 3.2.S "Drug Substance" of the CTD, all required information concerning the quality of the extract such as production and evaluation methods, stability as well as quality and origin of the plant material is fixed. If an extract is directly applied as active pharmaceutical ingredient of a phytopharmaceutical, a plant-specific documentation is prepared for the approval documents of the extract. If in contrast, the extract is only an intermediate product for the allocation of an isolated natural substance, which is used as an API or as educt for a partial synthesis, the required information is documented in a drug master file (DMF).

For food products, there is also a general German law named German Food and Feed Code (LFGB). Hazard analysis and critical control point (HACCP) is an important instrument for guarantee of quality and defense against risks. As contrary to pharmaceuticals, approval procedures are missing, the regulatory supervision has a much higher responsibility. There is also an increasing importance of guidelines by the European Union and the function of the European Food Safety Authority (EFSA) as appropriate central authority. In the United States, the FDA has comparable tasks. The described institutions and their sets of regulations are amended regarding quality by the food enterprises with enterprise-specific guidelines and policies, for example, concerning risks of allergies, stability, and transportation.

If the application of a plant extract and its products is not covered by pharmaceutical or food law, the regulatory frame is set by the laws and regulations for chemicals. National law is based on the Chemicals Act (ChemG) and the Ordinance on Hazardous Substances (GefStoffVO), which are in the European Union completed by the Guideline Registration, Evaluation, Authorisation and Restriction of Chemicals (REACH). A registration of plant extracts for the production of pharmaceuticals or food products is not required, because the already named rules and standards apply. Plant extracts used for cosmetics or medicinal products also do not need to be registered if the extraction is carried out with aqueous extraction agent, as steam distillation or through separation of water via heating (dry extraction). European Chemicals Agency (ECHA) is for this product sector a central European authority. Again, enterprise-specific guidelines regarding further processing and secondary products are of great importance for the production of plant extracts.

2.3 Systematic Apparatus and Process Design

A data-based and systematic approach for economically optimal design of solid–liquid extraction processes is indispensable. In principle, there are three methods that can be distinguished: DoE, graphical design, and physicochemical modeling.

Concerning process technological design of the extraction of ingredients from plant material, German groups primarily have published data on salvia, sugar beets, vanilla beans, and pepper. Extraction of salvia is discussed experimentally with aqueous two-phase systems compared to ethanol and ethanol/water mixtures [4, 5]. Furthermore, there are existing publications on extraction with other organic solvents and supercritical carbon dioxide [6]. Extraction of saccharose from sugar beets with water was investigated by means of DoE and physicochemical modeling, whereat operating parameters as well as apparative modifications were optimized [7]. Analyses for extraction of vanilla and pepper via physicochemical modeling and a combination of modeling and DoE have been published. The experiments mostly serve for determination of physicochemical model parameters and for model validation [8–10]. Further publications subsist with respect to supercritical extraction [11, 12] and the subsequent purification of the extracts [7, 13].

Besides the German research activities, there are further European and worldwide approaches for process technological design and optimization of phytoextraction. Mostly, process-intensifying methods such as ultrasound, microwaves or subcritical solvents, and the approach of “green extraction” are discussed [14, 15].

The focus of this publication is on the systematic design and optimization of solvent extraction.

2.3.1
Design of Experiments

DoE as an experimentally based design method is the most frequently applied method in literature. The dependence of parameters of potential influence such as

- Temperature
- Particle size
- Solvent flow rate and flow direction
- Choice of solvent
- Extraction time.

on target functions such as

- Yield
- Purity
- Space – Time – Yield
- Extraction kinetics and by this extraction time
- Apparative efforts
- Concentration of extract and dilution.

are to be determined specifically for each system. Moreover, the interactions between the single parameters can be identified by an appropriate experimental design. Thus, the optimal parameter set can be found for the process. The determined data are specific for the system and not transferable to further systems. Hence, for each raw material, the same experimental effort is necessary. Process design is therefore cost intensive and time intensive.

Opportunities reducing experimental effort and creating data, which are not specific to the system, are the graphical method and physicochemical modeling. Both approaches enable the effective and predictable design of extraction processes.

2.3.2
Graphical Calculation Methods

The calculation of extraction yield via equilibrium composition is shown in Figure 2.1. For design of the extraction both limiting cases are defined, between which the real case should be

1) A constant carrier phase fraction
2) A constant ratio between carrier phase and liquid phase.

A constant carrier phase fraction results if the sum of solvent fraction and target component fraction is constant. This means that the extracted mass of target component is replaced by the same amount of solvent (limiting case 1). If this is

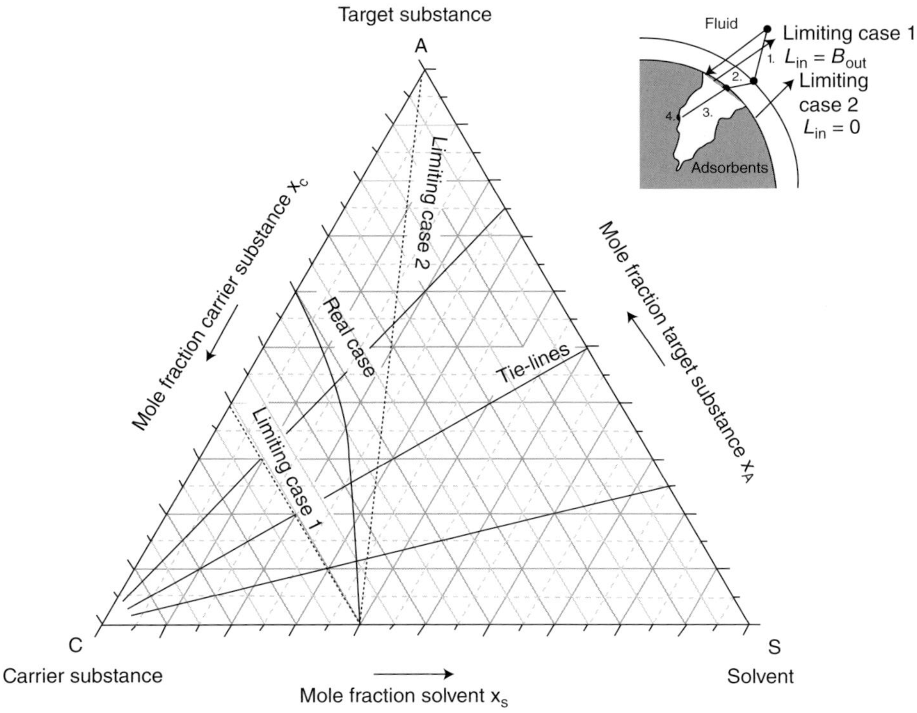

Figure 2.1 Extraction equilibria according to Mersmann *et al.* [16].

not exchanged, only the extraction of the target component from the raw material occurs, so the mass of liquid phase and of carrier matrix in the raw material and their proportion stay constant (limiting case 2) [16, 17].

In the current application, the equilibrium always exists between the extract phase, consisting of target component and solvent, and the solid phase containing the carrier matrix, the target component, and solvent. From this, it becomes obvious that by a change of the remaining solvent fraction, as for instance induced through pressing, the ratios of the three components and thus the equilibrium lines are shifted – which generally contradicts the definition of the equilibrium line via chemical potential.

An equilibrium line independent from the process can be obtained by closing mass balances and is a prerequisite for physicochemical modeling [8, 10]. In the next chapter, the determination of the equilibrium line independent from the process, as well as the determination of model parameters for description of fluid dynamics and mass transfer kinetics, is discussed.

2.3.3
Physicochemical Modeling

The model assumptions most present in literature are "shrinking core," "broken and intact cells," and "desorption." A summary of these models including the respective assumptions as well as examined systems has already been given extensively [18].

For a model-based design of solvent extraction, "desorption" is the most frequently applied approach. "Broken and intact cells" and "shrinking core" models are primarily used for supercritical carbon dioxide extraction. The foundation of the "desorption" approach is the "distributed plug flow" model. Hereby, for description of the extraction process, equilibrium for the target components between the raw material matrix and the liquid phase, extraction kinetics, and fluid dynamics are considered. Equation 2.1 shows the balance for the liquid phase, whereby mass transfer kinetics is included via the mass transfer coefficient k_f in the film.

$$\frac{\partial c_L(z,t)}{\partial t} = D_{ax} \cdot \frac{\partial^2 c_L(z,t)}{\partial z^2} - \frac{u_Z}{\varepsilon} \cdot \frac{\partial c_L(z,t)}{\partial z} - \frac{1-\varepsilon}{\varepsilon} \cdot k_f \cdot a_p \cdot (c_L(z,t) - c_P(z, r = r_P, t)) \tag{2.1}$$

Fluid dynamics is described as the sum of convection, defined via the internal flow velocity u_z, and dispersion, which is modeled with the axial dispersion coefficient D_{ax}.

The balance of the raw material phase is given in Equation 2.2 assuming a flow through packed bed. For a countercurrent process, along with mass transfer kinetics, the terms for convection and dispersion need to be included analogous to the liquid phase balance [7]. Pore diffusion is defined by the effective pore diffusion coefficient D_e.

$$\frac{\partial q(z,r,t)}{\partial t} = D_e \cdot \frac{1}{r^2} \cdot \frac{\partial}{\partial r}\left(r^2 \cdot \frac{\partial c_P(z,r,t)}{\partial r}\right) \tag{2.2}$$

The total loading q of a particle is composed of the solid-phase concentration bound onto the matrix c_S and liquid-phase concentration in the matrix pores c_P (see Equation 2.3). Describing the packed bed, there is a distinction between internal particle porosity and external porosity. Particle porosity ε_P gives the pore volume in relation to the total particle volume. External porosity is defined as voidage between the particles, being calculated as the quotient from the void volume related to the total volume of the apparatus.

$$q(z,r,t) = \varepsilon_P \cdot c_P(z,r,t) + (1-\varepsilon_P) \cdot c_S(z,r,t) \tag{2.3}$$

The effective pore diffusion coefficient in the matrix D_e as given in Equation 2.2 can be estimated from the binary diffusion coefficient D_{12}, particle porosity ε_P, constrictivity δ, and tortuosity τ according to Equation 2.4 [19, 20].

$$D_e = \frac{D_{12} \cdot \varepsilon_P \cdot \delta}{\tau} \tag{2.4}$$

An approximation to constrictivity of the components in the solid is given in Equation 2.5, wherein λ is the dimensionless pore diameter. This is defined as the relation between the molecules and the pore diameter and thus dependent on the diffusing component [10].

$$q(z,r,t) = \varepsilon_P \cdot c_P(z,r,t) + (1-\varepsilon_P) \cdot c_S(z,r,t) \tag{2.5}$$

For a description of the equilibrium line, Equation 2.6 is given as an example for the Langmuir approach. Equilibrium lines may also be of linear or anti-Langmuir type, depending on the system.

$$c_S = c_{max} \cdot k_H \cdot \frac{c_P}{1 + c_P \cdot k_H} \tag{2.6}$$

The introduced model parameters

1) Equilibrium
2) Mass transfer kinetics
3) Fluid dynamics.

are determined corresponding to literature in multistage maceration and percolation experiments [7, 10]. For predictive process design, model parameters need to be determined with a relative mass balance error not exceeding ±5% as shown by a sensitivity analysis [10]. Therefore, mass balances for

- Target and side components
- Carrier matrix
- Solvent
- Water brought into the process via solvent and raw material.

ought to be closed. It is only then that the error in model parameter determination and thus in subsequent process design can be minimized. The experimental setup for determination of model parameters has been described in the literature extensively [18]. Figure 2.2 shows the experimental setup. For closure of mass balance, the required analytics comprises along with HPLC and GC for determination of target and side components also dry weighing, toluene distillation,

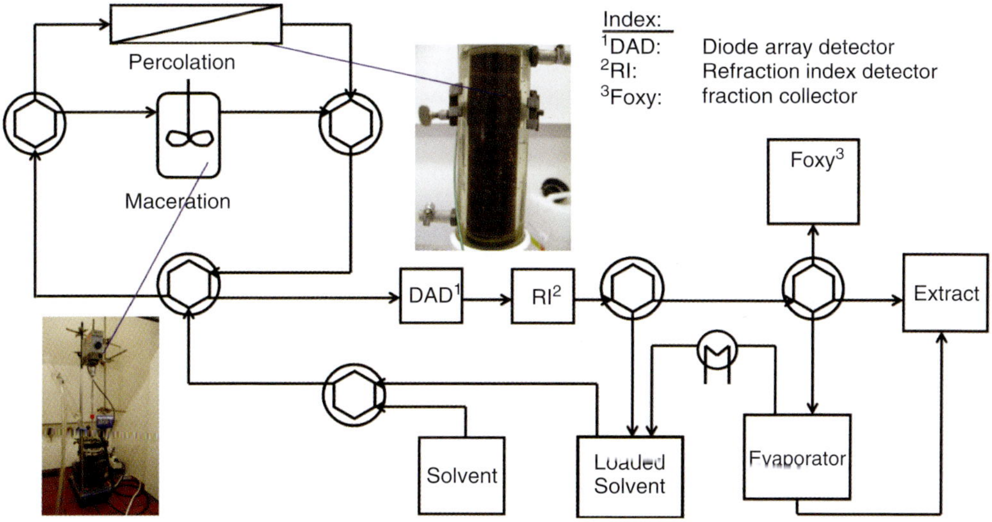

Figure 2.2 Standing apparatus for measurement of model parameters equilibrium, fluid dynamics, and mass transfer according to [18].

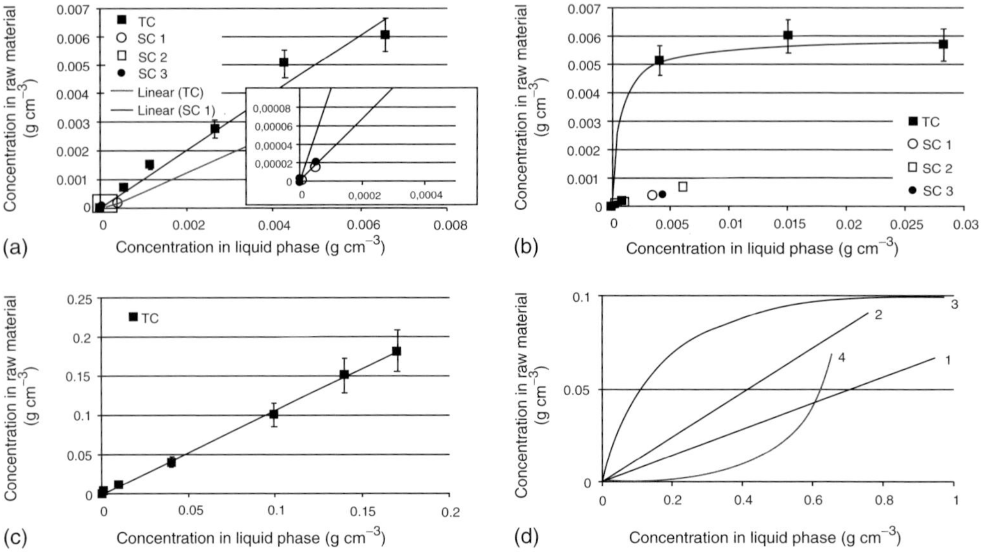

Figure 2.3 Equilibrium lines (a) vanilla beans [8, 10], (b) peppercorns [10], (c) sugar beets [7], and (d) schematic course of equilibrium lines – 1 and 2: linear, 3: Langmuir, 4: "anti-Langmuir" – TC: target component, SC: side component [21].

and Karl–Fischer titration for determination of water, solvent, and solid matrix, respectively.

The equilibrium for component i between the carrier matrix and the liquid phase is determined via multistage extraction. Thereby, it is assumed that the loading of the liquid phase in the pores and in the bulk are identical. Figure 2.3 shows equilibrium lines for saccharose from sugar beets [7] as well as target and side components from vanilla beans [8–10] and peppercorns [9, 10]. Hereby, different types of equilibrium lines become apparent. Vanilla beans and sugar beets exhibit a linear relationship between the equilibrium concentrations in liquid and solid phase. The equilibrium line for peppercorns follows the Langmuir course. In principle, an anti-Langmuir course is also possible, but this points at insufficient solvent capacity. The maximum concentration of the regarded component in the extraction agent is limited by capacity.

Further parameters to be determined are mass transfer kinetics quantified by film and pore diffusion coefficients and fluid dynamics or rather the axial dispersion coefficient. Therefore, methods for calculation and for experimental determination may be applied. Calculation methods provide an orienting value in early process development stages. Diffusion coefficients can be calculated according to [22] via the Schmidt and Sherwood dimensionless numbers in Equations 2.7 and 2.8.

$$\mathrm{Sc} = \frac{\eta_{\mathrm{L}}}{\rho_{\mathrm{L}} \cdot D_{12}} \tag{2.7}$$

$$\mathrm{Sh} = \frac{k_\mathrm{f} \cdot d_\mathrm{P}}{D_{12}} \tag{2.8}$$

Axial dispersion in a flow through packed bed can be estimated from the Péclet number in Equation 2.9 [22].

$$\mathrm{Pe} = \frac{d_\mathrm{P} \cdot u_\mathrm{z}}{\varepsilon \cdot D_\mathrm{ax}} \tag{2.9}$$

The dimensionless numbers are interrelated by empirical correlations. Equation 2.10 shows the relation between the Péclet number, which is defined as the relation between convective to diffusive mass transfer, and the Reynolds number (inertial force to viscous force, Equation 2.11). Equation 2.12 describes the relation at mass transfer between the Schmidt, Sherwood, and Reynolds dimensionless numbers [19].

$$\mathrm{Pe} = \frac{0.2}{\varepsilon} + \frac{0.11}{\varepsilon} \cdot (\varepsilon \cdot \mathrm{Re})^{0.48} \tag{2.10}$$

$$\mathrm{Re} = \frac{u_\mathrm{z} \cdot d_\mathrm{P} \cdot \rho_\mathrm{L}}{\varepsilon \cdot \eta_\mathrm{L}} \tag{2.11}$$

$$\mathrm{Sh} = 2 + 1.1 \cdot \mathrm{Sc}^{0.33} \cdot \mathrm{Re}^{0.6} \tag{2.12}$$

With increasing knowledge about the system, the calculated values may be replaced by real measured values and thus, prediction accuracy of the model rises. In contrast to DoE, there is the possibility to obtain values for the early process development stages on the basis of similarly behaving systems. Diffusion coefficients, for example, show similar values for comparable raw material matrix, target component properties, and solvents. A first characterization of systems related to this is given in Figure 2.4 according to [9]. Here, primarily the three factors, localization of ingredients, matrix properties, and humidity of the raw material are defined as raw material characteristics. Equilibrium between the phases and extraction kinetics is dependent on the system and fluid dynamics on the apparatus.

The current state of research is the quantification of these parameters. By this diffusion paths and barriers shall be identified. First approaches including botanic properties into process design are available [8–10, 23, 24], but until now, these provide mostly qualitative information.

2.3.4 Approaches for Description of Diffusion

According to Fick's first law, particle flow density J is proportional to the concentration gradient (Equation 2.13).

$$J = -D\frac{\partial c}{\partial x} \tag{2.13}$$

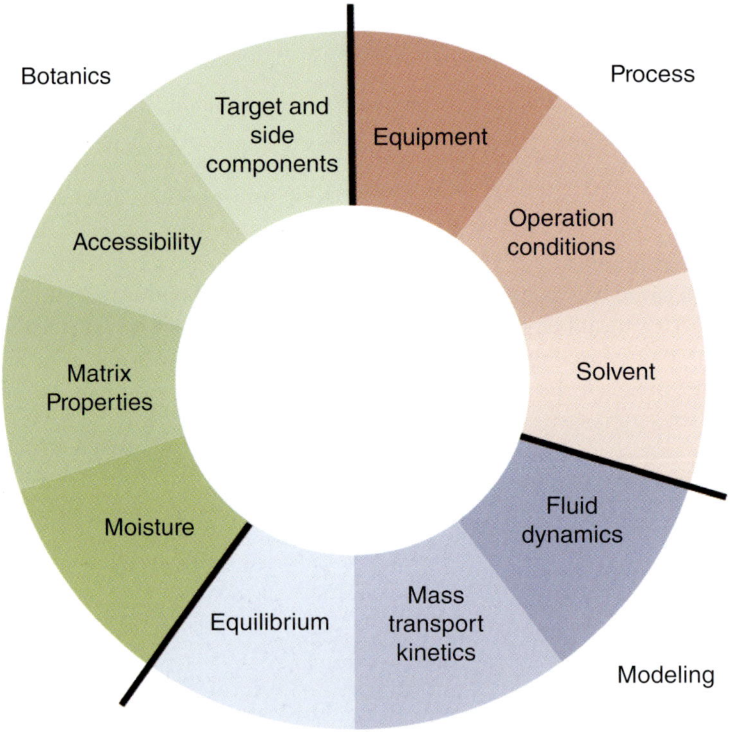

Figure 2.4 Classification of systems according to their characteristics [8].

Therein, Fick's diffusion coefficient D is the proportionality factor. Particle flow density thus makes a statement about how many particles diffuse per time through an area orthogonal to their moving direction. Fick's second law is defined as given in Equation 2.14, whereby both local and temporal concentration differences can be represented. In contrast to Fick's first law, which only regards temporal constant particle flow densities, Fick's second law is applicable for description of transient diffusion processes.

$$\frac{\partial c}{\partial t} = D\frac{\partial^2 c}{\partial x^2} \tag{2.14}$$

For porous spherical particles, an extended version of Fick's second law as given in Equation 2.15 may be applied.

$$\frac{\partial c}{\partial t} = D\left(\frac{\partial^2 c}{\partial x^2} + \frac{2}{x}\cdot\frac{\partial c}{\partial x}\right) \tag{2.15}$$

2.3.4.1 Maxwell-Stefan Approach

In general, Maxwell–Stefan diffusion considers the equilibrium between driving forces and drag forces in a system. Drag forces occur between the components of a system. Driving forces are, for example, gradients in potential, pressure, or concentration. The model is based on energy conservation during collision of molecules in the considered volume element. The amount of collisions correlates to the number of molecules. The general Maxwell–Stefan equation (2.16) is valid for calculation of the total force acting on molecules of type 1 in a binary system per volume element in a certain direction [25, 26]. This also applies to molecules of type 2 analogously.

$$d_1 = \frac{1}{P}\nabla p_1 = -\frac{x_1 \cdot x_2(u_1 - u_2)}{D_{12}} \tag{2.16}$$

Thereby, D_{12} is defined as the quotient of driving and drag forces or as the drag coefficient F_{12}. Mole fractions are given as x_i and velocities of molecules as u_i. For the isothermal and isobaric cases, Equation 2.16 can be transferred to the simplified version in Equation 2.17.

$$d_1 = \nabla x_1 = -\frac{x_1 \cdot x_2(u_1 - u_2)}{D_{12}} \tag{2.17}$$

For a multicomponent system, the model may be extended so that Equation 2.18 shows the general relation and the final form of the Maxwell–Stefan equation. Moreover, in Equation 2.18, the dependence of the acting total force on velocity and concentration of the respective molecules is replaced by a dependence on molar fluxes N.

$$d_i = \nabla x_1 = -\sum_{j=1}^{n} \frac{x_i \cdot N_j - x_j \cdot N_i}{c_t \cdot D_{ij}} \tag{2.18}$$

A transfer of Maxwell–Stefan equations from gaseous to liquid systems is possible, because the aggregation state is accounted for in the calculation of the Maxwell–Stefan diffusion coefficient and driving force according to Equation 2.19.

$$d_i = \frac{x_i}{R \cdot T}\nabla_{T,P}\mu \tag{2.19}$$

At diffusion of substances in porous systems, interaction effects between the molecules as well as between the molecules and the porous matrix occur, both hindering mass transfer significantly. Krishna and Wesselingh [26] showed that, for the example of diffusion in a catalyst pellet, Fick's law cannot describe the interactions and thus often leads to erroneous results.

The so-called "dusty gas model" is currently the most applied model for the description of diffusion processes in porous materials [27, 28]. For mass transfer in porous media, the following mechanisms are possible, depending on the relation of mean free path to pore diameter:

- Knudsen diffusion J_i^K
- Molecular diffusion J_i^M
- Convective flux J_i^V
- Surface diffusion J_i^S.

Knudsen diffusion occurs if the mean free path is larger than the pore diameter. Hence, there are more frequent collisions of a molecule with the wall than collisions of two molecules. In convective flux, the mixture is considered a continuum

Table 2.1 Characterization of porous matter according to [8, 29].

Pore size	Description	Mass transfer
<2 nm	Micropore	Activated transport
2–50 nm	Meso pore	Knudsen diffusion, capillary transport
>50 nm	Macropore	Molecular diffusion

moved by the pressure gradient. Thereby, collisions of two molecules dominate compared to collisions of molecules with a wall. In literature, the pores are classified into micro-, meso-, and macropores as shown in Table 2.1, for examination as to which of the diffusion mechanisms occurs [29].

Above a pore diameter of 50 nm, Knudsen diffusion can be neglected. Along with the named characterization, the influence of Knudsen diffusion may be evaluated by the dimensionless Knudsen number (Equation 2.20). λ is thereby the mean free path of the respective molecule and d_{P} the diameter of the pore. If Kn is substantially larger than 1 and thus the mean free path larger than the pore diameter, Knudsen diffusion occurs and has to be considered in the description of the diffusion.

$$\mathrm{Kn} = \frac{\lambda}{d_{\mathrm{P}}} \tag{2.20}$$

Molecular diffusion occurs if the molecules move because of to concentration gradients or other external forces. In this case, collisions between molecules dominate compared to collisions of molecules with a wall.

Surface diffusion differs basically from the other three mechanisms. Molecules, which have diffused and adsorbed to a surface, move laterally on the surface. The Maxwell–Stefan equation for molecular diffusion is extended by Knudsen diffusion to Equation 2.21.

$$-\sum_{j-1}^{n-1} \left(\Gamma_{ij} \cdot \nabla x_j\right) = -\sum_{j=1}^{n} \frac{x_j \cdot N_i - x_i \cdot N_j}{c_t \cdot D_{ij}^{\mathrm{eff}}} + \frac{N_i}{D_{i\mathrm{K}}^{\mathrm{eff}}} \tag{2.21}$$

D_{ij}^{eff} is thereby the effective Maxwell–Stefan diffusion coefficient accounting for porosity and tortuosity of the system and $D_{i\mathrm{K}}^{\mathrm{eff}}$ is the effective Knudsen diffusion coefficient, which incorporates the system properties described above. Γ is the thermodynamic factor describing the nonidealities of the system.

2.3.4.2 Calculation of Diffusion Coefficients

A method for calculation of the Maxwell–Stefan diffusion coefficients for a binary mixture is the so-called Vignes method [30–32]. Equation 2.22 shows the basis for two-component systems and Equation 2.23 for multicomponent systems. The Knudsen diffusion coefficient may be calculated from Equation 2.24 [25].

$$D_{\mathrm{MS}} = \left(\overset{\circ}{D}_{ij}\right)^{x_j} \cdot \left(\overset{\circ}{D}_{ji}\right)^{x_i} \tag{2.22}$$

$$D_{\mathrm{MS}} = \left(\overset{\circ}{D}_{ij}\right)^{(1+x_j-x_i)/2} \cdot \left(\overset{\circ}{D}_{ji}\right)^{(1+x_i-x_j)/2} \tag{2.23}$$

$$D_{i\mathrm{K}} = \frac{2}{3} r \sqrt{\frac{8\,R \cdot T}{\pi \cdot M}} \tag{2.24}$$

The effective diffusion coefficient accounts for porosity and tortuosity of the porous solid (Equation 2.25), with ε as porosity and τ as tortuosity [33, 34].

$$D_{ij}^{\mathrm{eff}} = \frac{\varepsilon}{\tau} \cdot D_{ij} \tag{2.25}$$

2.3.4.3 **Thermodynamic Factor**

The thermodynamic factor describes the nonideality of a system. This is, as given in Equation 2.26, defined via the Kronecker factor δ_{ij}, mole fractions, and activity coefficients.

$$\Gamma_{ij} = \delta_{ij} + x_i \cdot \left(\frac{\partial \ln\left(\gamma_{ij}\right)}{\partial x_j}\right)_{\mathrm{T,p,}\,\Sigma} \tag{2.26}$$

2.3.4.4 **Determination of Activity Coefficients**

Activity coefficients in nonideal multicomponent systems may be calculated by, for example, UNIFAC, NRTL, or UNIQUAC. Determination of activity coefficients is only possible by adjustment to experimental data. The UNIFAC model is based on a group contribution method. Activity coefficient is thereby composed of a combinatory part γ_i^{C} and a residual part γ_i^{R}. In the combinatory part, the geometric size of the molecule groups is considered, whereas the residual part accounts for the interactions. For simple systems such as water/ethanol, the simplified calculation via the approach by Taylor and Krishna [25] can be chosen.

2.3.4.5 **Proof of Principle**

Figure 2.5 shows the named parameters for the mixture ethanol/water above its composition. Figure 2.5a gives thereby the three approaches van Laar, Margules, and NRTL in comparison to data by Tyn and Claus [35]. The models NRTL and van Laar show the real course of Fick's diffusion coefficient with lowest deviations. According to the definition of the thermodynamic factor, which contains the derivative of the activity coefficient with respect to the mole fraction, the thermodynamic factor depends on the model. In Figure 2.5b, the thermodynamic factor is shown for the three models named. A comparison to experimental data is not possible. Up to now, no real course of the thermodynamic factor has been measured.

Figure 2.5c shows the course of Fick's diffusion coefficient, calculated via NRTL, in comparison to the Maxwell–Stefan diffusion coefficient. Maxwell–Stefan diffusion coefficient changes slightly above composition in relation to Fick's diffusion coefficient. Activity coefficients required for the models are determined by UNIFAC. Figure 2.5d depicts both courses.

The models according to Fick and Maxwell–Stefan can hereby be applied for the description of extraction of water from a porous solid matrix. In countercurrent flow to the extraction of water, the solvent, in this case, ethanol, diffuses into

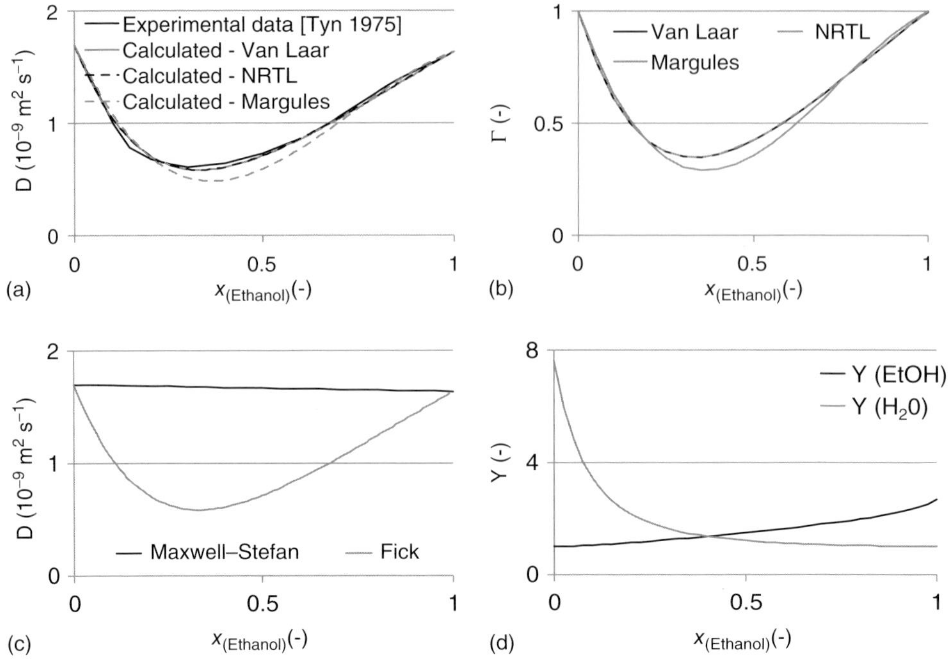

Figure 2.5 (a) Diffusion coefficients, (b) thermodynamic factor, (c) Maxwell–Stefan and Fick's diffusion coefficient, and (d) activity coefficient.

the matrix. Figure 2.6 shows the comparison of both approaches to experimental data (example vanillin). Thereby, it becomes apparent that the Maxwell–Stefan model describes the real case more exactly. Deviations of both models are especially recognized in sector II. Deviations may be explained by sources of error in model parameter determination, calculational as well as experimental. In sector III, there are uncertainties of measurement at closure of mass balances. The data in Figure 2.6 exhibit an error in mass balance of up to 4%. Via a sensitivity analysis, this error is quantified to be acceptable [36].

For diffusion in other porous media, such as catalysts and zeolite, the better description with the Maxwell–Stefan approach has already been thematized [35]. The extraction of water may thus be described hereby and serves for determination of a defined solvent composition in the solid. By this, the real case with variable equilibrium line, degree of swelling, or mass transfer kinetics can be depicted in the model depending on the current solvent mixture.

2.4 Model-Based Realization: Apparatus and Process Design

Model-based design by the "desorption approach" may be applied for extraction as a single unit operation and for total process design and optimization. In the

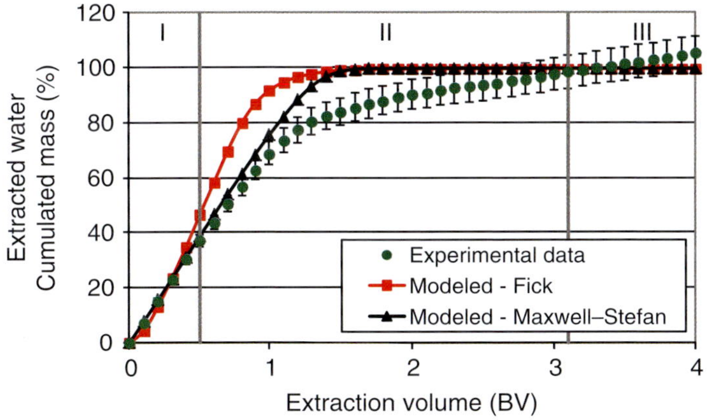

Figure 2.6 Comparison of Fick and Maxwell–Stefan approaches to experimental data [8].

following, some examples are given. Models need to be checked beforehand for their validity. For development of new processes, validity is to be checked by a variation of operating conditions such as flow velocity or particle size. Furthermore, scale-up behavior needs to be predicted.

If phytoextraction is considered as single unit operation, the process can be optimized with respect to different target variables:

- Yield
- Space–time–yield
- Purity and side-component spectrum
- Reaching the equilibrium and diluting the extract
- Extraction time
- Solvent consumption
- Apparative effort.

Further target variable such as energy consumption can be defined depending on the process. Figure 2.7 shows the comparison of the variables for the system vanillin from vanilla beans according to [8]. The maximum yield of vanillin is hereby determined from a five-stage maceration and finally a control percolation for verification of total extraction. Purity is defined as the ratio of vanillin to the sum of the five mainly obtained side components. Further ingredients have, because of their small mass, a very low impact on purity. For evaluation of apparative effort, solvent consumption, and extraction time, the five-stage maceration with control percolation serves as reference process. As outline for an industrial process, this is not applicable owing to high apparative effort, but exhibits a total extraction and thus yields of 100% for all ingredients. Process designs can thus be scaled by this.

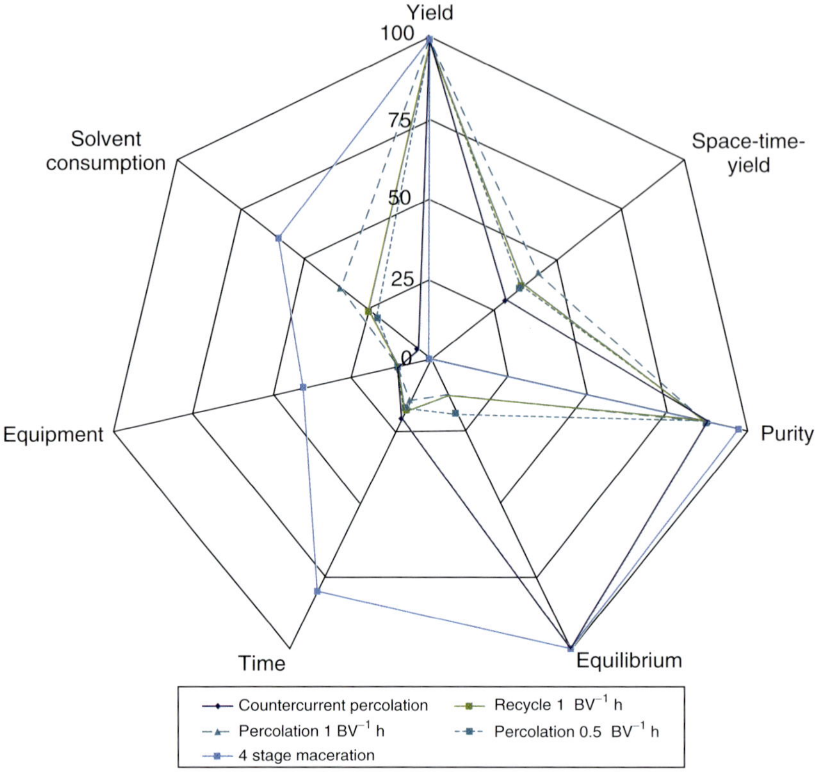

Figure 2.7 Model-based design: dependence of target variables on chosen process.

2.4.1 Quantification of Determining Factors

Priorities concerning the named variables are to be chosen specific for the product. For high-value products such as vanillin, yield is the determining factor, whereas solvent consumption and costs are secondary. Contrarily, for low-value products such as saccharose from sugar beets, solvent costs should be minimized. Solvent consumption and costs mostly increase significantly with increasing yield, so that yields close to 100% can only be achieved by a disproportionately high amount of solvent. Yield losses can thus increase the profitability of the process [8, 10].

Along with design of new extraction processes, physicochemical modeling of the current state is applicable for optimization of existing processes. Model validation is hereby not carried out by a variation of operating conditions but by prediction of the industrial reference process. Parameters equilibrium and mass transfer kinetics depending on the raw material may be determined in laboratory experiments. Fluid dynamics depends on the process and thus has to be determined in industrial scale.

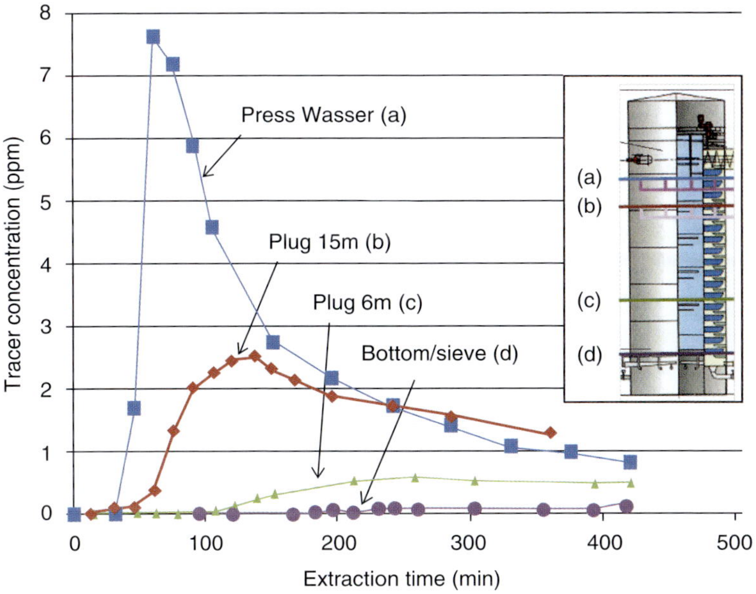

Figure 2.8 Tracer experiments of liquid phase [7].

2.4.2
Proof of Principle – Process Optimization

Figures 2.8 and 2.9 show for the optimization of a sugar extraction tower tracer experiments for determination of residence time and hence quantification of fluid dynamics exemplary for the liquid phase. Profiles are collected in four heights and the evaluated model is based analogous to literature. With the validated model, optimization of process parameters such as package density may be conducted.

2.4.3
Proof of Principle – Cost-Driven Decision

If phytoextraction is not only to be designed or optimized on its own but to be integrated into a total process, models of solid extraction need to be combined with models of other unit operations for purification. Development of purity over extraction time now becomes an important factor, and total process optima require a combined assessment of the unit operations involved. Here a decision depending on the product is required, whether an increase in yield, connected with elevated solvent consumption, increased extraction time, and lowered purity of the extract, and thus higher purification efforts results in an economically optimal process.

All previously listed parameters in the spider diagram finally result in specific separation costs, given as €/kg of product.

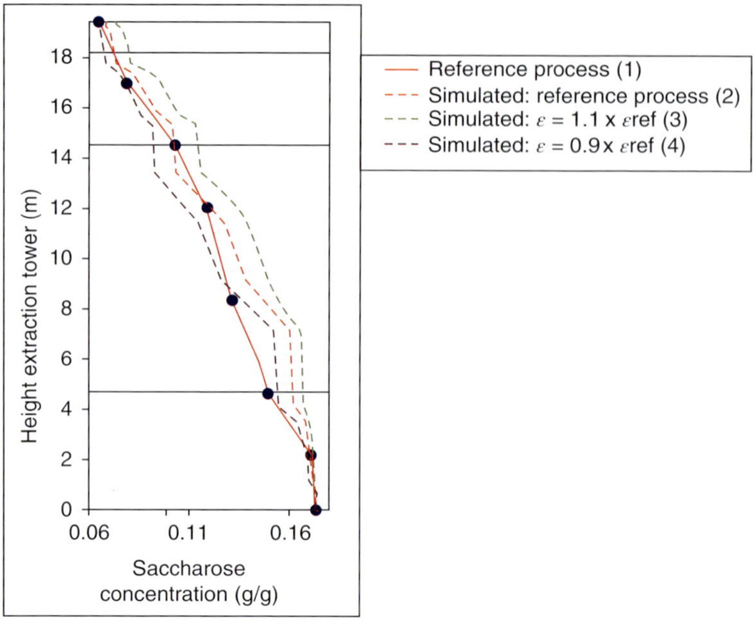

Figure 2.9 Optimization of packing density for the example of cossette phase [7].

The quality of an extract is primarily defined by its purity. To ensure highest quality demands in terms of TQM-understanding, reproducibility of this quality norm is of the highest importance. A proven reproducibility always guarantees defined degrees of identity, purity, and content.

High-value products are driven by product yield, whereas low-value products are mainly focused on solvent consumption.

Specific separation costs are essential for evaluation of economic manufacturing process parameters. Spider diagrams clearly visualize those cost drivers.

2.5
Extract Purification

In previous publications, the opportunities for a structured proceeding via data on chemical media and/or experimental investigation for plant extracts are discussed [13, 37, 38]. Figure 2.10 shows the arrangement and the proceeding schematically. The complex mixture under research is composed of structurally known and unknown substances, whereby unknown substances may be transferred to known ones by separation, for instance, by multidimensional chromatography and subsequent NMR identification. Therefore, samples and purification of the extract are required, which is linked to high costs or experimental effort.

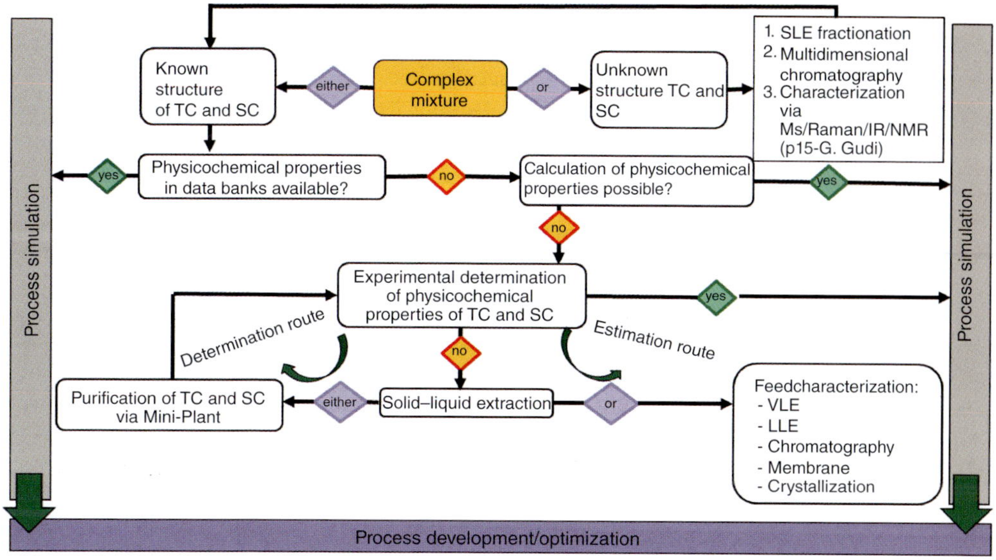

Figure 2.10 Systematic process design for complex mixtures according to [38].

Data on chemical media provide first indication for the choice of a unit operation or of a solvent and are required for calculation of various model parameters. For the unit operations regarded in this work, the choice of unit operation is from

- Solid extraction
- Liquid–liquid extraction
- Distillation
- Chromatography.

and the required data on chemical media of target and side components are primarily

- Molar mass
- Density
- Polarity, given as distribution equilibria between octanol and water
- Boiling temperature and evaporation enthalpy
- Melting temperature and enthalpy
- Isoelectric point.

In databases such as the Dortmund database (DDB) [39], Design Institute for Physical Properties (DIPPR) [40], or Reaxys [41], data on chemical media for plant ingredients are partly deposited. For process design, these data on pure chemicals can provide orienting values. For simple molecules such as alcohols or esters, the required data can be found within databases. Thereby, up to now, no standardized method for chemical data measurement is applied. Thus, statements on validity of the data are not to be made and data need to be checked on their applicability for calculation of model parameters and choice of unit operation. With increasing complexity of the molecules, the amount of available data decreases [13, 38].

Along with the usage of databases, chemical data may be determined by calculation methods. For simple molecules such as monoterpenes, group contribution methods can provide adaptable results. With increasing complexity, these models fail. Chemical data of many diterpenes can only be determined with significant error margin even with PC-SAFT or quantum mechanical calculation such as COSMO-RS [38]. A general statement as to which model calculates the data of a molecule of a certain class with lowest deviations from real values is not possible [38].

Process design for complex mixtures via data on pure chemicals is not constructive as the contained side components may have significant influence on separability. Especially for plant extracts, the solubility of the target component in water and organic solvents can be affected decisively by the side-component spectrum [8].

For process design of complex mixtures, an experimental determination of data results, whereby there is a distinction between data on chemical media and parameters required for the models applied in process design. For model-based process design, an experimental model parameter determination and model validation is performed, following the approach by Josch *et al.* [13].

Model parameters for the multicomponent mixture may be determined via standardized methods. Depending on modeling depth and hence considered effects, equilibrium, mass transfer kinetics, and fluid dynamics are the parameters that need to be determined primarily [37].

2.5.1
Modeling Approaches

Analogous to solid extraction, known and comparable systems can provide orienting values in the early stages of process development. Parameters such as diffusion coefficients and the axial dispersion coefficient may be estimated from correlations specific to the unit operation. Table 2.2 gives the model parameters of the respective unit operations.

Unit operations and total process connections are subsequently to be designed model based to get an optimal overall process. Hence process modeling is always combined with cost calculations [42]. The parameters liquid–liquid equilibrium, vapor–liquid equilibrium, molecular weight, and dissociation constant are in focus in the following studies (Figure 2.11).

2.5.2
Scale-Up and Mini-plant

Mini-plant technology has proved an efficient standard to reduce efforts and time in bulk and fine chemical process development [43, 44].

Table 2.2 Unit operations and specific model parameters according to [13].

Unit operation	Thermodynamic equilibrium for key components	Kinetics	Fluid dynamics	Cost calculation
Size exclusion chromatography	Molecular weight, size	Diffusion coefficient (bulk, film, pore, surface)	Axial dispersion (liquid)	Henry, capacity, productivity, dilution
Ion exchange chromatography	Sequence of elution (concentration of modifier and adsorbent), isoelectric point	Diffusion coefficient (bulk, film, pore, surface), isotherm for target component	Axial dispersion (liquid)	Henry, capacity, productivity, dilution
Reversed/normal phase chromatography	Sequence of elution (concentration of modifier and adsorbent), hydrophobicity	Diffusion coefficient (bulk, film, pore, surface), isotherm for target component	Axial dispersion (liquid)	Henry, capacity, productivity, dilution
Distillation	Vapor–liquid equilibrium in mixture at temperature and salt concentration	Mass transfer coefficient	Axial dispersion (liquid, gas)	Heat capacity of components and solvent, evaporation enthalpy of low boiling component/solvent
Liquid–liquid-extraction	Liquid–liquid equilibrium in mixture at pH, temperature, solvent, and salt concentration	Diffusion coefficient (bulk, film)	Axial dispersion (continuous and disperse phase)	Capacity, density, viscosity, surface tension
Membrane	Molecular weight, size	Diffusion coefficient, density of pure solvent	Axial dispersion (permeate, retentate)	Molecular weight cutoff, pressure–flow curve
Solid–liquid-extraction	Solid–liquid equilibrium in mixture at pH, temperature, solvent, and salt concentration	Diffusion coefficient (bulk, film, pore), solubility	Axial dispersion (liquid)	Capacity, solid–liquid equilibrium line, dilution

Products, for example, in nutrition, agrochemicals, cosmetics, flavors, and pharmaceuticals tend to become more complex in molecular structure. Besides classic synthesis routes, biotechnological fermentation broths or plant-based extracts make an increasing contribution to production-scale supply of value products. Broths and extracts are complex mixtures which have to be separated up to high product purity specifications. Literature research points out that less than

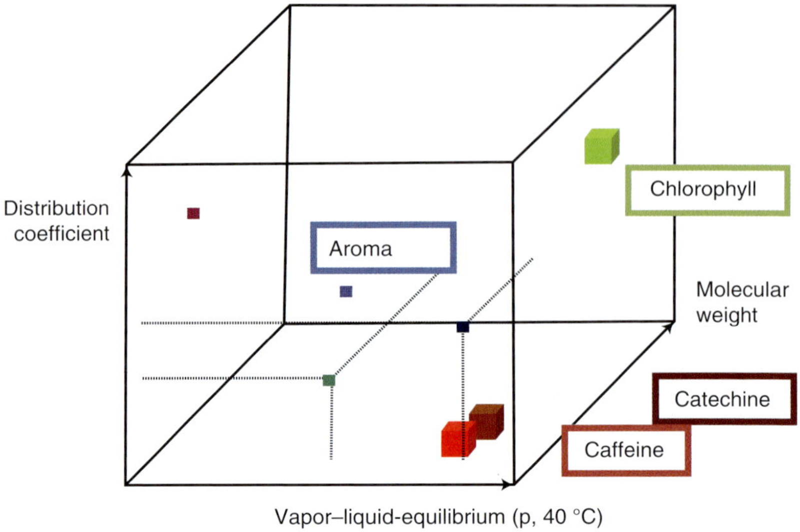

Figure 2.11 Separation properties.

5% of the physical properties of the ingredients can be found in databases, mixture properties even less. Therefore, chemical engineering approaches of process simulation lack basic input data. Experiments in miniaturized laboratory-scale measurement cells are necessary to determine model parameters directly from the original complex mixture. The experimental data is evaluated model based and the model parameters are determined with sufficient accuracy. Conceptual process design predicts potentially efficient process concepts [45]. These process alternatives are finally validated experimentally by mini-plant experiments and test amounts are generated for quality control.

A typical modern mini-plant for all unit operations in separation science is depicted in Figure 2.12.

The objectives of research are the development of methods for process modeling combined with an efficient experimental model parameter determination of complex mixtures and a final sound experimental validation of process modeling and data.

The mini-/micro-plant is designed and configured to be operated with real industrial complex mixtures under ATEX and CE-near, TÜV certified. GMP near DQ, OQ, PQ has been aimed at.

Distillation, liquid–liquid extraction with scrub and re-extraction, solid extraction by maceration or percolation, precipitation/crystallization, membrane technology, and chromatography in batch and SMB/MCC mode can be combined. More special equipment is rented and integrated project based, if necessary.

Further equipment integration into the plant is focused on micro-separation equipment starting with distillation and extraction. The basic platform is an Ehrfeld-Bayer Technology Services standard device setup. Ideas for chromatography and membrane units tend to be more long term. Micro-structured devices

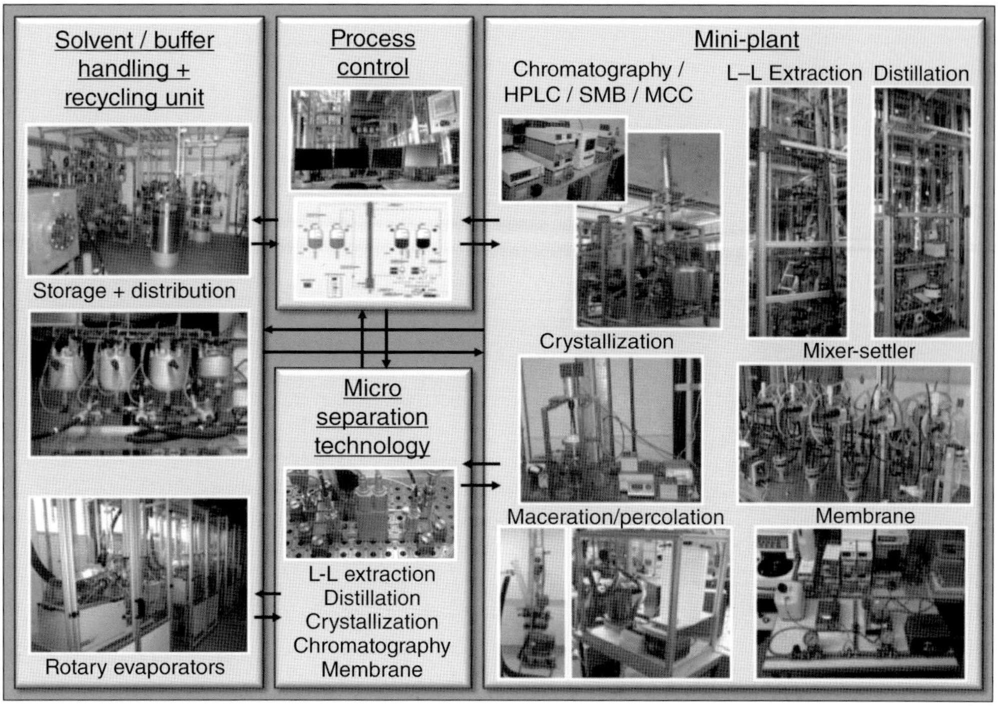

Figure 2.12 Mini- and micro-plant technology at institute for separation and process technology, Clausthal University of Technology.

are built by cooperation partners and characterized at the institute. The objective is to change equipment easily and to compare performance of mini- and micro-devices directly in process operation.

The plant-operation range is about $1-2\,l\,h^{-1}$. A buffer and solvent preparation and handling unit with about $11\,m^3$ total vessel volume is available. Recycling of solvents is done with the aid of the distillation column or three automated, continuous rotary evaporators. Each unit operation could be operated stand-alone or in any process sequence combined by flexible dry-coupled tubing. The plant has been designed, manufactured, and built including the process control system, which was configured at the institute. Modifications caused by research and project requirements are now efficiently possible. Basic operational analytics are done at the institute. A close cooperation with the Institute of Organic Chemistry (Prof. Kaufmann and Prof. Schmidt) provides GC/HPLC-MS and NMR.

All the unit operations installed are characterized concerning fluid dynamics and separation performance with IUPAC/EFCE standard test systems, which allow a sound error analysis and interpretation of data of the complex mixtures generated.

The distillation columns feature an inner diameter of 50 mm with 6 m overall height and 3 m active packing height. Between every 0.5 m section, a measurement

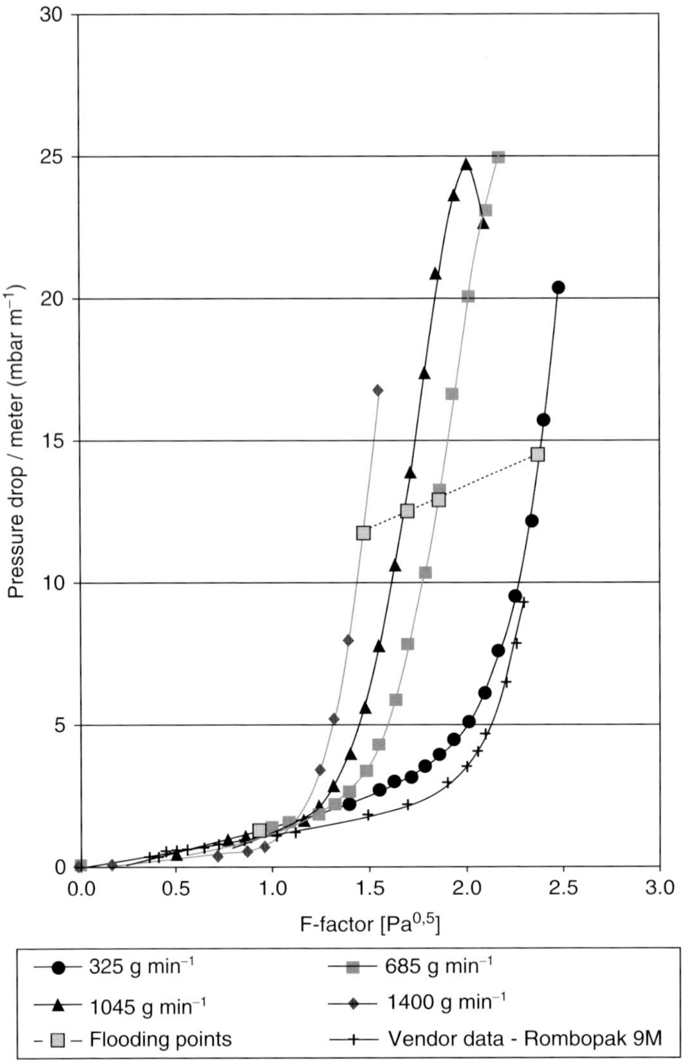

Figure 2.13 Diagram of pressure drop per meter over F-factor for different liquid flow rates [46].

port for collection of pressure, temperature, and sample data is introduced. The currently used packings are Kühni/Sulzer Chemtech Rombopak® 9 M and 12 M. A vacuum pump enables vacuum distillation experiments, a switchable column bottom with either batch or continuous process operation. Besides feed, bottom, and distillate heating/cooling, a coat surrounds the column as insulation and compensates for heat losses.

According to common procedure, first basic hydrodynamics such as loading and flooding point are determined using a water–air system. Figures 2.13 and 2.14

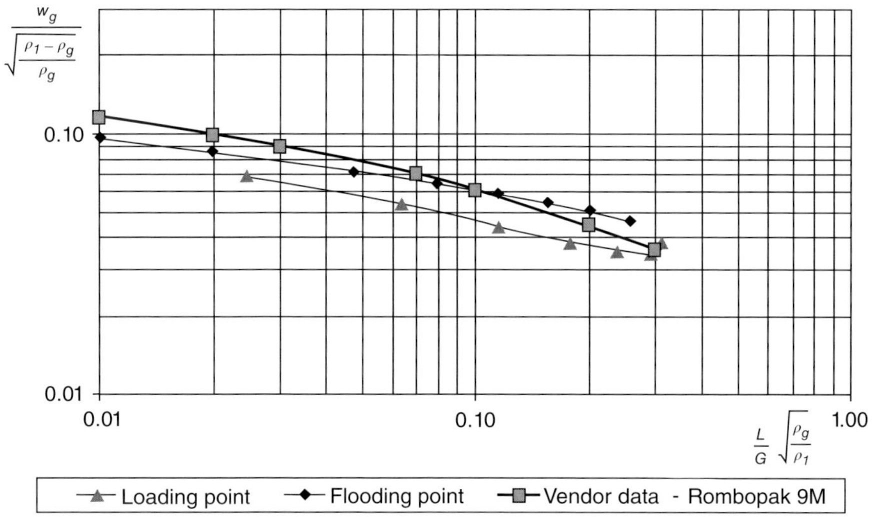

Figure 2.14 Diagram of loading factor over stream parameter [46].

display initial results for pressure drop per height and flooding points respectively in comparison to packing vendor data [44]. Although displaying a small difference, the results for the given column height and diameter are in the same order of magnitude as released vendor data for which no liquid flow rate has been given.

Determination of separation performance is carried out employing the EFCE test system chlorobenzene/ethylbenzene, which is rated for expected 10–40 separation stages and an operation pressure of 30–1013 mbar [47].

The results of the separation test have been evaluated using the Fenske equation and give a packed column separation stage number of 18 (data not displayed). For the given column, this yields an HETP of 0.16, which corresponds to the vendor information for this packing [48].

The mini-plant includes three units for liquid–liquid extraction research besides a single droplet measurement cell and a 200-stage Craig extractor. The first unit is a 10-stage mixer–settler–battery with approximately 180 ml volume per stage and, if required, coalescence aids such as Rhodius coalescence nets.

In addition, two extraction columns are available. Both have an inner diameter of 26 mm, an insulating jacket, and an overall height of 5 m. Settling areas have been designed with a 50 mm diameter each. Hence, the total volume of the column is nearly 4 l. Active countercurrent height of the columns is 3.5 m. Both general types of energy input are available: Currently, one column is filled with a structured Montz B1-750 packing and pulsation energy input. The other contains a stirred Kühni-ECR geometry. Steady-state residence time distributions for light and heavy phases are determined via inline tracer detection and first estimates of droplet size distribution over column height by analysis of pictures taken at different column sections.

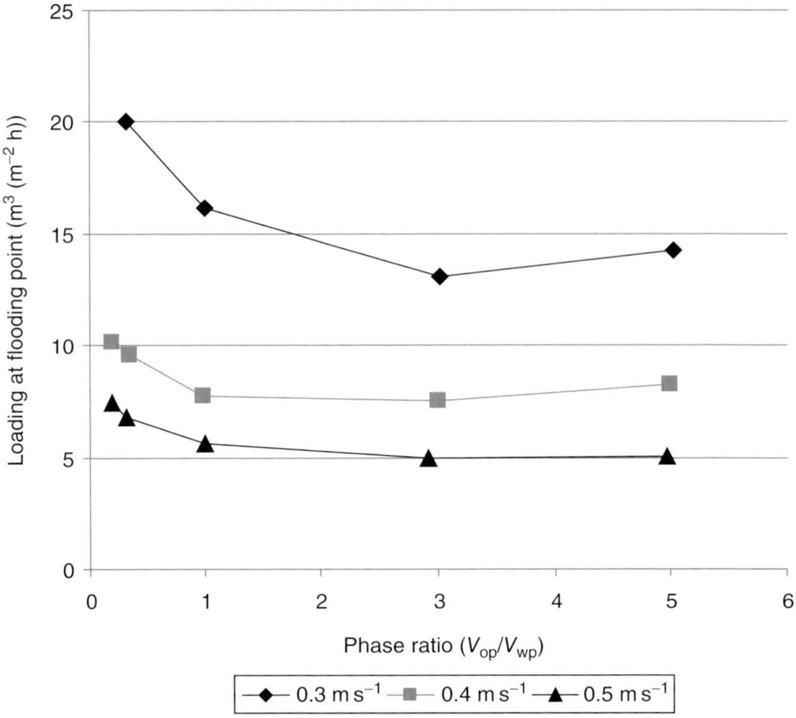

Figure 2.15 Diagram of flooding point loading at different phase ratios and tip speeds.

Depending on the applied test systems, such as the EFCE systems toluene/acetone/water or butylacetate/acetone/water [49], and process operation parameters such as energy input, phase ratio, and flow rates (in relation to flooding point) sets of data are collected. Figure 2.15 displays exemplary first results for operation points of the stirred column with a butylacetate/water-system without mass transfer. Compared to larger scale columns, a lower loading at flooding point can be observed. Apart from increased wall effects, the optimization of this aspect especially targets shaft-bearing rings which have proven to be the most common seed point for flooding.

About 4 NTP/m are gained in RDC-extraction, 5 NTP/m in the extraction pulsed packed column.

Membrane and chromatography equipment have already been well characterized [50–52].

2.6
Total Process Development and Design

- Cultivation (GACP) and harvesting instead of wild collection is described in Chapter 3 in detail by Hansjörg Hagels. The major objective of all these activities

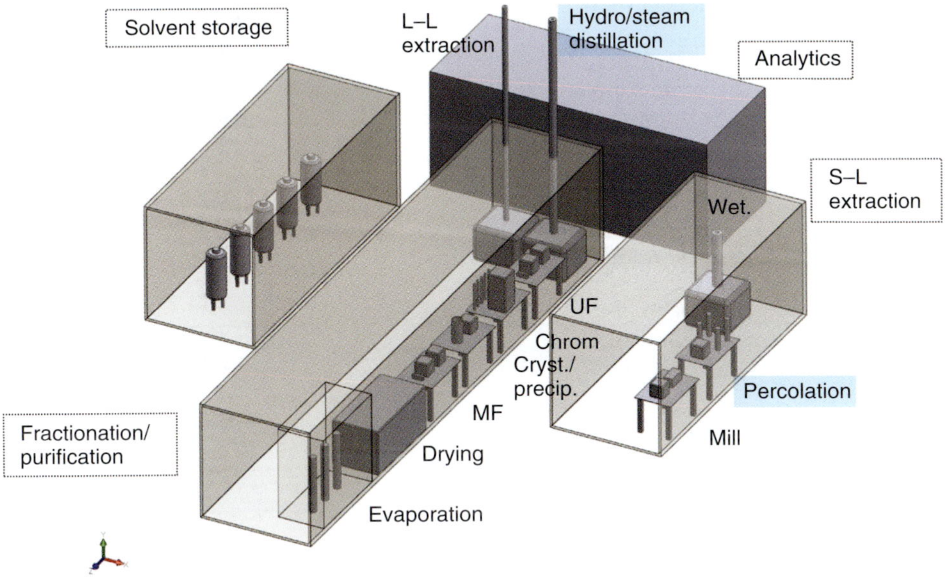

Figure 2.16 Production plant concept for phytochemicals in a container [55].

is to harmonize the ingredient composition for any further robust manufacturing, especially purification.

- The general rationale for development of separation strategies are already described in detail in Chapter 5 of the book on fluid process technology by Strube *et al.* [53] and Chapter 5 in [24].

Therefore, finally, innovative manufacturing strategies are needed.

In addition, as the era of blockbuster therapeutics cherished by most pharma companies in the past has most probably come to an end, stratified medicine is looked upon as the new hope. Provided that the stratification issue works out, there is higher potential for clinical success, but by definition, going along with low volume manufacturing without economy-of-scale potentials. Here, the proposed concept of continuous, nevertheless flexible manufacturing proves significant potential also for small and medium entities (SME) [54].

These concepts have also significant application potential in traditional batch productions, as seen for phytochemicals, nutraceuticals, food, cosmetics, and flavors [22].

If these manufacturing chains are transferred into continuous operation, small footprint container-type plants with high productivity and product flexibility result as shown in Figure 2.16.

This approach is even feasible for small-scale products in 100 kg/a scale, as analyzed in this study. Typical cost reductions are in a magnitude of factor of 3–10 in CAPEX and TOTEX, as compared in Figure 2.17.

Batch to conti

Product per year	(kg)	1.852	1.852	Large
Drug-amount	(kg/batch)	300	1.5	Large
TC-amount	(%)	4%	4%	Large
CapEX	(T€)	5.884	1.620	Large
Product per year	(kg)	185	185	Small
Drug-amount	(kg/batch)	30	0.15	Small
TC-amount	(%)	4%	4%	Small
CapEX	(T€)	2.231	872	Small

Figure 2.17 Cost comparison of batch and continuous manufacturing at two scales.

As shown, investment (CAPEX) of less than €1 million enables SMEs to operate in a fully integrated mode, including manufacturing with acceptable financial risk and therefore access to the necessary financing. Such a production concept permits SMEs to have much more control over their value chain and to compete in all markets accessible for them while protecting their research as well as their business better and longer.

The size of a unit operation shrinks below conventional pilot plant size, sometimes even close to standard laboratory size equipment, which often may allow equipment selection from existing product catalogs with pricing impacted by "economy of scale" due to the large unit numbers produced and sold to the laboratory market. Specific equipment development can be kept to the minimum, if available and proven technologies can be combined intelligently. Further, process control can be developed cost-effectively by reconfiguration of standard industrial systems such as Siemens S7.

Looking, for example, at the development of the phytochemicals market, it could be foreseen that existing products defined as standard extracts will continue to have their share on the market, although the amounts of raw material needed for that are quite small and business growth potentials are moderate at best. However, there is enough room for new and innovative products, if they are filed on an evidence basis, are supplied as pure compounds, and therefore have a much higher value. In case studies conducted, typical ROIs (returns on investment) are in a range of 1–3 years. Detailed experimental studies are still necessary to show, that an up to 10-fold COG (cost of good) reduction potential is feasible under robust operation conditions. For this, however, funding activities have to be initiated, as none of the SME companies active in those fields could generate and carry out such a study alone.

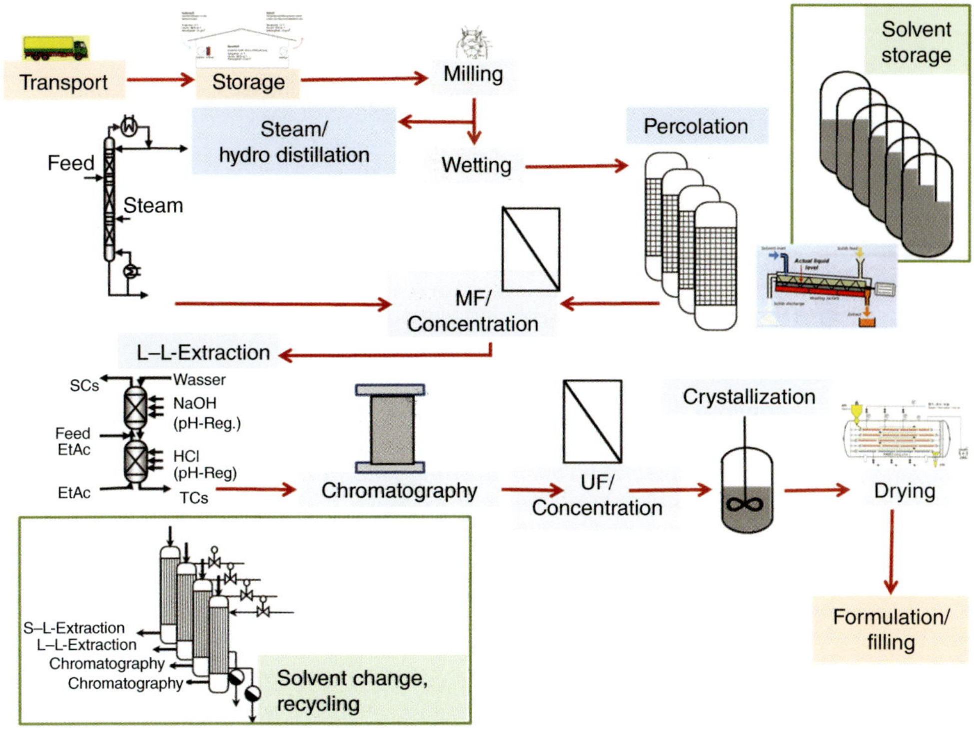

Figure 2.18 TPTP – phytochemical manufacturing flowchart.

The proposed approach includes purification technologies to gain, in addition to the existing standard extracts, new pure substance products with higher value on new market segments.

Figure 2.18 summarizes the general flowchart for these manufacturing approaches, including solvent recycling. Compounds are fractionated and purified by distillation or solid–liquid extraction. The raw material is preferably harvested and extracted fresh in order to save drying and maintain maximum-value compound concentration and costs. Process efficiency could be put in place by continuous harvesting from greenhouses or plant cell fermentations. The energetic manufacturing concept is optimized and closed with the aid of renewable resources. Therefore, this novel integrated manufacturing concept could be considered as “traceless plant and traceless production (TPTP)” [55].

2.7 Conclusions and Summary

Analyzing trends in engineering of chemical pharmaceutical manufacturing plants reveal relevant changes of boundary conditions on a global level.

Regionally varying market growth and energy supply combined with globally increasing competitiveness point toward

- higher flexibility in process and plant design and corresponding set up;
- enhanced process and cost efficiencies;
- variable capacities to adjust for demand.

While various options have been intensively researched, most of which have yet to prove their practical value, some additional variables have been assessed and compiled, resulting in the following suggestions:

- platform methods instead of platform processes;
- support process development and fast track filing by model-based QbD approaches early on;
- continuous operation instead of batch processing by small-scale flexible decentralized production in containers;
- novel operator or CMO manufacturing concepts to add value by selling not only equipment but knowledge, know how, and support, thereby;
- making high-end manufacturing technology applicable worldwide, including
  - automated production operation by skilled personnel
  - use of automation and continuous operation for higher manufacturing robustness, thereby improved product quality.

In order to propose efficient manufacturing strategies for value-added products in a regulated environment, efficient engineering and production concepts are more a realistic dream than a nightmare. Nevertheless, while broad studies have shown the theoretical feasibility, the experimental proof is still to be delivered. Most of the technologies applied do already exist, however their combination is innovative, therefore the risks appear to be moderate. Challenging is the consequent transfer into manufacturing. As no single company is capable to solve that on its own, different consortia with national funding are in discussion at the moment.

Acknowledgments

The authors from Clausthal University of Technology would like to thank especially Christoph Helling, Iraj Koudous, Steffen Zobel, and Martin Lucke for their contribution and enthusiasm in supporting the tedious work on those topics.

References

1. Tegtmeier, M. (2012) Pflanzenextraktion: schlüsseltechnologie zur nachhaltigen Nutzung von Bio-Ressourcen. Plant extraction: key technology for sustained use of bio-resources. *Chem. Ing. Tech.*, **84** (6), 880–882.
2. Bart, H.-J. *et al.* (2005) Konzeptpapier der Fachgruppe "Phytoextrakte – Produkte und Prozesse", DECHEMA.
3. Bart, H.-J. *et al.* (2012) Positionspapier der Fachgruppe

Phytoextrakte – Produkte und Prozesse, DECHEMA.

4. Schneider, P., Bischoff, F., Müller, U., Bart, H.-J., Schlitter, K., and Jordan, V. (2011) Plant extraction with aqueous two-phase systems. *Chem. Eng. Technol.*, **34** (3), 452–458.
5. Schneider, P., Hosseiny, S.S., Bischoff, F., Müller, U., Bart, H.-J., Schlitter, K., and Jordan, V. (2011) Surfactant mediated extraction of triterpenes and their direct HPLC analysis from the micellar system. *Instrum. Sci. Technol.*, **39** (5), 407–418.
6. Bart, H.-J. and Schmidt, M. (2007) Feststoffextraktion. *Chem. Ing. Tech.*, **79** (5), 663–667.
7. Both, S., Eggersglüß, J., Lehnberger, A., Schulz, T., Schulze, T., and Strube, J. (2013) Optimizing established processes like sugar extraction from sugar beets – design of experiments versus physicochemical modeling. *Chem. Eng. Technol.*, **36** (12), 2125–2136.
8. Both, S., Koudous, I., Jenelten, U., and Strube, J. (2014) Model-based equipment-design for plant-based extraction processes – considering botanic and thermodynamic aspects. *C.R. Chim.*, **17** (3), 187–196.
9. Kassing, M. (2012) Process development for plant-based extract production. Dissertation. Clausthal University of Technology, Shaker Verlag, Aachen.
10. Kassing, M., Jenelten, U., Schenk, J., Hänsch, R., and Strube, J. (2012) Combination of rigorous and statistical modeling for process development of plant-based extractions based on mass balances and biological aspects. *Chem. Eng. Technol.*, **35** (1), 109–132.
11. Brunner, G. (2010) Applications of supercritical fluids. *Annu. Rev. Chem. Biomol. Eng.*, **1** (1), 321–342.
12. Laurent, A., Lack, E., Gamse, T., Marr, R. (2001): In: (eds A. Bertucco, G. Vetter) *Industrial Chemistry Library: High Pressure Process Technology: Fundamentals and Applications*, vol. 9, Elsevier, pp. 351–403.
13. Josch, J.-P., Both, S., and Strube, J. (2012) Characterization of feed properties for conceptual process design involving complex mixtures, such as natural extracts. *Food Nutr. Sci.*, **3** (6), 836–850.
14. Chemat, F. (2011) *Éco-Extraction du Végétal*, Dunod, Paris.
15. Chemat, F., Albert-Vian, M., and Fabiano-Tixier, A.S. (2013) Program Book of Abstracts, International Congress on "Green Extraction of Natural Products", Avignon, France, April 16–17, 2013, Green Extraction of Natural Products, *www.univ-avignon.fr/fr/minisite/minigenp2013/accueil.html* (accessed 13 August 2014).
16. Mersmann, A., Kind, M., and Stichlmair, J. (2005) *Thermische Verfahrenstechnik. Grundlagen und Methoden*, 2. Aufl, Springer-Verlag (VDI-Buch).
17. Harker, J.H., Backhurst, J.R., and Richardson, J.F. (2002) *Particle Technology and Separation Processes*, Chemical Engineering, vol. 2, 5 Aufl, Butterworth-Heinemann.
18. Kassing, M., Jenelten, U., Schenk, J., and Strube, J. (2010) A new approach for process development of plant-based extraction processes. *Chem. Eng. Technol.*, **33** (3), 377–387.
19. Ndocko Ndocko, E., Bäcker, W., and Strube, J. (2008) Process design method for manufacturing of natural compounds and related molecules. *Sep. Sci. Technol.*, **43** (3), 642–670.
20. Wilke, C.R. and Chang, P. (1955) Correlation of diffusion coefficients in dilute solutions. *AIChE J.*, **1** (2), 264–270.
21. Both, S., Tegtmeier, M., and Strube, J. (2014) Produktion von Pflanzenextrakten – Von Evolution zur Revolution. *Z Arznei- Gewurzpfla.*, **19** (4), 169–177.
22. Herbert, V.G. (2002) *Verfahrensentwicklung: Von der ersten Idee zur chemischen Produktionsanlage*, Wiley-VCH Verlag GmbH, Weinheim.
23. Allaf, T., Tomao, V., Ruiz, K., and Chemat, F. (2013) Instant controlled pressure drop technology and ultrasound assisted extraction for sequential extraction of essential oil and antioxidants. *Ultrason. Sonochem.*, **20** (1), 239–246.
24. Bart, H.-J. and Pilz, S. (eds) (Hg.) (2011) *Industrial Scale Natural Products Extraction*, Wiley-VCH Verlag GmbH & Co. KgaA.

25. Taylor, R. and Krishna, R. (1993) *Multicomponent Mass Transfer*, John Wiley & Sons, Inc., New York.
26. Krishna, R. and Wesselingh, J.A. (1997) The Maxwell-Stefan approach to mass transfer. *Chem. Eng. Sci.*, **52** (6), 861–911.
27. Krishna, R. (1987) A simplified procedure for the solution of the dusty-gas model equations for steady-state transport in non-reacting systems. *Chem. Eng. J.*, **35** (2), 75–81.
28. Mason, E.A. and Malinauskas, A.P. (1983) *Gas Transport in Porous Media: The Dusty-Gas Model*, Elsevier Science Publishers B. V., Amsterdam.
29. Dörfler, H.-D. (2002) *Grenzflächen und kolloid-disperse systeme*, Springer-Verlag, Berlin.
30. Caldwell, C.S. and Babb, A.L. (1956) Diffusion in ideal binary liquid mixtures. *J. Phys. Chem.*, **60** (1), 51–56.
31. Danner, R.P. and Daubert, T.E. (1983) *Manual for Predicting Chemical Process Design Data*, AIChE, New York.
32. Vignes, A. (1966) Diffusion in binary solutions. Variation of diffusion coefficient with composition. *Ind. Eng. Chem. Fundam.*, **5** (2), 189–199.
33. Brenner, H. and Gaydos, L.J. (1977) The constrained brownian movement of spherical particles in cylindrical pores of comparable radius: models of the diffusive and convective transport of solute molecules in membranes and porous media. *J. Colloid Interface Sci.*, **58** (2), 312–356.
34. Guiochon, G., Shirazi, S.G., and Katti, A.M. (1994) *Fundamentals of Preparative and Nonlinear Chromatography*, Academic Press, Boston, MA.
35. Tyn, M.T. and Calus, W.F. (1975) Temperature and concentration dependence of mutual diffusion coefficients of some binary liquid systems. *J. Chem. Eng. Data*, **20** (3), 310–316.
36. Borrmann, C. (2012) Methode zur Auslegung von integrierten Downstream Processing Verfahren am Beispiel der hydrophoben Interaktions- und Ionenaustauschchromatographie. Dissertation. Clausthal University of Technology, Shaker Verlag, Aachen.
37. Josch, J.P. (2012) Charakterisierung von Rohstoffgemischen für die Verfahrensentwicklung zur Produktreinigung aus komplexen Matrizes. Dissertation. Institut für Thermische Verfahrens- und Prozesstechnik, TU Clausthal.
38. Koudous, I., Both, S., Gudi, G., Schulz, H., and Strube, J. (2014) Process design based on physicochemical properties for the example of obtaining valuable products from plant-based extracts. *C.R. Chim.*, **17** (3), 218–231.
39. *http://http://www.ddbst.com/online.html* (accessed January 2014), 657–705.
40. DIPPR *http://dippr.byu.edu/students/* (accessed January 2014).
41. ELSEVIER *https://www.reaxys.com* (accessed January 2014).
42. Both, S., Helling, C., Namyslo, J., Kaufmann, D., Rother, B., Harling, H., and Strube, J. (2013) Resource-efficient process technology for energy plants. *Chem. Ing. Tech.*, **85** (8), 1282–1289.
43. Deibele, L. and Dohrn, R. (2006) *Miniplant-Technik in der Prozessindustrie*, Wiley-VCH Verlag GmbH, Weinheim.
44. Goedecke, R. (2006) *Fluidverfahrenstechnik. Grundlagen, Methodik, Technik, Praxis*, Wiley-VCH Verlag GmbH, Weinheim.
45. Strube, J., Grote, F., Josch, J.P., and Ditz, R. (2011) Process development and design of downstream processes. *Chem. Ing. Tech.*, **83** (7), 1044–1065.
46. Helling, C., Fröhlich, H., Eggersglüß, J., and Strube, J. (2012) Fundamentals towards a modular microstructured production plant. *Chem. Ing. Tech.*, **84** (6), 892–904.
47. Onken, U. and Arlt, W. (1990) *Recommended Test Mixtures for Distillation Columns*, 2nd edn, Institute of Chemical Engineers, Rugby.
48. Sulzer Chemtech Ltd (2010) Structured Packings for Distillation, Absorption and Reactive Distillation, Winterthur, *www.sulzerchemtech.com/portaldata/11/Resources//brochures/mtt/Structured_Packings_April_2010.pdf* (accessed 19 August 2014).

49. Misek, T., Berger, R., and Schröter, J. (1985) *Standard Test Systems for Liquid Extraction*, 2nd edn, Institute of Chemical Engineers, Rugby.
50. Borrmann, C., Helling, C., Lohrmann, M., Sommerfeld, S., and Strube, J. (2011) Phenomena and modeling of hydrophobic interaction chromatography. *Sep. Sci. Technol.*, **46** (8), 1289–1305.
51. Grote, F., Fröhlich, H., and Strube, J. (2011) Integration of ultrafiltration unit operations in biotechnology process design. *Chem. Eng. Technol.*, **34** (5), 673–687.
52. Grote, F., Froehlich, H., and Strube, J. (2012) Integration of reverse-osmosis unit operations in biotechnology process design. *Chem. Eng. Technol.*, **35** (1), 191–197.
53. Strube, J., Schulte, M., Arlt, W. (2006): Chromatographie, in *Fluidverfahrenstechnik. Grundlagen, Methodik, Technik, Praxis.* (ed. R. Goedecke) Wiley-VCH Verlag GmbH, Weinheim.
54. Strube, J., Grote, F., and Ditz, R. (2012) in *Bioprocess Design and Production Technology for the Future* (ed G. Subramanian), Wiley-VCH Verlag GmbH, Weinheim, 657–705.
55. Strube, J., Ditz, R., Fröhlich, H., Köster, D., Grützner, T., Koch, J., and Schütte, R. (2014) Efficient engineering and production concepts for products in regulated environments – dream or nightmare? *Chem. Ing. Tech.*, **86** (5), 687–694.

3 Tailor-Made Production of Plants for Green Extraction

Hansjoerg Hagels

3.1 Introduction

Herbal raw material is the key structure-forming basis for natural products and, therefore, the basis of their quality. It is the foundation of the product – the phytoextract. Key quality aspects are

1) the content and composition of active compounds;
2) the matrix of the herbal material (which may differ very much depending on growing conditions and postharvest treatment);
3) the purity (residual pesticides, heavy metals, mycotoxins, poly-aromatic hydrocarbons (PAH), etc.).

Many producers underestimate this fact or do not even take it into account and are then not aware of subsequent problems caused by unspecified quality-relevant attributes that lead to deviations in extraction, quality control, and quality assurance processes.

The supply of herbal raw materials can be carried out in two completely different ways:

1) Cultivation
2) Wild collection.

Generally, cultivation is preferred, as this allows the producer (farmer) to control the growth conditions much better than in wild collection. However, the establishment of plant cultivation is very time consuming for the majority of plants and may, depending on the requirements of the plant, take several decades.

The first step is the evaluation of the biodiversity of the species: as many as possible different accessions of the plant are collected in the original biotopes and, if possible, from botanical gardens and other plant-preserving institutions. These accessions are evaluated regarding growth conditions, biomass production,

Green Extraction of Natural Products: Theory and Practice, First Edition.
Edited by Farid Chemat and Jochen Strube.

biosynthesis of the substance/substance group of interest and substances to be avoided, such as the hepatotoxic pyrrolizidine alkaloids in butterbur *Petasites hybridus* [1] or the bitter taste of steviosids in *Stevia rebaudiana* extracts [2]. It is preferable to start directly with material that does not contain unwanted compounds rather than eliminating them by technical means downstream.

Domestication aims to find suitable accessions that are able to survive even in monocultures. Various soils and fertilizations are tested to learn about optimal growth conditions.

Then the cultivation itself is developed and continuously optimized with regard to sowing, fertilization, and weed and pest control – and potentially, even irrigation. In parallel, the harvest and postharvest treatment is established. If possible, technology developed for other crops is applied, modifications are engineered, or, in cases where unique technologies are required, devices are developed.

Most medicinal and aromatic plants (MAPs), however, are still collected in the wild [3]. The main reason is the need for small quantities only – the requirement for most of the species is less than a few tonnes per year, so that cost-intensive investment in cultivation development does not make sense.

Some plants cannot be cultivated in an economically reasonable way, as they need particular growing conditions that cannot be achieved in agricultural processes. These can be symbioses with other plants or microorganisms that cannot be verified in large-scale agriculture.

3.2 Sustainable Processes

Irrespective of the sourcing method, cultivation and wild collection should be carried out on the basis of sustainable approaches. The word "sustainability" became an established term with the Brundtland Commission of the United Nations in March 1987 "Sustainable development is development that meets the needs of the present without compromising the ability of future generations to meet their own needs" [4].

Sustainability consists of three "pillars" that contribute to the overall process (see Figure 3.1) [5] – if just one pillar is weak, then the entire system becomes instable:

1) *Social sustainability* is the ability of a social system (family, community, or organization) to function with a level of well-being (which should be defined). Topics such as poverty, injustice, poor educational systems, and war demonstrate social unsustainability.
2) *Environmental sustainability* is the ability of the environment to provide a level of environmental quality and to assure that no species becomes endangered or even extinct.
3) *Economic sustainability* is the ability of an economy to assure a defined level of economic production.

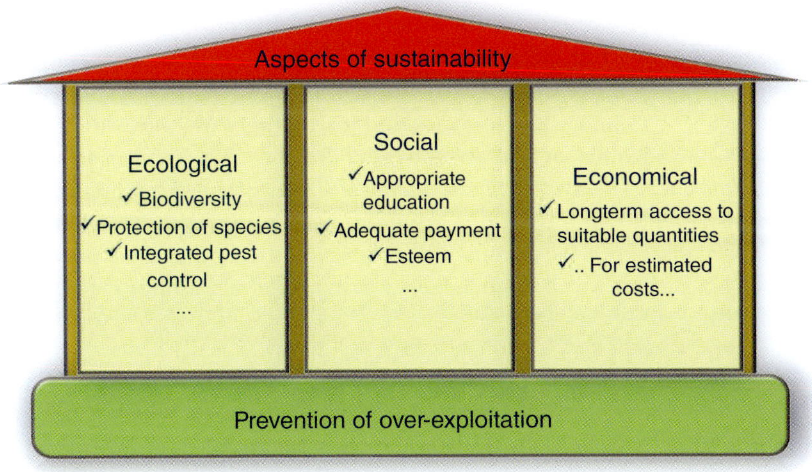

Figure 3.1 The sustainability house.

3.2.1
Social Sustainability

All people involved in production have a precarious income; the organization of the work should be well known. Wild collection processes in particular are not easy to supervise: interim dealers should be controlled regarding their share of the overall income and regarding the payment history of their collectors. Child labor must be excluded. To obtain a picture of all these aspects, repeated audits should be carried out. Another crucial topic is the education aspect: in cultivation processes, every employee needs to have a minimum level of education about the tasks within his responsibility. Workers who drive tractors or harvesters should be aware of potential risks, but also about general requirements regarding the operation and maintenance of the equipment for which they are responsible. Pest control workers should be suitably qualified. General topics include the appropriate use of fertilizers to ensure an optimal supply of nutrients, but also so as to avoid overfertilization. Overfertilization has interactions with the two other pillars, as it leads to dysfunctions within the adjacent biotopes and generates avoidable investments.

Training requirements for wild collectors should not be underestimated: staff should know exactly how to behave in the wilderness. This implies having a detailed understanding of how to handle waste, how to deal with sanitary issues, to protect both nature in general and the crop of interest, in particular. Furthermore, they must be experienced in the local botany, for example, how to handle poisonous plants – a critical topic when the crop of interest is poisonous. They must also be familiar with local wild animals and know how to behave if they encounter one. The collector must know the plant species exactly and should be able to distinguish it from ones that are closely related or look similar. There are some typical examples of mix-ups of wild collection food plants and poisonous species even in central Europe, such as the leaves of bear's garlic (*Allium ursinum*, used as a spice comparable to garlic) with lily of the valley (*Convallaria majalis*, a medicinal plant containing cardiac glycosides) [6]. Further examples are a mix-up of the roots and rhizomes of gentian (*Gentiana lutea*), which are used as bitters, with white hellebore (*Veratrum album*, also a cardiac glycosides containing plant) [7]. These examples demonstrate how important it is that collectors have the necessary botanical knowledge, especially when considering that the biodiversity in many regions of the world is much higher than in central Europe.

3.2.2
Environmental Sustainability

Cultivation should not cause the transformation of original habitats to cultivated land, but only be undertaken by converting it from farmland for another crop. Unfortunately, there are many negative examples, some of which should be mentioned: the decision to use plant-based renewable energies more intensively led to increased demand for cropland in several regions of the world. In Indonesia, for example, several thousand square miles of rain forest have been converted to cultivated land for the production of palm oil [8]. Many ecosystems are sensitive to influences and might lose their viability through land use and change forever. The treatment of the soil in the long term is crucial for the sustainable development of our civilization overall [9]. The cultivation of land should not have a negative impact on the ecosystem of the adjacent land. One example is overfertilization of farmland, which obviously has a negative impact on the soil, but also damages air and water quality. Environmental damage caused by nitrogen fertilization has been well known for decades and led, in the European Union, as far back as 1991, to a Council directive on nitrogen concentrations in drinking water [10]. Some EU member states meanwhile frequently calculate what are known as "farm-gate balances": the input of nitrogen entering the farm in fertilizer, feedstuff, and so on, is tracked, as is the output in produce (crops, milk, meat, and eggs). The remainder is assumed to remain on the land [11]. Meanwhile, chemical analytics and geo-systems have developed sophisticated systems that allow farmers to archive analytical results of soil analysis in geo-mapping systems, with which they can control the application of fertilizer to the particular needs of each block on the entire farm.

The control of wild collection regarding sustainability is much more complicated. It starts with ensuring compliance with international legislation: the Convention on International Trade in Endangered Species of Wild Fauna and Flora (CITES) entered into force in 1975 [12]. It is an international agreement between governments that endeavor to ensure that international trade in specimens of wild animals and plants does not threaten their survival. Frequently updated annexes to the convention contain lists of species

1) that are threatened with extinction;
2) whose trade must be controlled in order to avoid endangering them;
3) that are protected in at least one of the member states for which CITES parties should support control.

Companies intending to commercialize a product derived from a plant that is listed in one of the annexes should contact the responsible local authorities to find ways of ensuring sustainable production of the product.

Further international legislation comprises the "Convention on Biological Diversity" (CBD). It has three main objectives:

1) the conservation of biological diversity;
2) the sustainable use of the components of biological diversity;
3) the fair and equitable sharing of the benefits arising from the utilization of genetic resources.

The CBD was initiated by the United Nations Environment Programme (UNEP) and entered into force on 29 December 1993 [13]. This convention must also be taken into account if it is planned to commercialize a particular crop.

The Nagoya Protocol (2010) pursues the third goal of the CBD by giving clear guidance on how to ensure the fair and equitable sharing of the benefits arising from the utilization of a genetic resource [14].

3.2.3 Economic Sustainability

This means that an economy can support a defined level of economic production indefinitely in the long term, without funding or subsidies. It is very closely linked to social sustainability, as it interacts directly with the income of workers. The poverty threshold or poverty line is defined as "the minimum level of income deemed necessary to achieve an adequate standard of living in a given country." Work is sustainable only when workers earn enough money to support themselves.

3.3 Production Technology

3.3.1 Choice of Cultivation Location

The two essential environmental criteria are the soil and the climate.

Soil is characterized by its mineral composition, such as structure, texture, organic matter, pH value, and nutrients. Owing to the fact that this composition is so essential to the overall agricultural use of properties, soil quality rating systems have been developed in order to evaluate the potential of each particular area. The “World Reference Base for Soil Resources” (WRB) of the Food and Agriculture Organization (FAO) [15] is a globally applicable tool for classifying general characteristics of soils. A more functionally focused method is what is known as the “Muencheberg Soil Quality Rating” (M-SQR), which leads to comparative crop yield estimation calculations [16].

Climate is defined as the weather prevailing in an area for a relatively long period of time , which means it is a kind of average weather. The local climate is composed of light (duration and intensity), air humidity, precipitation, air temperature, and wind. While weather describes the current situation, climate is a statistically based description of the average weather (including volatility) of the area. By studying the local climate, the vendor or lessor of a property learns about potential risks and challenges of newly acquired land in an unfamiliar location. The weather forecast informs the farmer about the challenges for the next few days or weeks; the climate describes what weather events can be expected to happen in the long term. Another important factor to keep in mind is climate change: locations that are currently used for the cultivation or even wild collection of a crop might become extinct or at least endangered in the long run as a result of the change. Agriculture is always a long-term business and involves intensive financial investments. This is why the consideration of climate change is a key factor for decision making. The Intergovernmental Panel on Climate Change (IPCC), established by the UNEP and the World Meteorological Organization (WMO), is organized under the umbrella of the United Nations (UN). It frequently updates global reports on climate change. National and local official bodies provide data on regional climate development, as this topic is crucial for the feeding of the world.

Books can be filled with the complexity of the interaction between soil and weather, climate, and climate change (e.g., [17]). The current biodiversity is the result of millions and millions of years of evolution. For the last several thousand years, human beings have interfered with this evolution by growing their essential crops. And for the last several hundred years, they have unintentionally but intensively influenced evolution by the huge change in the way how they use the land.

Keeping in mind that nature is the result of evolution, it should be noted that each and every species has its needs based on the biotope where it has developed. This is why the basis of systematic cultivation always starts with the description of the natural habitat of the plant.

The cultivation starts with similar conditions to those of the original biotope of the plant. This is why soil and climate, including the length of day and night and the seasons, should be comparable to those of the origin of the plant. In later stages of cultivation, these factors may be altered. An important aid is the use of the “Koeppen and Geiger” [18] climate classifications maps. When comparing the historical climate data (Figure 3.2) observed during 1975–2000 (a) with those in scenario A1FI 2000–2025 (b), one can determine where lower or higher changes can be expected.

Soil analysis for contaminants is also a fundamental tool for the evaluation of the quality of a property. Heavy metals, for example, may exist in the soil as a natural resource; in most cases, however, they are found as contamination from

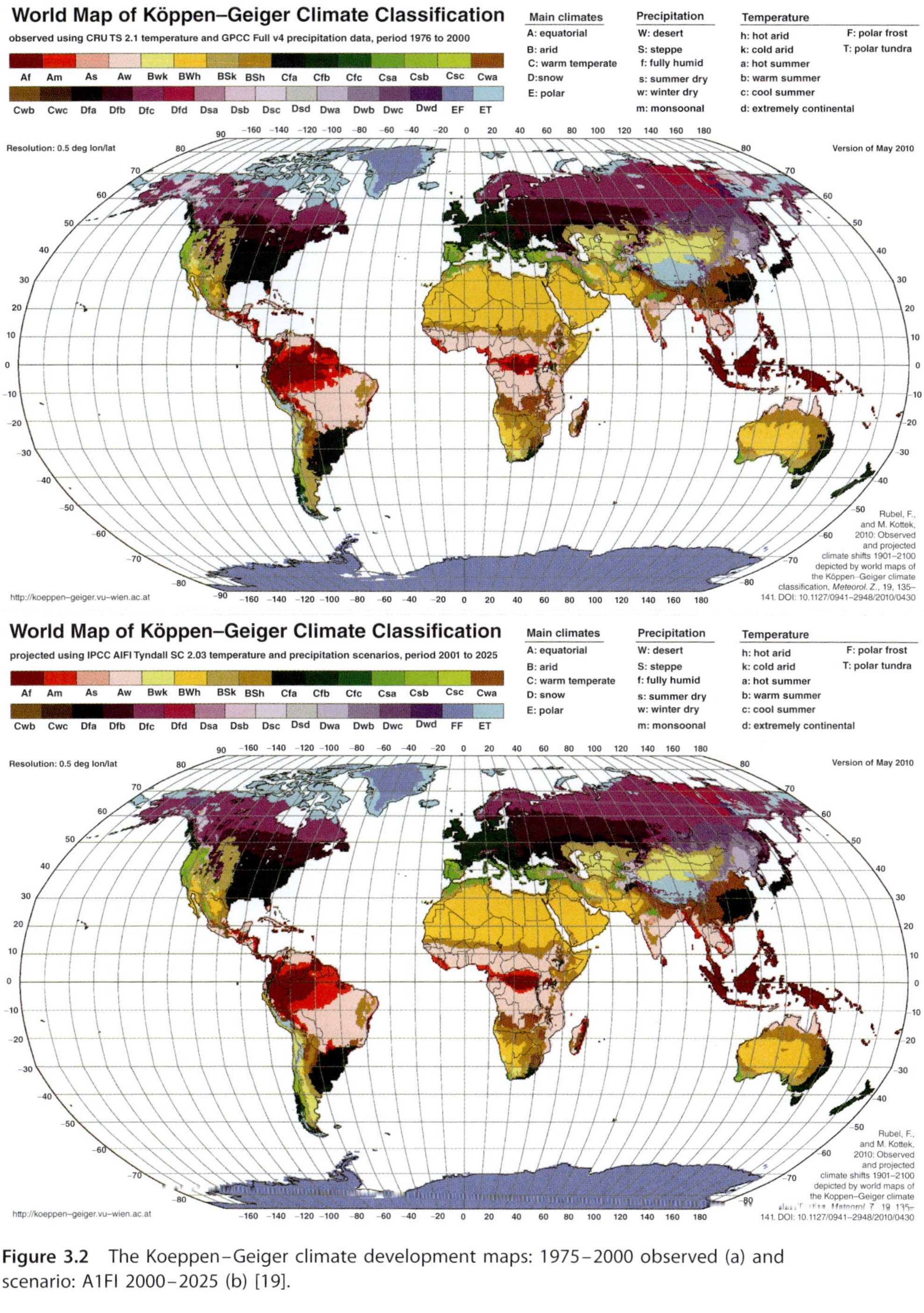

Figure 3.2 The Koeppen–Geiger climate development maps: 1975–2000 observed (a) and scenario: A1FI 2000–2025 (b) [19].

previous human activities such as mining or chemical industries. When leasing or buying a property, an investigation of the former land use is highly advisable. The key heavy metals to be tested for are at a minimum: cadmium, lead, mercury, and arsenic (see Section 3.5 on quality criteria); in some cases, even barium, zinc, and copper play a role. Depending on the previous land use, tests for further substance groups such as those below make sense:

- coal ash deposits, containing PAH, a group of potentially carcinogenic substances;
- sewage sludge, containing a variety of contaminants such as heavy metals, pesticides, pharmaceuticals, and even organisms;
- pesticides and herbicides, a huge variety of different, mainly organic substances which were used intensively in the past (e.g., Dichlordiphenyltrichlorethan (DDT) in the fight against malaria, Agent Orange in the Vietnam War).

In any case, the clearance of unaltered natural habitats to cultivate crops should be avoided, as global land use is already intensive, and the remaining natural environment should be protected.

3.3.2
Crop Rotation

Crop rotation means frequently changing the type of crop on the same agricultural area. It is a practice that has been carried out for thousands of years and has the following benefits for the soil:

- replenishment of nitrogen by green manure or the cultivation of legumes (e.g., alfalfa, beans, clover, lentils, soya beans, peanuts, and peas), which live in symbiosis with nitrogen-fixing bacteria, called rhizobia, in the root nodules of their rooting system;
- mitigation of the accumulation of plant-specific pathogens;
- restructuring of the soil texture by rotating shallow-rooted with deep-rooted crops.

Agronomic crop rotation between feeding crops and medicinal or aromatic plants usually does not lead to major problems, as most of the species belong to plant families that differ from prevalent arable crops. Because of this, they disrupt the prevailing disease and pest pressure. They enrich crop rotation for the given reasons, and in some cases even lead to an increase in organic matter or decrease of soil pests such as nematodes. Furthermore, perennial crops reduce erosion and the washing out of nutrients. In erosion-susceptible areas, there are certain cultivation techniques such as reduced or even zero tillage. They ensure a permanently covered surface that can withstand heavy rainfall, strong wind, and even prevent the evaporation of water. Weed control for these techniques is, however, a challenge. A more labor-intensive but even more erosion-preventing technique is to leave the stubble [20].

In firms specialized in MAPs, however, long crop intervals can result in a significant crop rotation problem. The main issue is increased disease and pest pressure. To avoid such calamities, the following approaches, which are known from normal agriculture, are applied:

- Intervals between crops of the same plant family
- Rotation between leaf, root (rhizome), and grain crops
- Choice of a suitable preceding crop
- In cases of extreme susceptibility, leaving areas fallow or cultivation of green manure to enrich the soil.

Some crops, for example, rhizome crops such as *Agropyron repens* or *Silybium marianum*, may cause problems for the subsequent crop, as their vegetative organs are persistent, and they may germinate even years later. Manual removal of these rhizomes is crucial, as the use of chemical herbicides should be avoided at all costs.

The choice of the immediately preceding crop is directly dependent on the requirements of the subsequent crop. For example, species that require a fine seed bed should not directly follow those crops that leave coarse plant residue, such as maize. The bound nitrogen from leguminous plants should be used by a following crop that requires a high amount of nitrogen.

3.3.3 **Fertilization**

Fertilization in agriculture is the application of additional nutrients to the soil or directly to the crop. The composition and concentration of the application is dependent on the quality of the soil and the requirements of the crop. Knowledge of these factors is the foundation for customized fertilization. Systematic soil analysis to identify nutrients, contamination, the composition, and characteristics such as the acidity or pH level is essential.

It is estimated that fertilization and achievements of the industrial revolution account for 30–50% of global crop yields [21]. In addition, fertilizers are an important industry, which is expected to reach sales of approximately US $195 billion in 2019 [22]. MAPs are costly overall in terms of labor, which is why measures including the application of suitable fertilizers are taken to achieve optimal yields.

Fertilizers can be inorganic or organic.

All the main inorganic elements play a role in fertilization; their proportions vary:

- The macronutrients (concentrations in dried plant material (without water) between 0.15% and 6.0%) are nitrogen (N), potassium (K), phosphorus (P), magnesium (Mg), calcium (Ca), and sulfur (S)
- The micronutrients (concentrations in dried plant material (without water) from 0.15 to 400 ppm) are: boron (B), manganese (Mn), iron (Fe), zinc (Zn), chlorine (Cl), copper (Cu), molybdenum (Mo), and nickel (Ni).

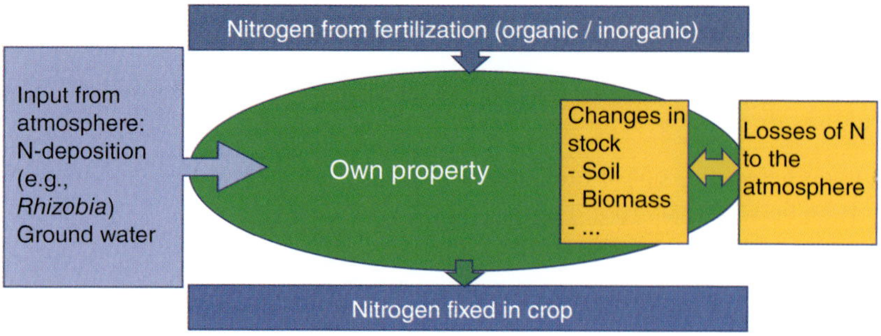

Figure 3.3 Farm gate cycle amended from [24].

Justus von Liebig defined the law of the minimum in the 1840s [23]. This states that crop yield is proportional to the concentration of the most limiting nutrient. If this nutrient is supplied to the crop, yields improve to the point where it competes with another nutrient. This demonstrates how important knowledge about the requirements of the crop and about the soil is.

As with nearly everything in nature, even nutrients are needed to a varying extent, which means that the agronomist should identify the limiting nutrients in order to know which concentrations can be sufficiently substituted. However, the agronomist should not apply excessively high concentrations, as these would otherwise cause senseless expenditure, but even more importantly, ecological damage as well. Meanwhile, scientists are striving to establish systems to achieve what are known as farm gate balances: the quotient between nutrients introduced into the soil divided by the nutrients utilized by the crop should be approximately "one." Such calculations allow assumptions to be made regarding the appropriate level of fertilization. However, the cycle of nutrients in the soil is very complex, as it does not depend only on interaction between the soil, added nutrients, and the crop. Figure 3.3 illustrates the interactions within the nitrogen cycle on a farm.

With the exception of nitrogen, which is chiefly converted chemically from atmospheric nitrogen to nitrogen salts, all nutrients are mining products, and consequently often contain other minerals. Even after purification steps, contamination with heavy metals cannot be completely excluded. If such mineral fertilizers are applied to a field over a long period of time, these heavy metals can accumulate there and cause critical concentrations in the crop.

Inorganic fertilizers tend to be released directly by the first leachings/rainfalls, which can then cause a harmful interaction such as nitrogen burn, when excessively high concentrations are applied to the plants. This led to the development of controlled-release fertilizers – formulations of the required inorganic substances with soluble polymers or coated with thermoplastics or sulfur. Such instead of these formulations lead to a slow release of the fertilizer, that is, a successive release after each leaching.

Inorganic fertilizers are regulated in the European Union [25] in order to guarantee effectiveness, but also to prevent adverse effects on the environment or the health of humans, plants, and animals when used according to the instructions.

Organic fertilization can be classified into the following sources:

- Naturally occurring mineral deposits such as saltpeter (containing potassium and nitrate as a salt), limestone (a calcium carbonate source chiefly used to adjust soil pH values), phosphate rock, and bone meal (high phosphate concentrations)
- Animal- and human-sourced materials such as manure, chicken litter, and guano
- Plant-sourced materials such as compost and seaweed, but also crop rotation with cover crops such as legumes, which are able to fix nitrogen from atmospheric nitrogen (N_2) in ammonia (NH_4^+) in the soil by symbiosis with bacteria called rhizobia [26].

Many of these sources provide organic matter and micronutrients, microorganisms, and a depository for carbon dioxide and thus enrich the soil with a kind of soil biodiversity. All of these organic fertilizers contain minerals mentioned in the section on inorganic fertilizers; however, their concentration in the organic matrix is not exactly quantified. This is why organic fertilization cannot be applied as systematically as inorganic fertilizers to adjust an imbalance in the soil; the release of its nutrients in the soil is low [27].

However, some biowastes can even support the growth of harmful soil pathogens that cause problems for the crops themselves and consumers (e.g., *Escherichia coli* or *Salmonella sp.* on vegetables). Many organic fertilizers such as sewage sludge contain contamination with heavy metals and organic chemicals such as active pharmaceutical ingredients, pesticides, and herbicides, making application of the fertilizers problematic [28].

On the one hand, fertilization is the essential catalyst of the green revolution, which has led to much more efficient agriculture for feeding the continuously growing population of the world. On the other hand, it must be applied in smart, appropriate ways, as the world is already experiencing a number of problems, such as the following:

- *In soil*: eutrophication of oceans, blue baby syndrome, heavy metal, and persistent organic pollutants;
- *In the atmosphere*: greenhouse gases:
  - Methane emissions caused by ammonia-containing fertilizers (e.g., from rice paddy fields)
  - Laughing gas (N_2O) generally through nitrogen fertilization
  - Carbon dioxide by decomposition of compost and manure on/in the soil.

Organic farming with its various branches strives to keep the complex cycles between crop, soil, and atmosphere in a sustainable balance – an aim that ensures the most respectful treatment of nature.

3.3.4 Organic Farming

Organic farming has a long and diverse history: since the start of agriculture, different methods have been developed to benefit from ecological interactions such as crop rotation or forest gardening. However, even hundreds of years ago, people brought about the downfall of their own culture by unsustainable agriculture; a well-known example is the deforestation of the Easter Islands [29].

There is no clear definition for organic farming, but almost all organic farming systems apply alternative farming methods such as mechanical cultivation including weeding, green, and animal manures, crop rotation, and integrated pest management. The aim is to avoid harmful substances, such as synthetic pesticides and herbicides, mined fertilizers, and other mechanistic farming approaches. The industrialization of agriculture, which started at the end of the nineteenth century, generated a massive boost in crop yield. The core steps were the mechanization and automation of field operations and the application of mineral fertilizers driven by inventions such as the Haber–Bosch process, where atmospheric nitrogen is converted into ammonia. In parallel with this rapid agricultural industrialization, opposition grew within the agronomic community. People such as Sir Albert Howard, who worked together with his wife, Gabrielle, in Bengal, intensively studied traditional Indian agricultural methods. They came to the conclusion that some of these were superior to the conventional way. In 1940, they published their book "An Agricultural Testament" [30], which remains a kind of bible for organic farming. Sir Albert Howard is regarded as the father of organic farming.

Rudolf Steiner, the creator of the philosophy of anthroposophy, gave his "Lectures on Agriculture" in 1925 [31] – the start of biodynamic agriculture.

Meanwhile, a large variety of different approaches to organic farming techniques exist. In this book, only some prominent ones are mentioned, such as the following:

- *Biodynamic farming* has a very comprehensive modus operandi. It has a holistic view of ecological interrelations between plant growth, soil fertility, and livestock care. It is embedded in Rudolf Steiner's (1861–1925) spiritual and mystical anthroposophic movement. In common with general organic farming, it is against the use of artificial chemicals. It differs by the fact that it treats the handling of the soil, plants, and animals as an integral system. Biodynamic farmers raise domestic varieties and local breeds, and they promote local distribution systems. The whole process is organized using an astrological sowing and planting calendar. There are several certification agencies for biodynamic products; the most important one is the Demeter International group [32].
- *Natural farming* is also called *Fukuoka farming*, named after its founder, the Japanese philosopher Masanobu Fukuoka (1913–2008), is an ecological farming approach. It is also called *do-nothing farming*, as it claims to avoid the need for manufactured inputs and equipment. The approach is to exploit the complexity of the living organisms in each particular ecosystem. Natural farming is

embedded in Fukuoka's philosophy of achieving the overall goal of "the cultivation and perfection of human beings" [33]. Fukuoka identified five core principles for natural farming: (i) no tillage, (ii) no fertilizer, (iii) no pesticides (or herbicides), (iv) no weeding, and (v) no pruning. Although developed in the western Japanese area of Shikoku, a subtropical region, his method has been used even in India (*rishi kheti*) [34] and parts of Africa. The soil is kept covered with vegetation while it is cultivated. Sometimes, further beneficial weeds, such as alfalfa or other herbaceous legumes, are added to enrich the soil. Weeds are never removed from the soil but cut at most. The cut material is left on the ground to become compost and add organic material to the soil. The farming system integrates even livestock – typical examples are free-range chicken or carp and ducks in rice fields.

- *French intensive farming and bio-intensive farming* is claimed to be as intensive as possible (highest crop yield possible) while maintaining and even improving soil fertility. In a closed system, the highest possible sustainability is striven for. Many aspects of the techniques involved in bio-intensive farming are derived from the ancient Mayans, Chinese, and Greeks. It was Alan Chadwick (1909–1980) who merged biodynamic farming with these techniques – and even French intensive gardening [35]. Strongly influenced by Rudolf Steiner (he had been a student of his), Chadwick created an integrated system involving the cultivation of four adjacent areas of topsoil and subsoil, a method also known as *double-digging*. The aim is to establish aerated and raised beds, carry out compost planting, intensive planting, and companion planting, use open-pollinated seeds, and achieve a carefully balanced planting ratio of about 60% carbon-rich crops (to produce compost), about 30% energy (calorie)-rich crops (food production), and an optional 10% in cash crops (for sale) [36].
- *No-till farming, also known as zero tillage or direct drilling* means not disturbing the soil through tillage while growing crops or pasture from year to year. The soil's surface is kept intact to ensure the storage of water entering the soil and the retention and circulation of nutrients so as to reduce erosion caused by rainfall and wind. While ploughing applications turn the whole subsurface ecosystem upside down, zero tillage preserves the amount and variety of life in the soil, thus improving biological fertility and soil resilience. The zero-tillage pioneer, Edward Faulkner, established the first process in the 1940s in the United States [37]. Successors such as Herbert Bartz in Paraná, Brazil, however, combined it with massive applications of herbicides to automate the process [38].

Given the diversity within the different approaches, it is obvious that private consumers as well as industrial customers did not have comprehensive transparency about the quality of the products produced. This led to the establishment of intergovernmental standards.

The European Council of Agricultural Ministers adopted Council regulation (EEC) no. 2092/91 [39] on organic farming and the corresponding labeling of agricultural products and foods back in 1991. At that time, the implementation of this regulation had been part of the reform of the then EEC common agricultural policy for the 15 member states. Initially, it was only applicable to plant products;

provisions for animal products were introduced later. Genetically modified organisms either for use in production or as products had been generally excluded from organic production.

As several amendments had been made since 1991, the regulation became very comprehensive and was not easy to understand, which is why it was completely revised in 2007 as Council regulation (EC) no. 834/2007 [40].

Council regulation (EC) no. 834/2007 now sets common minimum standards for produce within the entire European Union as well as for imported produce. The member states as well as private organizations are still entitled to enact their own additional, stricter standards, some of which are mentioned above. Producers that fulfill the requirements can label their produce with the EU organic logo [41].

In the cosmetics industry, the "Ecocert" standard [42] defines and promotes quality levels that are higher than legal minimum requirements and ensures the utilization of natural and organically produced ingredients. The standard, ECOCERT 2002, has been developed by a consortium of experts, consumers, and members of the responsible authorities. The standard is officially recognized, and its implementation is supervised by the authorities. The French-based organization has certified more than 8000 different products globally.

The following is the fundamental policy of Ecocert:

1) Ingredients should be produced from renewable sources and produced with environmentally friendly processes. At least 95% of the ingredients should be of natural origin and the production process should be monitored.
2) Depending on the chosen labeling, the use of organically farmed ingredients is as follows:
   a. For bio-cosmetics, at least 95% of the herbal ingredients of the formulation and 10% of the overall ingredients have to be produced according to organic farming requirements
   b. For natural cosmetics, at least 50% of the herbal ingredients of the formulation and 5% of the overall ingredients have to be produced according to organic farming requirements
3) A transparent declaration giving clear information to the customer.

3.4 Seed and Seed Stock

3.4.1 Breeding

Plant breeding is the agricultural discipline that genetically optimizes the characteristics of the cultivated plant in order to obtain better quality. Since agriculture began in the late Neolithic period, humans have practiced breeding, as the dwellers optimized their crops by preferring to take seeds from those individual plants with the most desired quality – selection. This is why major crops such as

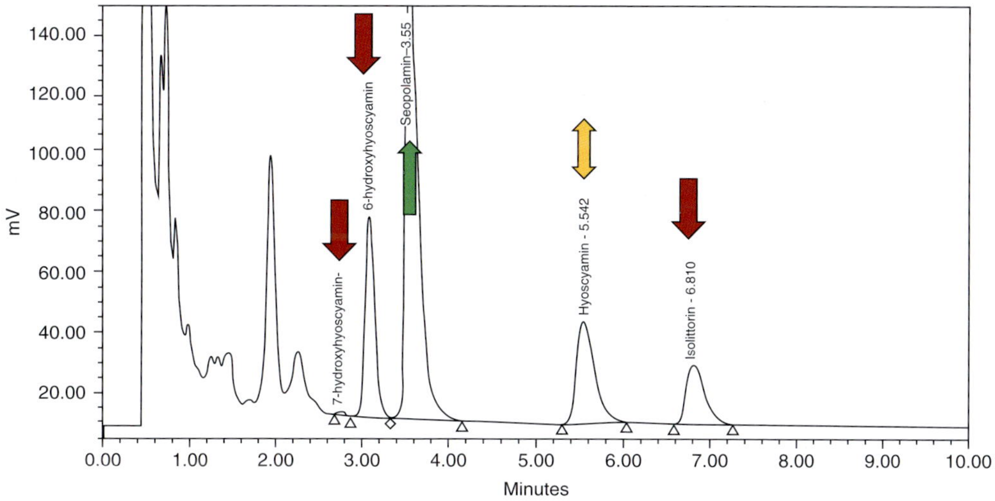

Figure 3.4 High Performance Liquid Chromatography-fingerprint chromatogram (HPLC-fingerprint chromatogram) of tropane alkaloids in a *Duboisia* hybrid sample.

wheat, rye, corn, potatoes, rice, and so on, can look back to a breeding history of thousands of years. Breeding became much more focused when Gregor Mendel identified the "laws of inheritance" in 1865. Today, breeding on major crops is mainly done by professional plant breeding institutions organized as enterprises or as government institutions, associations, or research centers.

Breeding ambitions can be driven by various intentions, such as the following:

1) yield increase (e.g., content of secondary compounds, biomass per time vegetation period) or even optimizing the composition of secondary compounds within one species; in Figure 3.4, the challenge to increase the content of the tropane alkaloid scopolamine and reduction of content of other tropane alkaloids in an HPLC fingerprint chromatogram is visualized;
2) adaption to various climatic conditions such as drought, wind, and heat tolerance (e.g., when the plant is going to be cultivated in another area or because of climate change reasons);
3) optimization of its tolerance or resistance to certain pests;
4) improvement of harvesting efficiency.

Breeding programs are set up mostly to obtain not only one but several of these ambitions.

Breeding approaches can be distinguished between classical breeding and genetic engineering. While classical breeding targets to reach the intended breeding aims by classical techniques, genetic engineering uses knowledge about the functionality of certain genes that are known to be responsible for the breeding aim and transfers these into the plant by inserting them into the host plant. As genetic engineering is an absolute no-go criterion in organic farming, it is not described in further detail.

Before starting a breeding program, it is strictly recommended to do an economic evaluation of the potential benefits of the breeding aim, because breeding, depending on the applied strategy, can be very costly and, in any case, lasts several years. This is why it is worth carrying out breeding only for those crops that are needed for the long term and are required in a significant quantity. If only small quantities are needed and the life cycle of the product made from it cannot be predicted to be longer than at least two decades, the establishment of a sustainable wild collection is in many cases more efficient.

Breeding should not be started unless breeding aims are concretely specified to avoid confusion during the execution of the program.

The following are the major activities of selective breeding [43]:

1) *Collection of variation*: as many as possible different accessions with differing characteristics should be collected from nature (considering the CITES and CBD) or from botanical gardens or other germplasm pools. The various characteristics should be described as thoroughly as possible.
2) *Selection*: is systematically carried out by comparing the collected accessions.
   a. *Positive mass selection*: those plants with the desired characteristics are used for crossing. Each individual is inspected thoroughly to observe the phenotype. Meanwhile, certain morphological characteristics are selected with automated camera systems. The result can either be released directly (4) or crossed (3) in a combined breeding concept.
   b. *Negative mass selection*: those plants with undesirable characteristics are eliminated from the population.
   c. *Combination breeding*: both steps are merged so that in the first generation, plants of interesting characteristics are detected (genotypes), which can be brought together by crossing in the next generation.
3) *Crossing*: self-pollinating selected plants can commonly cease blooming. For cross-pollinating plants, the pollen of other individuals of a selected population either cease blooming together or, more targeted, the pollen of a desired individual can be used to pollinate the whole population.
   In combination breeding, the desired characteristics of selected plants are attempted to be fused by bringing the pollen of the one to the stigma of the other – by this the most targeted pollination is secured.
4) *Release*: the seeds of bred fruits are purified and cultivated to obtain the next generation with merged characteristics. The cultivation should be carried out under conditions similar to large-scale cultivation to ensure that in later steps even the effects of the environment are considered. When the plants are in the development state in which conclusions about their performance can be made, they can be selected for future use.
5) *Multiplication*: the selected plant is only one given individual or a small population that needs to be multiplied to allow large-scale cultivation. The multiplication can be done by either of the following methods of production:

a. *Vegetatively* – clonally propagated plants having the same genetics as the selected plant itself. Many woody plants are propagated by cuttings. In a greenhouse/nursery, shoots of the selected plants are taken and are treated to produce roots. After the roots have robustly established, the young plants can be planted outdoors. Non-ligneous plants can be vegetatively propagated by various *in vitro* techniques.
b. *Generatively* – sexual propagation by the production of seeds. To ensure genetically homogenous seeds, crossing techniques as mentioned above (2) are applied.

6) *Distribution of the new variety*: the bred result is obtained after several selection generations and, depending on the propagation cycle of the species, might last for decades, as breeding depends on crossing, and crossing depends on flowers. This is why breeding is costly, and it can be of interest to protect intellectual property. For this, the plant breeders' rights have been established. With these rights, the owner of the bred plant can either license the variety to others or become exclusive marketer of the variety. Requirements for the application for plant breeders' rights are (i) the variety must have not yet been commercialized, and (ii) it must be stable, distinct, new, and uniform. Stability must be proven by obtaining the same phenotype for several generations, which means that the characteristics are genetically fixed. In order to prove distinction, it needs to have botanical characteristics that have not been described before, such as another shape, color, or height [44].

The following breeding technologies are more focused:

Heterosis breeding: breeding program over several generations, as cross pollinators are pollinated in a targeted manner until nearly homozygous plants are obtained – the alleles of the DNA are nearly the same. Through heterosis breeding, increased yields as well as tolerance and resistance against climatic conditions and pests are achievable. It is essential to maintain mother blocks or kinds of conservation breeding, because generative propagation of heterosis-bred seeds leads to inbreeding depression.

Hybrid breeding: the results of heterosis breeding programs – inbred lines – are crossed. The next generation has the so-called "heterosis effect": they are on average more vigorous than the parent generation. Furthermore, positive characteristics of the parent generation are combined. Seed production must be carried out comparable to that needed in heterosis breeding.

Haploid breeding [45]: using unripe pollen, haploid plants (with only one DNA strand) are developed. These are treated with colchicine or similar substances, which causes the resulting double-haploid (homozygous) plant to fail to form daughter cells = mitosis inhibition.

Mutation breeding: various kinds of stress, such as X-rays, heat or cold shock and chemical mutagens are applied to the plant material to obtain mutants. Only a very small portion of the treated material can be successfully used for the following breeding steps, as the DNA is changed in an untargeted manner by the stress.

3.4.2
Seed

Commercial seed can be prepared for the market with various treatments [46]:

- *Normal seed*: from selected plants are harvested, dried, purified, and stored under defined conditions.
- *Calibrated drill seed*: normal seed is sieved to obtain distinct seed size classes in order to conduct mechanized drilling efficiently.
- *Graded seed*: seeds with a low specific weight are separated. This is done to increase average germination capacity by eliminating those seeds having the lowest vigor.
- *Incrusted seed*: seeds are covered with fungicides, insecticides, nutrients, natural extracts, or other actives by incrustation. They are often even covered with a colored abrasion-resistant layer.
- *Pelleted seed*: seeds are covered by a composition of growth-stimulating pest repellent formulation comparable to incrusted seed, with the difference that the shape is homogenously formed for high-quality mechanized drilling.

The quality requirements for major food and feed are regulated by the European Union [47]. For minor crops typical quality criteria are foreign matter (not more than between 2% and 5%, depending on the seed) odor, loss on drying, color, and thousand seed weight (TSW) in grams.

In organic farming, the use of seeds produced organically by agricultural operations is crucial. To ease acquisition of such seeds, EU member states keep an "organic seed" database [48].

3.4.3
Vegetative Propagation

For some plants, vegetative (clonal) propagation is economically more attractive than seed production. Especially woody plants are propagated efficiently by cuttings (see Section 3.4.1 Breeding). For some MAPs, vegetative propagation is obligatory, as they are hybrids so that generative propagation is either impossible or leads to backcrossing results. For example, peppermint is a hybrid of *M. aquatica* × *M. spicata*, in which *M. spicata* is a hybrid of *M. rotundifolia* and *M. longifolia*. The production of seeds would lead to following generation of heterogeneous plants with characteristics of the original species with different compositions of natural compounds. It is, therefore, cultivated by stolons [49]. Garlic is not able to even produce seeds and is, therefore, propagated by cloves [50].

3.4.4 Stock Maintenance

3.4.4.1 Diseases

Plant pathogens belong to the following groups of organisms [51]:

1) *Bacteria*: most of the plant-associated bacteria are saprophytic and do not harm them. However, some of them damage plants by producing toxins. *Agrobacterium* has an influence on the plant by its own production of the plant hormones "auxines" and causing tumors. *Erwinia* produces cell wall–degrading enzymes which cause soft rot. There is no large-scale chemical control available. *Erwinia* is controllable with sprayed solutions of hydrated lime. Generally, coppicing of the affected areas and observant avoidance of carryover is the most effective measure.
2) *Viruses, viroids, and virus like organisms*: they exist in several types and as a massive number of different species. Most of them are asymptomatic or cause only minor crop yield loss. Of economic importance are virus infections in perennial plants such as fruit trees. They need vectors, organisms which transmit them: insects or aphids, but even fungi, nematodes, or protozoa are such vectors. If they can be controlled, the impact of viruses is reduced to a minimum.
3) *Fungi*: these reproduce both sexually and asexually and propagate by spores that can be transported for long distances by wind or water. Some are even soilborne, which means that they can stay in the soil for a long time until they reinfect a plant. Conventional agriculture deploys chemical fungicides for control, but struggles with emerging resistant strains. In organic agriculture, hygiene regimes are essential. Cultivation is carried out in a manner such that affected leaves are disposed of. Soil is covered with mulch to prevent soilborne fungi from reaching the plant. If irrigated, plants are watered only at the stem; the leaves remain dry. Woody plants are frequently pruned to ensure ventilation. The use of copper as a fungicide in organic agriculture is controversial, as copper ions have comparatively high ecotoxicity [52].
4) *Oomycetes*: these are fungus-like organisms. Their most prominent member is *Phytophthora.* There is no chemical treatment for outdoor cultivation available; it can be controlled only by keeping affected blocks fallow, by allowing no stagnant moisture, and by disinfection of equipment that is used.
5) *Parasitic plants*: the most prominent parasitic plant is mistletoe, which is itself used as a medicinal plant for the treatment of cancer.
6) *Nematodes*: these wormlike animals live in the soil, mainly in subtropical, and tropical regions, and some parasitize plant roots, which causes crop yield loss. Potato cyst nematodes are the most prominent example and cause €100 million worth of loss every year. The best applicable measure for controlling them is crop rotation with those crops that are not favorable for nematodes.

3.4.5
Pests

Pests in this context comprise higher animals that cause harm to plants. These comprise the following:

1) *Insects*: several species feed on plants and directly damage crops in the field or when stored (e.g., caterpillars, aphids, larvae, grasshoppers, and crickets). Others cause indirect damage by spreading viral diseases of crop plants, for example, sucking insects (even grasshoppers and aphids). Several chemical insecticides have been developed to control infestation. Insecticides can be divided into contact and systemic-acting substances. Contact insecticides are applied by spaying; they remain on the surface of the treated plants (Cypermetrine, Permetrine, etc.). Systemic insecticides are applied to seeds, to soil, or even sprayed onto plants. They are distributed through the whole plant during growth (Dimethoate, Imidacloprid, Clotianidine, etc.). While residues of contact insecticides can be diluted by washing, the systemically absorbed ones remain in the crop. Undifferentiated insecticide application even harms beneficial insects such as honeybees. In the European Union, some insecticides of the highly bee-toxic neonicotinoid group were banned for this reason in 2013 [53]. Many insects can be controlled by deploying beneficials – organisms that have a positive impact on the growing process. They are, among others, insects, bacteria, nematodes, or even plants or fungi, and they improve maintenance of soil health, pest control, and pollination. A prominent example is the aphid-eating ladybug [54]. Many cultivated plants profit from companion planting, such as carrots with onion [55]. *Bacillus thuringiensis* is a bacterial beneficial that affects butterflies.
2) *Snails*: these feed on various crops. In conventional agriculture, they are controlled using chemicals such as Methiocarb and Thiodicarb – nerve-toxic substances. Conventional agriculture prefers planting snail-deflecting crops such as spinach, cress, or mustard around the crop or to quench them by growing sage, thyme, chamomile, or lovage. Liquid manure made from tomato leaves is another alternative; the establishment of snail-feeding animals such as hedgehogs, moles, shrews, and birds such as blackbirds supports control.

3.4.5.1 Weed Control

The agronomist strives to provide the best conditions for his plant and because of this pays attention to the competition of his plants with those that are not intentionally planted. Historically, weed control had been carried out by weeding, ploughing, tilling, and other mechanical means or by covering with weed mats or films. Chemical herbicides have been developed to selectively harm undesired plants while not affecting the crop plant. Organic agriculture applies mechanical methods with smart crop rotation regimes.

3.4.6 Harvesting Technology

For the major crops, many harvesting technologies have been developed and optimized efficiently. Devices are available for each and every plant organ, flowers, leaves, herbs, roots, rhizomes, and grains, most of them capable of directly separating the crop from foreign matter. For most cultivated MAPs, such devices are adapted to the morphology of their particular crop organs. Generally, equipment that is used should be maintained and be cleaned before use to avoid cross-contamination with the residue of the crop that had been harvested earlier. Lubricants should be wiped off; stains should be removed. The quality of the crop depends on the harvesting process and also the choice of the right time: many secondary compounds are known to peak in a certain vegetative phase of the plant. For example, anthocyanins in red wine decline shortly before leaf drop, and the content of epicuticular flavonoids in birch leaves wanes during the summer [56].

3.4.7 Purification of Harvest

Even though harvesting technology is engineered to harvest the crop as selectively as possible, foreign matter cannot be prevented completely. For most harvesting procedures, a further purification step is therefore crucial. Depending on the crop, sieving, classification, threshing, or manual elimination of unwanted material is carried out. Roots and rhizomes are washed to eliminate remains of the soil. Purification can be carried out on fresh or on dried material. If purification is carried out on fresh material, the process should be accomplished as soon as possible to avoid fermentation and bacterial growth on the fresh crop. Roots and rhizomes are washed before drying.

3.4.8 Mechanical Treatment

The crop is then either directly brought into a preserved condition, which can be done by drying or freezing; or, it is comminuted to obtain the optimal size for extraction. The comminution itself is done either by crushing (flowers, leaves, and herbs) or by cutting (leaves, barks, roots, and rhizomes).

3.4.9 Thermal Treatment

The drying process is a further postharvest treatment process step that is also critical to the quality of the product. If dried at too low a temperature, fermentation can start or the crop can be adulterated by microbial growth. If dried at too high a temperature, natural compounds such as enzymes or secondary compounds can degrade, or the texture of the crop might change because of the Maillard reaction.

The drying process is carried out until the crop is visibly dry. Herbs and leaves are crushed. They have, on average, a loss on drying of less than 10%, which is necessary to avoid microbial propagation during storage.

3.4.9.1 Natural Drying

Especially wild collected plants are still dried without a technical aid. They are put in thin layers on a clean surface and then aerated in the shade until the crop is visibly dry. Hygiene and avoidance of direct light are critical to preserve optimal quality.

3.4.9.2 Artificial Drying

The main artificial drying technologies are tumble dryers, tray dryers, and belt (or layer) dryers, all of them having their pros and cons. Heat is produced by burning gas or other combustible media, using heat exchangers to prevent direct contact between the exhaust gases and the crop. The ecologically most-appreciated drying technology is the use of artificial dryers with solar energy supply.

3.5 Quality Criteria

3.5.1 Quality Management

Several organizations have developed quality management approaches for the production of aromatic as well as medicinal plants. The most prominent are the following:

1) The WHO (World Health Organization) "Guidelines on Good Agricultural and Collection Practices (GACP) for *Medicinal Plants*" [57]
2) The EUROPAM (European Herbs Growers Association) "Guidelines for Good Agricultural and Wild Collection Practice of *Medicinal and Aromatic Plants* (GACP-MAP)" [58]
3) The EMA (European Medicines Agency) "Guideline on Good Agricultural and Collection Practice (GACP) for *Starting Materials of Herbal Origin*" [59].

The focus of the three guidelines is on differing products. Whereas the EMA's and the WHO's guidelines focus on plants for medicinal plants only, the EUROPAM guidelines include even aromatic plants.

As the EMA GACP guide [59] is the strictest one, it is exemplarily described by summarizing the various chapters:

1) *Introduction*: the purpose of the guide is to set out minimum requirements for the establishment of an appropriate quality management system, as the quality of the raw material has a massive impact on the quality of the products made from it.

2) *General*: the document is intended to address specific aspects of the processes and aims to ensure the identification of critical quality-relevant process steps needed to achieve good quality. Furthermore, it sets out minimum standards for consumer safety regarding hygiene and the handling of the crops to ensure that consumers are not adversely affected. All partners involved in the production – growers, collectors, traders, and processors – should comply with these considerations.
3) *Quality assurance*: agreements between the involved partners should comply with the legal requirements and should be laid down in written form.
4) *Personnel and education*: the processing procedures should comply with legal requirements, and staff should be trained regarding the processes and hygiene responsibilities. The welfare of the staff as well as its health (e.g., protection from toxic or allergenic plants) should be ensured – this includes the allocation of adequate protective clothing (social responsibility). People with open wounds or infections or suffering from infectious diseases should be excluded from areas with open handling of the goods. Growers and collectors need further suitable training, for example, on how to clearly identify the species, but also on how and when to harvest in a sustainable manner. They need training for the execution of the appropriate documentation. If they drive vehicles such as tractors or harvesters or apply chemicals such as pesticides, they need sufficient training. Generally, people involved in cultivation or collection need appropriate knowledge on the protection of the environment and the conservation of plant species.
5) *Buildings and facilities*: these should be clean, well-ventilated, and separated from housing livestock. They must provide adequate protection of the goods against birds, rodents, and domestic animals and a well-maintained pest control program should be in place. The goods should be appropriately packaged and stored on pallets to prevent cross-contamination and adulteration.
6) *Equipment*: this should be clean, well maintained and, if needed, regularly calibrated. Surfaces with direct product contact must be cleaned after use to prevent cross-contamination.
7) *Documentation*: each and every quality-relevant production step, such as irrigation, fertilization, pesticide application, trimming, harvesting, and postharvest treatment, should be documented. Records for the documentation of quality-influencing events such as extreme weather conditions should be recorded. Geographical locations should be documented, and the batch records should be established in a way that traceability from the area where the plants were grown to the packaging of the dried material is ensured. All agreements between the involved parties (staff, traders, and customers) should be documented, preferably as contracts.
8) *Seeds and propagation material*: the origin of the materials used should be documented in a traceable way in terms of genus, species, variety/cultivar/chemotype, and origin. Starting materials should be free of pests and diseases.

9) *Cultivation*: the cultivation should not have an impact on the environment. Appropriate crop rotation should be implemented. Soil contaminated with sludge, heavy metals, or other residues shall not be used. The application of chemicals should be reduced to a minimum. Manure should be composted and without human feces. Fertilizers should be applied strictly according to their specifications. Irrigation should be done with a suitable water quality and according to the requirement of the species, and any pesticide, herbicide, or other chemical application should be used as little as possible and strictly according to legislation and the specifications of the producer.
10) *Collection*: this has to be carried out carefully to keep the natural habitats as untouched as possible. In all cases, the legal requirements including the CITES requirements must be fulfilled.
11) *Harvest*: this should be carried out at the time that provides the optimal quality for the product. Damaged plants must be excluded, and harvest should be executed under optimal weather conditions: potential contamination with wet soil, dew, rain, humidity, and so on, should be avoided. If avoidance is impossible, processes that reduce potential risks such cleaning and drying soon after harvest should be in place. Generally, harvested material should be brought into a preserved condition (by drying, freezing, direct distillation, etc.) to prevent fermentation or growth of microbes. All equipment that comes in contact with the crop should be as clean as possible. If possible, contact of the crop with soil should be avoided. Incidental co-harvest of other plants such as toxic weeds should be excluded.
12) *Primary processing*: this includes washing, cutting before drying, fumigation, freezing, distillation, drying and should be carried out immediately after drying. The packing step should be carried out directly after the preservation step. In between, the crop should be protected from direct sunlight, rainfall, insects, and so on. In the case of open-air drying, a strong focus must be on the implementation of strict hygiene rules. Direct light must be excluded. The crop should be spread in thin layers and air circulation must be ensured. In the case of other drying technologies, the drying process should be adapted to the specific requirements of the plant. Records must be maintained and archived. All materials should be inspected and most likely also sieved (in well-maintained sieves) to eliminate foreign matter.
13) *Packaging*: this should be carried out in preferably new materials. If packaging materials are recycled, they should be cleaned in suitable ways. The packaging materials should be stored in clean and dry places separately from other products and inaccessible to livestock, domestic animals, and pests.
14) *Storage and distribution*: warehouses should protect the material from direct light and should be well ventilated, without extreme temperature fluctuations. Fresh material should be stored between 1 and 5 °C, while frozen should be stored below −18 °C. When bulk transport is carried out, measures should be implemented to avoid mold formation, such as

suitable aeration. If fumigation has to be applied, it must be done only by certified experts and with registered chemicals. Any application must be documented.

3.5.2
Quality Control

Herbal raw materials used for the production of pharmaceutical products (e.g., phytoextracts and natural compounds) should comply with the general monograph "Herbal drugs" (1433) of the European Pharmacopoeia [60].

For each species, a detailed description of the appearance and the organoleptic properties is required.

To verify the identity, macroscopic and microscopic examinations (2.8.23) are required; chromatographic tests such as thin-layer, high-performance liquid or gas chromatography are also used.

Purity is tested by performing a test for foreign matter (2.8.2), which should not exceed 2%. The test for total ash (2.4.16) or ash soluble in hydrochloric acid (2.8.1) is carried out to verify how much soil has entered the product. Loss on drying (2.2.32) or water content (2.2.13) gives information about the drying process. The maximum content is different for various materials. It should, on average, not be more than 12%. High water content is not in the interests of the buyer, as it reduces the amount of extractable matter. In concentrations greater than 12%, it even promotes the growth of microbes, which can lead to degradation of the ingredients and produce toxins such as mycotoxins (2.8.16, 2.8.23). A test for the heavy metals (2.4.27) of cadmium, lead, and mercury is obligatory. A test for other heavy metals should be considered if they are known to exist in high concentrations in the soil or if the plant has a tendency to accumulate particular heavy metals. Pesticide and herbicide residues should not exceed the limits listed in the Pharmacopoeia (2.8.13) and EU directive 396/2005. Even radioactivity should be measured during the qualification of a new supply. Microbial contamination (5.1.8) must be tested; the acceptable limits are specified and depend on the subsequent production steps. For dried herbs directly used as herbal teas, the limits are stricter than for those used as raw materials for extraction.

As samples should be representative for a whole batch, it is essential for the robustness of the analytical results that the sampling process is also meaningfully specified (2.8.20). If a batch consists of fewer than three containers, samples of each container are taken and homogenized. If there are three containers or more, sampling is carried out on $n^* = \sqrt{N} + 1$ containers. Within each container, sampling is carried out in the upper part, the middle part, and the lower part. There are further requirements such as the definition of the mass of each sample and minimum weights depending on the plant organ (roots, barks, rhizomes, and stems >500 g; leaves, flowers, seeds, and fruits >250 g; fragmented plant parts >125 g).

Glossary and Abbreviations

DDT	Dichlordiphenyltrichlorethan an isecticide with a very long persistemnce
EMA	European Medicines Agency
EUROPAM	European Herbs Growers Association
FAO	Food and Agriculture Organization of the United Nations
GACP	Good Agricultural and Collection Practices
HPLC	High Performance Liquid Chromatography
MAP	Medicinal and Aromatic Plants
WHO	World Health Organization

References

1. Chizzola, R. (1993) The main pyrollizidine alkaloids of petasis hybridus: variation within and between populations. ISHS Acta Horticulturae 333: WOCMAP I - Medicinal and Aromatic Plants Conference: Part 1 of 4, November 1993
2. Yadav, A.K., Singh, S., Dhyani, D., and Ahuja, P.S. (2011) A review on the improvement of stevia [Stevia rebaudiana (Bertoni)]. *Can. J. Plant Sci.*, **91** (1), 1–27.
3. ISSC-MAP – Bundesamt für Naturschutz *http://www.bfn.de/fileadmin/MDB/documents/service/skript195.pdf* (accessed 29 March 2014).
4. United Nations General Assembly (1987) Report of the World Commission on Environment and Development: Our Common Future, Transmitted to the General Assembly as an Annex to Document A/42/427 - Development and International Co-operation: Environment; Our Common Future, Chapter 2: Towards Sustainable Development; Paragraph 1. United Nations General Assembly, 20 March 1987, *http://www.un-documents.net/ocf-02.htm* (accessed 29 March 2014).
5. Thwink The Three Pillars of Sustainability, *http://www.thwink.org/sustain/glossary/ThreePillarsOfSustainability.htm* (accessed 29 March 2013)
6. BfR (2005) Risk of Mix-up With Bear's Garlic, 10/2005, *http://www.bfr.bund.de/en/press_information/2005/10/risk_of_mix_up_with_bears_garlic-6228.html* (accessed 14 March 2014).
7. Grobosch, T., Binscheck, T., Martens, F., and Lampe, D. (2008) Accidental intoxication with *Veratrum album*. *J. Anal. Toxicol.*, **32**, 768–773.
8. Smith, J. (2010) *Biofuels and the Globalization of Risk*, Zed Books Ltd, London.
9. David, R. (2007) *Montgomery: Dirt – The Erosion of Civilizations*, University of California Press, Berkeley.
10. EUR-Lex Council Directive 91/676/EEC of 12 December 1991 Concerning the Protection of Waters Against Pollution Caused by Nitrates From Agricultural Sources, *http://eur-lex.europa.eu/LexUriServ/LexUriServ.do?uri=CELEX:31991L0676:en:NOT* (accessed 29 December 2013).
11. European Community. Nitrogen in Agriculture, *http://ec.europa.eu/agriculture/envir/report/en/nitro_en/report.htm* (accessed 29 December 2013).
12. CITES *http://www.cites.org/eng/disc/E-Text.pdf* (accessed 29 December 2013).
13. CBD *http://www.cbd.int/intro/default.shtml* (accessed 29 December 2013).
14. CBD The Nagoya Protocol, *https://www.cbd.int/abs/* (accessed 19 March 2014).
15. WRB (2006) World Reference Base for Soil Resources 2006, A Framework for International Classification, Correlation and Communication. World Soil

Resources Reports 103, FAO, Rome, p. 145.
16. Mueller, L. *et al.* (2010) Assessing agricultural soil quality on a global scale. 19th World Congress of Soil Science, Soil Solutions for a Changing World, Brisbane, Australia, August 1–6, 2010
17. McKibben, P. (2012) *The Global Warming Reader: A Century of Writing About Climate Change*, Penguin Press, 432 p.
18. World Maps of Köppen-Geiger Climate Classification *http://koeppen-geiger.vu-wien.ac.at/shifts.htm* (accessed 29 March 2014)
19. Rubel, F. and Kottek, M. (2010) Observed and projected climate shifts 1901–2100 depicted by world maps of the Köppen – Geiger climate classification. *Meteorol. Z.*, **19**, 135–141.
20. Unger, P.W. and McCalla, T.M. (1989) Conservation tillage systems. *Adv. Agron.*, **33**, 2–53.
21. Stewart, W.M., Dibb, D.W., Johnston, A.E., and Smyth, T.J. (2005) The contribution of commercial fertilizer nutrients to food production. *Agron. J.*, **97**, 1–6.
22. Ceresana (2013) Market Study: Fertilizers – World, May 2013, *http://www.ceresana.com/en/market-studies/agriculture/fertilizers-world/* (accessed 12 January 2014).
23. Brock, W.H. (1999) *Justus von Liebig*, translated by Siebeneicher, G.E. p. 27, Vieweg&Teubner, Braunschweig.
24. Pau Vall, M. and Vidal, C. Nitrogen in Agriculture in: 'Agriculture and Environment', *http://ec.europa.eu/agriculture/envir/report/en/nitro_en/report.htm* (accessed 11 January 2014).
25. Europa Summaries of Legislation Regulation (EC) No 2003/2003 of the European Parliament and of the Council of 13 October 2003 Relating to Fertilisers, *http://europa.eu/legislation_summaries/food_safety/contamination_environmental_factors/l21278_en.htm* (accessed 5 January 2014).
26. Maeder, P., Fliebach, A., Dubois, D., Gunst, L., Fried, P., and Niggli, U. (2002) Soil fertility and biodiversity in organic farming. *Science*, **296** (5573), 1694–1697.
27. Birkhofer, K., Bezemer, T.M., Bloem, J., Bonkowski, M., Christensen, S., Dubois, D., Ekelund, F., Fliessbach, A., Gunst, L., Hedlund, K., Maeder, P., Mikola, J., Robin, C., Setälä, H., Tatin-Froux, F., Van der Putten, W., and Scheu, S. (2008) Long-term organic farming fosters below and above ground biota, implications for soil quality, biological control and productivity. *Soil Biol. Biochem.*, **40** (9), 2297–2308.
28. European Community Use of Sewage Sludge in Agriculture, *http://ec.europa.eu/environment/waste/sludge/* (accessed 03 November 2014).
29. Dangerfield, W. (2007) The Mystery of Easter Island. Smithsonian Magazine (Apr. 1).
30. Howard, A. (1943) *An Agricultural Testament*, Oxford University Press, Oxford.
31. Kirchmann, H. and Bergström, L. (eds) (2008) *Organic Crop Production – Ambitions and Limitations*, Springer-Verlag, Berlin.
32. Demeter International *http://en.wikipedia.org/wiki/Demeter_International* (accessed 18 January 2014).
33. Hanley, P. (1990) Agriculture: a fundamental principle. *J. Bahá'í Stud.*, **3** Fukuoka, M., (1978). *One Straw Revolution.* Emmaus: Rodale Press (1), *http://www.google.de/url?sa=t&rct=j&q=&esrc=s&frm=1&source=web&cd=1&cad=rja&ved=0CCwQFjAA&url=http%3A%2F%2Fwww.bahai-studies.ca%2Fjournal%2Ffiles%2Fjbs%2F3.1%2520Hanley.pdf&ei=d2jjUvijF8PctAbO-YGwAw&usg=AFQjCNHCCUEwVmSgeTbWRqhcKqEI_QHNvw* (accessed 11 August 2014).
34. Tegta, M. (2010) Masanobu Fukuoka: The Man Who Did Nothing. DNA Daily News and Analysis (Aug. 22, 2010), Mumbai, India (25 Jan. 2014).
35. Alan Chadwick Biography on the Alan Chadwick Homepage, *http://alan-chadwick.org/html%20pages/chronology.html* (accessed 25 January 2014)
36. Jeavons, J. (2004) *How to Grow More Vegetables: And Fruits, Nuts, Berries, Grains and Other Crops Than You Ever Thought Possible on Less Land Than You*

Can Imagine, 6th edn, Ten Speed Press, Berkeley.
37. Faulkner, E. (1945) *Ploughman's Folly*, M. Joseph Ltd Press.
38. Derpsch, R. *http://www.rolf-derpsch.com/en/no-till/historical-review/*. (accessed 03 Nov 2014) A Short History of No-Till, *http://www.rolf-derpsch.com/notill.htm*
39. EN Council regulation (EEC) no. 2092/91 of 24 June 1991 on organic production of agricultural products and indications referring thereto on agricultural products and foodstuffs, *Off. J. Eur. Union*, **198**, 1, *http://eur-lex.europa.eu/LexUriServ/LexUriServ.do?uri=CONSLEG:1991R2092:20080514:EN:PDF* (accessed 11 August 2014).
40. EN Regulations Council regulation (EC) no. 834/2007 of 28 June 2007 on organic production and labelling of organic products and repealing Regulation (EEC) no. 2092/91, *Off. J. Eur. Union*, **189**, 1–23 *http://eur-lex.europa.eu/LexUriServ/LexUriServ.do?uri=OJ:L:2007:189:0001:0023:EN:PDF* (accessed 11 August 2014).
41. European Commission The EU Organic Logo, *http://ec.europa.eu/agriculture/organic/downloads/logo/index_en.htm* (accessed 11 August 2014).
42. ECOCERT *http://www.ecocert.com/en* (accessed 15 March 2014).
43. Diepenbrock, W., Ellmer, F., and Leon, J. (2012) *Ackerbau, Pflanzenbau und Pflanzenzüchtung*, 3rd edn, UTB Verlag Stuttgart.
44. CIOPORA *http://www.ciopora.org/* (accessed 30 March 2014).
45. Bundesministerium für Bildung und Forschung Biosicherheit: Gentechnik – Pflanzen – Umwelt, Haploidenzüchtung, *http://www.biosicherheit.de/lexikon/854.haploidenzuechtung.html* (accessed 30 March 2014).
46. Dacher, M. and Pelzmann, H. (1999) *Arznei- und Gewürzpflanzen*, 2nd edn, AV-Verlag, Kloster Neuburg, pp. 59–60.
47. EU (2004) Council directive 2004/117/EC: Marketing of plant seed amended as regards of examinations carried out under official supervision and equivalence of seed produced in third countries. *Off. J. Eur. Union*, **198**, 1, *http://eur-lex.europa.eu/LexUriServ/LexUriServ.do?uri=OJ:L:2005:014:0018:0033:EN:PDF* (accessed 30 March 2014).
48. European Commission Organic Seed Databases of the EU, *http://ec.europa.eu/agriculture/organic/eu-policy/eu-rules-on-production/seeds-database/index_en.htm* (accessed 30 March 2014).
49. Kooperation Phytopharmaka Arzneipflanzenlexikon der Kooperation Phytopharmaka – Monographie Pfefferminze, *http://www.koop-phyto.org/arzneipflanzenlexikon/pfefferminze.php* (accessed 30 March 2014).
50. Bayerische Landesanstalt für Weinbau und Gartenbau Kulturanleitung Knoblauch, *http://www.lwg.bayern.de/gartenakademie/infoschriften/gemuese/linkurl_0_7_0_9.pdf* (acccessed 30 March 2014)
51. Schloesser, E. (1997) *Allgemeine Phytopathologie*, Thieme Verlag Stuttgart.
52. Heibertshausen, D., Braus, O., Langen, G., Kogl, K.-H., Bleyer, G., Kassemeyer, H.-H., Loskill, B., Maier, K., Maixner, M., and Berkelmann-Löhnertz, B. (2010) *Kupferminimierung im ökologischen Rebschutz, Deutsches Weinbau Jahrbuch 2010*, Verlag Ulmer, Stuttgart.
53. European Commision Bees and Pesticides: Commission Goes Ahead with Plan to Better Protect Bees, *http://ec.europa.eu/food/animal/liveanimals/bees/neonicotinoids_en.htm* (accessed 31 March 2014)
54. Hodek, I., Honek, A., and van Emden, H.F. (2012) *Ecology and Behaviour of the Ladybird Beetles*, Wiley-Blackwell, Hoboken, NJ.
55. Mateeva, A., Ivanova, M., and Vassileva, M. (2002) Effect of intercropping on the population density of pests in some vegetables. *Acta Hortic.*, **579**, 507–511.
56. Valkama, E., Salminen, J., Koricheva, A., and Pihlaja, K. (2004) Changes in leaf trichomes and epicuticular

flavonoids during leaf development in three birch Taxa. *Ann. Bot.*, **94**, 233–242.
57. WHO (2003) Guidelines on Good Agricultural and Collection Practices (GACP) for Medicinal Plants, Geneva, *http://apps.who.int/medicinedocs/en/d/Js4928e/* (accessed 22 March 2014).
58. EUROPAM (2010) Guidelines for Good Agricultural and Wild Collection Practice of Medicinal and Aromatic Plants (GACP-MAP), Brussels, *http://www.europam.net/documents/gacp/EUROPAM_GACP_MAP_8.0.pdf* (accessed 22 March 2014).
59. EMEA GACP Guideline of the EMA, *http://www.ema.europa.eu/docs/en_GB/document_library/Scientific_guideline/2009/09/WC500003362.pdf* (accesed 15 March 2014).
60. European Pharmacopoeia *http://online.edqm.eu/EN/entry.htm* (accessed 15 March 2014).

Further Reading

European Commission EU Organic Farming Legislation *http://ec.europa.eu/agriculture/organic/eu-policy/legislation_en* (accessed 14 August 2014).

European Commission EU Organic Farming News, *http://ec.europa.eu/agriculture/organic/news_en* (accessed 14 August 2014).

EN EU directive on Traditional Herbal Medical Products, *http://eur-lex.europa.eu/LexUriServ/LexUriServ.do?uri=OJ:L:2004:136:0085:0090:en:PDF* (accessed 14 August 2014).

European Union EU Organic Farming, *http://europa.eu/geninfo/query/resultaction.jsp?ResultCount=25&Collection=EuropaFull&ResultMaxDocs=200&SourceQueryText=scadplus01&qtype=simple&DefaultLG=en&ResultTemplate=%2Fresult_fr.jsp&page=1&QueryText=Organic+farming&y=10&x=12* (accessed 14 August 2014).

4
Mass Transfer Enhancement for Solid–Liquid Extractions

Simon Both, Jochen Strube, and Giancarlo Cravatto

4.1
Introduction

The economic impact of plant-based extracts especially for food, nutraceuticals, cosmetics, flavors/fragrances, and pharmaceutical industry has grown in the last years owing to the consumer demand for products from natural sources. The market for herbal extracts from products like balm, green tea, or blueberry alone amounted to about €6.7 billion in Europe and about €17.5 billion worldwide in 2003. According to the FAO (Food and Agriculture Organization of the United Nations), the annual growth rates for food, food supplements, and drugs from plant-based sources are projected at 6–8% [1]. To manage this growing demand for plant-based extracts, the development and industrialization of new processes and ingredients have to be accelerated. Existing manufacturing processes need to be optimized with respect to, for example, energy, solvent consumption, and solvent selection. Companies in this global and fast-growing market with many new, short life-cycle products need to adapt their development and production processes to decrease the time to market and production cost of new ingredients.

Especially in the field of green extraction aiming at new sustainable extraction processes, the enhancement of extraction processes and thus new and innovative technologies, process intensification, agro-solvents, and energy saving are of major importance. The following are examples of enhancement of extraction processes:

- Microwave extraction
- Ultrasound extraction
- High pressure extraction
- Pulse electric field
- Instantaneous decompression
- Flash extraction.

In this chapter, an overview on traditional as well as new enhanced extraction methods is presented. Case report examples, in which different techniques have

Green Extraction of Natural Products: Theory and Practice, First Edition.
Edited by Farid Chemat and Jochen Strube.

been compared, show in detail some efficient extraction procedures from lab to pilot scale. The extraction of black tea, clove buds, and oily microalgae clearly shows how innovative techniques and improved methods can compete with the traditional procedures and how sustainable and green extraction processes should be designed.

4.2
State of the Art Solid-Liquid Extraction

Solid–liquid extraction is one of the oldest known processes to isolate substances from plants. Excavation findings in Mesopotamia prove that extracts for medical and cosmetic applications were being manufactured as early as 3500 BC. Solid–liquid extraction is still of high importance and is one of the most applied unit operations in process technology. Solid–liquid extraction is the selective release, washing, or leaching of substances from solids using liquid solvents. In contrast to other separation processes, complex multicomponent mixtures are obtained from solid–liquid extraction processes. One or more subsequent purification steps are required to achieve an enriched botanical extract or a pure substance.

During all processes, solvent recovery is of great economic importance. In Figure 4.1, the method of a solid–liquid extraction is presented in a simplified block diagram. The solid raw material is mixed with a solvent and extracted. Subsequently, the solid and liquid phases are separated, and with different purification steps, the target compound is produced from the extract. The solid–liquid extraction kinetics is mainly influenced by parameters such as solvent type and pH, solid–liquid ratio, particle size, and temperature. The solvent selection is matched on the target component to influence selectivity in the right direction. Hence, solvent polarity and pH are important for attaining a high yield with low degradation of the target component.

Solid–liquid extraction covers a variety of processes such as the extraction of oils and fats from oilseeds, essential oils from herbs, sugar from beet and sugar cane as well as pharma- and nutraceuticals from medicinal plants. Making tea and coffee from ground raw material with hot water is an everyday example for a solid–liquid extraction process [2, 3]. Although such processes are widely applied in many fields, automation, mass transfer efficiency, and economics still require optimization [4]. The state of the art for plant-based extraction processes regarding equipment and modeling has been determined by a survey of scientific literature. Other industrial sectors such as mining, paper, or oil industry using solid–liquid extraction have been excluded from the literature survey, because the focus is only on plant-based extractions. Patents have been checked and taken into account, but mainly refer to maceration approaches.

The scientific literature search for solid–liquid extraction or, for example, leaching, lixiviation, decoction, and elution shows almost 75 000 "hits." Excluding all

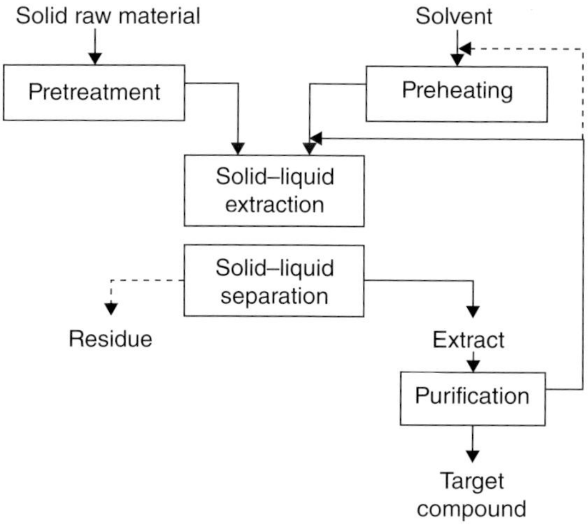

Figure 4.1 Process scheme solid–liquid extraction [2].

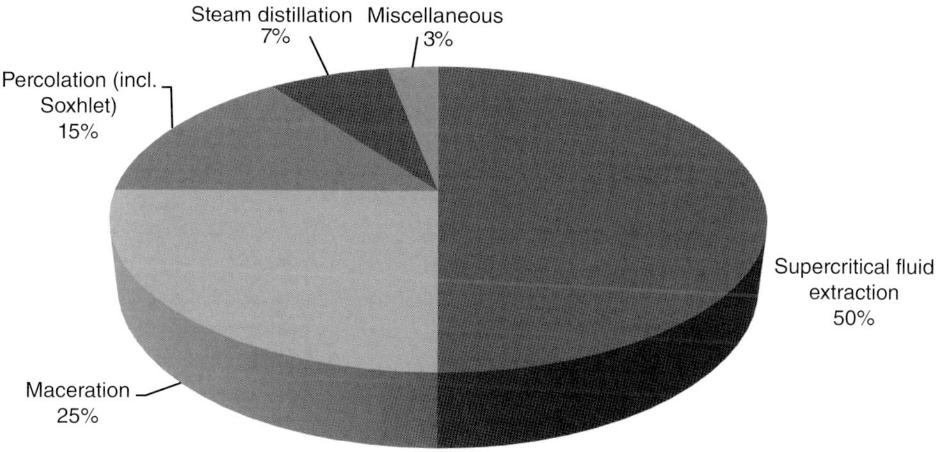

Figure 4.2 Literature survey on solid–liquid extraction.

articles dealing with the extraction of heavy metals from soil via plants, about 1500 articles remain, which finally leads to 300 articles. The articles dealing with plant-based extractions can be divided into several groups regarding the extraction method as can be seen in Figure 4.2. In about 45% of the published articles, the usage of supercritical fluid extraction mostly for pharmaceutical and food processing is described [5–7]. Plant-based extractions with solvents account for about 20% of the articles. Solvent extraction of plant material also includes pressurized liquid extraction (PLE) and accelerated solvent extraction (ASE™ – trade mark of Dionex), which apply pressure in the extraction cell to achieve temperatures of

Figure 4.3 Industrial ultrasonic batch reactors for plant extraction. (Kind permission of Martin-Bauer Italia, Nichelino, Turin – IT.)

Figure 4.4 Internal view of an industrial ultrasonic reactor. (Kind permission of Martin-Bauer Italia, Nichelino, Turin – IT.)

100–200 °C with organic solvents [8–10]. Around 10% of the articles deal with microwave-assisted extraction (MAE) [11, 12] and another 10% with ultrasound-assisted extraction (UAE) [13, 14]. Although scarcely reported in the scientific literature, since the 1990s, industrial applications are available with batch reactors from 100 up to 500 l as shown in Figures 4.3 and 4.4.

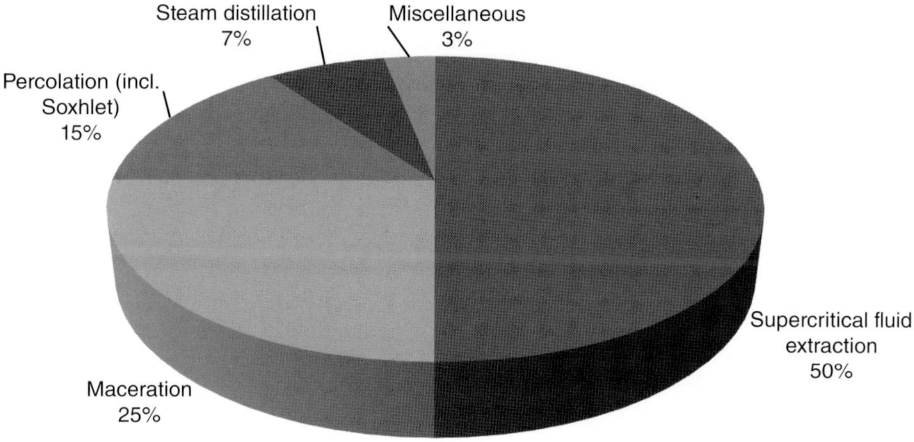

Figure 4.5 Literature survey on solid–liquid extraction equipments.

The scientific literature search for solid–liquid extraction equipment and apparatuses shows more than 16 000 "hits." Excluding all articles dealing with the extraction of heavy metals from soil via plants, there are about 800 articles, finally leading to 30 articles. Dividing the articles dealing with equipment for plant-based extraction into different groups ends up with the following split (Figure 4.5): about 50% of the articles are dedicated to supercritical fluid extractors, 25% discuss macerations, 15% cover percolation, and about 7% are focused on steam- and hydrodistillation.

4.2.1
Batch Processes

The most common apparatus for plant-based solvent extractions is maceration carried out in a stirred vessel equipped with a temperature control system. The solid–liquid extraction in a macerator usually ends up at an equilibrium state after a specific time depending on the solid–liquid partition coefficient, while diffusion and mass transfer coefficients determine the time until equilibrium is reached. Therefore, maceration processes are typically multistaged to reach an almost complete leaching of the raw material with the addition of fresh solvent. Usually macerators are equipped with a filter at the vessel bottom to realize a solid–liquid separation in the extractor. In Figure 4.6, a typical standard maceration equipment is shown. The raw material is placed on sieve plates and moved with the solvent. During the extraction, stirrers keep the mixture in motion until the desired concentration is reached, following which the liquid is drained. The solid remains in the tank reactor and a further extraction step can be added if necessary. In the setup shown here, it is possible to flush out the remaining solvent by means of indirect heating with steam, which is advantageous because of safety reasons and waste reduction. Nowadays, maceration extractors feature working volumes of up to 6000 l [15–17].

(a) (b)

Figure 4.6 Macerator from E&E [18] and percolation equipment from GEA Niro [19].

4.2.2 Continuous Processes

Continuous processes are mainly used for large-scale single-product plants such as in oilseed industry. Vegetable oils especially, as also some other substances that feature high oil content can be obtained by mechanical expulsion followed by centrifugation. Mechanical expulsion is the more economical method compared to solvent extraction processes, but at best about 5 wt% remains in the raw material if completely pressed. Owing to undesired heating of the solid cake by pressing and possible deterioration of the oil, a residual oil content of 15–20 wt% is usual in industry. The remaining oil in the micelle is obtained by extraction with light hydrocarbons such as hexane [3, 20].

Continuous solvent extractors feature relatively small dimensions and compact design. The process design can differ in the flow direction of solvent and raw material: if both are transported in the same direction, the process design is cocurrent and if they are transported toward each other, the process design is countercurrent. In percolating solid–liquid extractors, the process design is pseudo countercurrent, because the raw material is fixed while the solvent percolates through it. High concentrations of the target compounds and a low residual content of solute in the residue can be achieved via countercurrent process design, which allows a considerable reduction in solvent usage. Therefore, extraction times are shorter than those of batch methods and this has a beneficial effect on the quality of the end products [3]. Literature discusses a large number of solid–liquid extractor designs, but most of them have either not become widely accepted or they are not used, they are obsolete, or they have been used only for a short time period [21].

Today, there are a few companies on the market manufacturing continuous countercurrent extractors, most of them targeting oilseed extraction (Table 4.1):

Table 4.1 Overview of solid–liquid extractors [45].

Extractors			Capacity (t/d)	Yield (–)	Solid-liquid ratio (–)	Residence time (min)	Equilibrium stages (–)	Corn size (mm)
BMA, Germany	BMA tower extractor	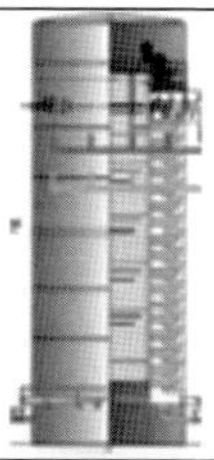	4 000–17 000	99	3–5	90–150	n/a	5–100
Crown Iron Works, USA	Model III™ percolation type		Up to 12 000	80–99	0.8–8	30–180	n/a	1–20
	Model IV™ Immersion type	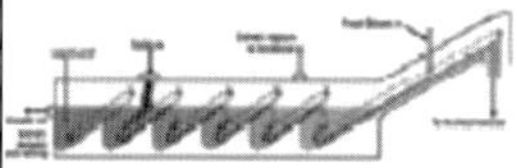	Up to 800	80–99	0.8–8	30–300	n/a	1–20
	Model V™ (speciality materials) percolation type	n/a	Up to 800	80–99	0.8–8	30–300	n/a	1–20
De Smet, Belgium	LM™ extractor	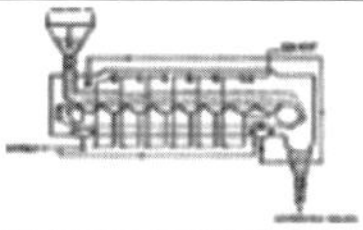	500–5 000	99	Approx. 1	60–120	5–10	0.3–15
	Reflex™ extractor		500-12 000	99	Approx 1	60–120	5–10	0.3–15
GEA Niro, Denmark	Contex™ extractor	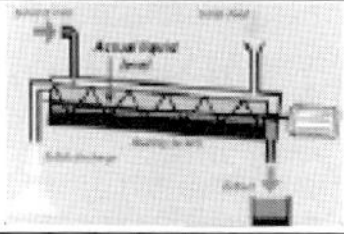	12–24	>90	Approx. 6	30–120	3–5	0.9–50
Harburg-Freudenberger, Germany	Carousel™ extractor	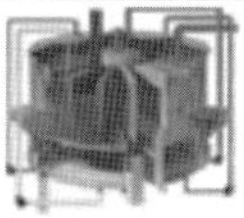	50–5 000	92–98	0.85–1	30–90	n/a	n/a
Lurgi, Germany	Sliding Cell™ extractor		100–5 000	98–99	n/a	n/a	8–12	0.5–20

BMA (Braunschweigische Maschinenbauanstalt AG) with the tower extractor, Crown Iron Works with the Model III™, Model IV™, and ModelV™ extractor, GEA Niro, and Vatron Mau with a screw extractor, De Smet with the REFLEX™ and LM™ extractor, Harburg-Freudenberger with the Carousel™ extractor, and Lurgi with the Sliding Cell™ extractor [22, 23].

BMA manufactures the BMA tower reactor for extracting sugar from beet cossettes and has the largest capacity of all available continuous solid–liquid extractors with up to 17 000 tpd. The cossette/juice mixture is delivered into the bottom of the extraction tower and discharged with screw conveyors at the top of the tower, while the extraction liquid (fresh water and press water at 80 °C) flows countercurrent to the cossettes. Well-positioned conveying blades and stops facilitate gentle conveyance of the cossettes through the extraction area [24].

The Model III™ and Model V™ extractors from Crown Iron Works in Roseville (USA) are continuous-loop shallow-bed extractors and differ in capacity according to the physical properties of the raw material. The Model III™ extractor shows capacities of up to 12 000 tpd. Material is fed into the extractor through the inlet hopper located in the top section of the extractor. The internal chain assembly carries the material across the length of the top section while a constant washing by several stage pumps takes place. In the tail section of the extractor, two important functions are automatically performed. First, the material is slowly turned over, ensuring that the bed is inverted when it leaves the tail section. Second, micelle added to the top section travels with the material, creating a soaking section which facilitates the removal of residual oil in the bottom section of the extractor. In the bottom section, the material is washed by successively leaner stages of miscella, until the last stage where pure solvent is used to remove the last remaining oil. The bed is raised up a shallow ramp in the drainage section, which ensures that no excess solvent can overflow into the discharge. The Model IV™ extractor is an immersion-style extractor with capacities similar to those of the Model V™ extractor and is designed for granular and powdery products which are too fine for percolation. In the Model IV™ extractor, a conveyor system draws the material along the smooth extractor bottom where it is totally immersed in solvent flowing from the opposite direction [20].

GEA Niro/Denmark and Vatron Mau/France build screw extractors for continuous countercurrent extraction. The GEA Niro CONTEX™ from GEA Niro is a test unit for scale-up of industry-sized continuous extractors. The solid is transported upward by two screws, while the miscella, which can be heated up to 95 °C, flows as countercurrent by gravity. The typical feed capacity is between 5 and 20 $kg\,h^{-1}$ (12–24 tpd for production scale) and the residence times range from 0.5 to 2 h [25]. The Vatron Mau extractor consists of several compartments, each equipped with a pump to recirculate the micelle and create a countercurrent flow. The feed capacity is similar to that of the Niro CONTEX™ and residence times of between 1 and 6 h can be realized [26]. The LM™ extractor from De Smet in Zaventem/Belgium is a continuous percolation extractor with capacities of up to 5000 tpd. The feed material is conveyed on a perforated belt. The height of the material bed is regulated by an adjustable slide. The solvent is sprinkled from the top on the length of the belt until the extraction residue is discharged through a discharge gate. The solvent percolating through the belt into catch basins is pumped to the top of a subsequent section of the bed, as countercurrent to the direction of solids' movement [3, 21, 27]. The Carousel™ extractor from Harburg-Freudenberger in Hamburg/Germany works as a countercurrent percolation extractor with an airtight casing and a slowly rotating cell wheel. The cell wheel continuously conveys the ground raw material from the entering shaft over a fixed sieve tray to the dropout shaft. In the meantime, the raw material is sprinkled with solvent from above. The solvent percolates through the raw material and the sieve trays until it is collected in the extract chambers. Before ejection of the filter cake, a product-specific washing step can be implemented. Depending on the throughput, the carousel extractor can be built with one or more stacked cell wheels to realize capacities from 50 to 5000 tpd [28]. The REFLEX™ extractor from De Smet applies the same principles with capacities of up to 12 000 tpd [27].

4.2.3
Hydro- and Steam Distillation

Water or water steam distillation is an often applied technology to produce substances with low thermal stress. The basic principle is that almost insoluble substances form low-boiling azeotropes, whose boiling point is below the boiling point of the higher volatile component.

The general scheme of water steam distillation is shown in Figure 4.7. In a distillation still pot, in addition to the preloaded lower boiling compounds, water steam is blown in. The azeotrope evaporates and is transferred through the vapors to the

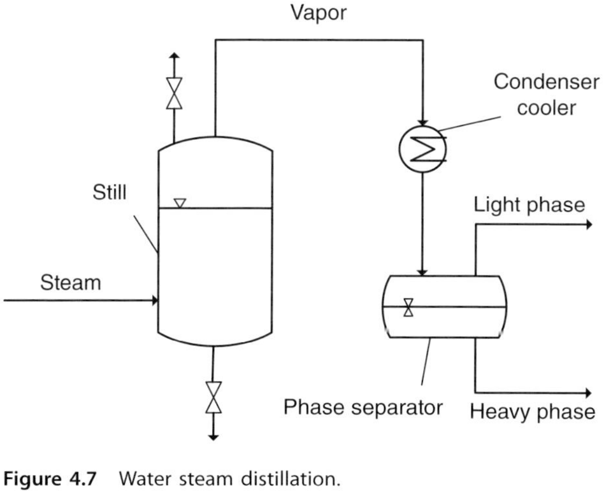

Figure 4.7 Water steam distillation.

condenser. There, the steam is liquefied again and flows into the phase separator vessel, where the liquid separates into specific lower density value component (organic phase) and process water.

In general the system pressure within a distillation is described with aid of the Raoult's law, which is given in Equation 4.1.

$$p = x_1 p^*_1 + x_2 p^*_2 \tag{4.1}$$

Here, p^*_1 and p^*_2 are saturation steam pressures of both components and x_1 and x_2 are their mole fractions. Often, in a good approximation, a very low solubility could be assumed, so that the substances involved are not mixed within the vapor phase. Thereby, both mole ratios are equal to one. The resulting total pressure in the case of a binary mixture is the sum of single saturation pressures p_i^*, as described in Equation 4.2.

$$p = p^*_1 + p^*_2 \tag{4.2}$$

At atmospheric pressure, the boiling point of the mixture is below 100 °C, which enables the already mentioned processing of temperature-sensitive compounds. For some binary systems, the resulting boiling temperatures can be directly read from the so-called Hausbrand diagram. In Figure 4.8, this is depicted for a pressure of 40 and 101 kPa.

If, instead of any supporting steam, water is directly provided, than the process is known as *hydrodistillation*. The theoretical fundamentals are identical to steam distillation.

The described separation principle is applied to the production of essential oil from plant-based raw material. Because of the very low mutual solubility of oil and water, a low boiling azeotrope results, allowing for the production of essential oils at low temperatures.

While in industry, large plants are operated, in a preparative scale, the distillation equipment must be used according to the European Pharmacopeia. This device is shown in Figure 4.9.

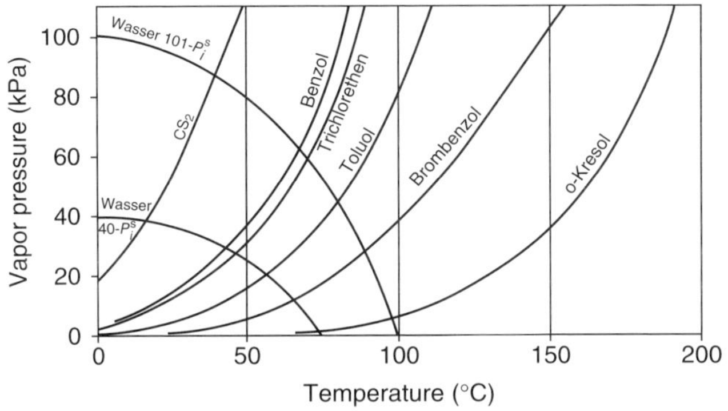

Figure 4.8 Hausbrand diagram according to [29].

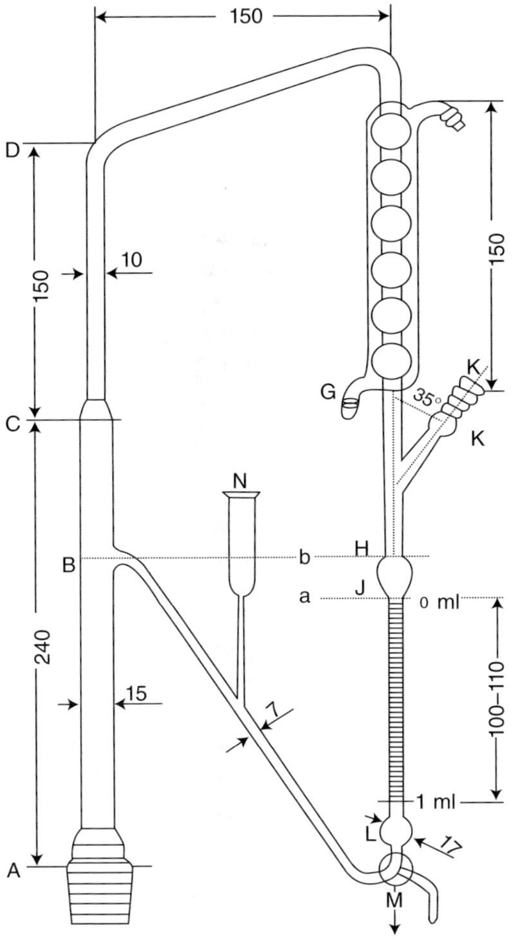

Figure 4.9 Water steam distillation according to [30].

4.2.4 Alembic Distillation

Distillation is the traditional method for separating substances of different volatility by evaporation and condensation within alembics. In this field, besides chemists and originally alchemists, the spirits industry has a centuries-old experience in the production of high-quality beverages by alembic distillation of fermented vegetables and fruits. The simple alembic requires well-defined conditions to maximize the spirit's quality and minimize toxic substances such as methanol and acetaldehyde. Recently, a study of dynamic modeling and simulation of alembic distillation described useful models to enhance alembic operations, to increase energy efficiency, and improve spirit safety and flavor [31].

Figure 4.10 Fourteenth century: the "Rosenhut" distillation apparatus. Reproduced from [32].

Herodotus and Pliny the Elder described the use of alembic distillation (Figure 4.10) to recover essential oils from plant material (see Section 2.3).

A modern version of the alembic is the microwave (MW) Clevenger apparatus introduced by Chemat [33], who developed a series of dedicated apparatuses produced by Milestone (Bergamo – Italy) for fast microwave-assisted distillation of essential oils.

4.2.5 Mechanical Expression (Extrusion)

Vegetable oils especially and some other substances featuring high oil content can be obtained by mechanical expulsion followed by centrifugation. It is the more economic method compared to solvent extraction processes, but at best about 5 wt% remains in the raw material if completely pressed. In general, the oil can be isolated from the seeds by three main processes: hard or full pressing, solvent extraction of prepressed seeds, and direct solvent extraction [34]. Hard pressing involves squeezing out as much oil as possible. In prepressing, only the easily extractable oil is pressed out before subjecting the partially de-oiled material to more complete extraction with a solvent. Direct solvent extraction without prepressing became possible with the development of an adapted expander technology. For soybean oil, this direct solvent extraction is widely applied as a large-scale process. In the choice of the seed extraction process, several factors play a role, such as oil content and market value, allowable oil residue in the meal without affecting product costs, and environmental issues generated by solvent handling.

High-quality oils for nutritional and cosmetic applications can be obtained using modern screw presses and purifying the oil only by sedimentation or filtration. The mechanical expression of oil from oilseeds has relatively low initial and operational costs and produces uncontaminated oils [35]. The production of outstanding cold-pressed oils requires however good know-how and attention during all the production chainwork. Improper raw material has to be discarded. Further critical points are the correct harvesting time, storage under controlled moisture and temperature, pressing of the raw material with a last-generation screw press, and fast purification of the resulting oil. Generally, from a batch of seeds, the prepressed oils and solvent-extracted oils from the residual press cake after prepressing are mixed and called *crude oil*, which will undergo the refining process. However, a preference for greener pressed oils, obtained without any solvent, might make it economically interesting to treat and commercialize both fractions separately. The pressed oil is free of solvent residues, and as any refining treatments are avoided, it is expected to have a higher overall quality. On the other hand, it is known that solvent extraction yields higher levels of phospholipids than simple pressing. Cold-pressed oil may retain a greater level of nutritionally valuable compounds such as phytosterols and tocopherols, which can, in turn, enhance the oxidative stability of the oil [36]. The oxidative stability of oil mainly depends on the fatty acid (FA) composition and the presence of antioxidative compounds as well as pro-oxidants.

Recently, three oils with outstanding organoleptic properties obtained with screw presses (Figure 4.11) were analyzed and compared to the conventional hexane extraction. The extraction yield, although inferior to the Soxhlet method, ranged from 18% for sapote oil, 9% for baobab oil to 3% for wheat germ oil [37].

The continuous process with seed explosion via fast extruder heating (100–105 °C) is one of the fastest ways to recover oil, although a regular flow is hampered by several technical requirements (Figure 4.12). Among them, feed material speed and seed moisture strongly affect the regular flow of the process.

Today, products of mechanical expression are of major importance in the food and cosmetic industries. During any extrusion process, the treatment of the material consists of mixing, mass kneading, heating and shearing, and finally extrusion through a die appropriately designed to form and dry the product under expansion. Different types of presses can be used depending on the purpose (expeller, expander, and twin-screw extruder) and on the capacity range ($3\,kg\,h^{-1}$ up to $150\,t\,day^{-1}$). At the laboratory scale, many authors have highlighted the effect of operating parameters (screw rotation speed, temperature, and back pressure) and raw material (seed species, variety, water content, and pretreatment) on process performance (oil yield and press capacity). In addition to classic oleaginous species, other plant species have been investigated with this solvent-free technique to obtain vegetable oils of nutritional and pharmaceutical interest [38].

Figure 4.11 Continuous screw press (30 kg seeds/h).

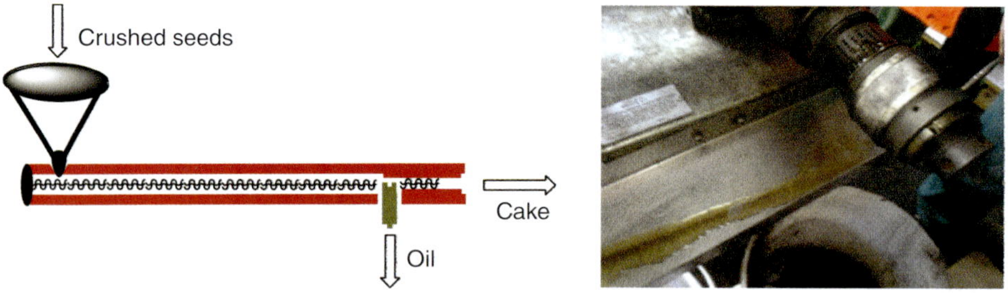

Figure 4.12 Oil delivery from seed extrusion. (Kind permission of Parodi Nutra srl, Campomorone – Italy.)

4.3
Enhancement of Solid–Liquid Extraction Processes

4.3.1
Microwave-Assisted Extraction (MAE)

In the last decade, MAE has been widely studied and successfully applied in solid–liquid extraction of plant material [39]. Despite the fact that classic methods (maceration, Soxhlet extraction, etc.) present several drawbacks, modern science is still far from fully replacing them with MAE or other nonconventional techniques. Microwave radiation is currently used for the rapid extraction of several classes of bioactive compounds, phytonutrients, functional food ingredients, and pharma-active substances from biomass. The main advantages of MAE are: reduced solvent consumption, minimal sample manipulation, shorter operation times, and good selectivity, recovery yields, and reproducibility. The method enables up to 10–15 samples to be run in a single extraction and so gives rise to high sample throughput. The fulfillment of GLP (good laboratory practice) and the easy automatization make this technique suitable for pharmaceutical applications.

In absence of wall effects and temperature gradient, the dielectric volumetric heating reduces thermal degradations. It has been reported that MAE of phenolics in water is not as efficient as conventional methods, because water has a higher dielectric constant and a lower dissipation factor than other solvents [40]. In MAE, it is better to use solvents with both a high dielectric constant and a high dissipation factor. Extractability also depends on the type of plant material extracted and the solvents used for the extraction. In the presence of polar molecules or ionic species, MAE provides rapid heating which leads to collisions with the surrounding molecules and so does not need to be carried out at high pressure. Power and extraction time for natural products are in the range of 25–750 W and 30 s to 10 min, respectively. MAE has been used for the extraction of polyphenolics from a number of plant sources, such as tea leaves, flax seeds, radix, and vanilla, among others. MAE causes the compounds of interest to desorb from the plant matrix because the free water molecules present in the gland and vascular systems are heated. By this, localized heating and an expansion occur, during which plant cell walls are ruptured, allowing the extracted molecules to flow toward the organic solvent. The effect of microwave (MW) energy is strongly dependent on the dielectric susceptibility of both the solvent and solid plant matrix. Most of the time, the sample is immersed in a single solvent or mixture of solvents that absorb MW energy strongly. Owing to the elevated temperature, the solvent's ability to penetrate the matrix increases. Another recent approach includes a fast preheating of fresh (or rehydrated) plant material in flow MW reactors, followed by proper extraction that is in most cases faster and more efficient. A system with an Archimedean screw is depicted in Figure 4.13.

The main disadvantages of MAE are its relatively high capital cost and possible need to filter the sample if fine particles are used for the extraction of compounds.

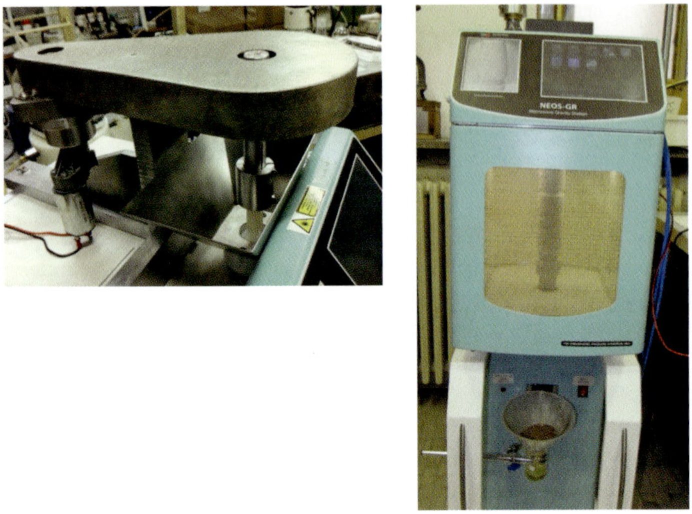

Figure 4.13 Flow Archimedian screw for MW preheating of seeds or other matrices.

Figure 4.14 SynthWAVE reactor (Milestone/MLS).

The last generation of equipments for highly efficient MAE, abandoned the concept of an oven, moving from sealed vessels to closed cavities (Figure 4.14). Being equipped with independent multiple gas inlets, these reactors enable modified atmosphere either with inert gas (N_2 or Ar) or even a reduced pressure in the presence of N_2 and a minimum amount of H_2. The latter conditions avoid any risk of oxidative degradation making the subcritical water extraction easy to handle.

A new microwave solvent-free extraction technique called *microwave hydrodiffusion and gravity* (MHG) has been used for extraction of essential oils, colors, and antioxidants. This technique is carried out in an original "upside down" microwave alembic combining microwave heating and earth gravity at atmospheric pressure. On the basis of a relatively simple principle, this method also involves placing the plant material in a microwave reactor, without adding any solvent or water.

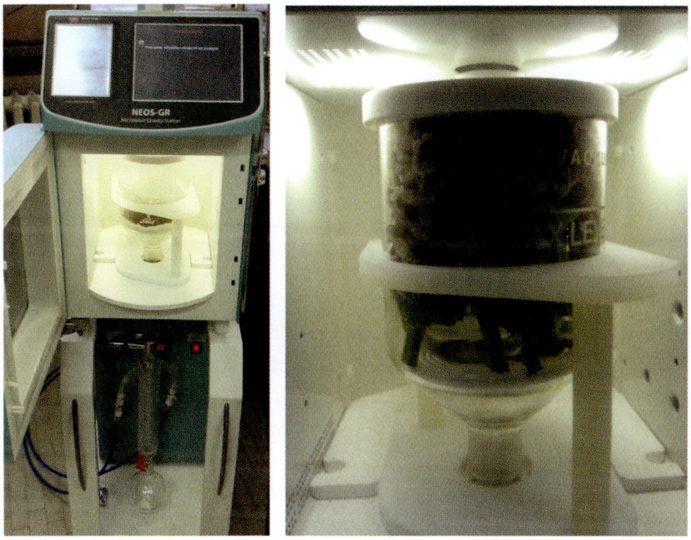

Figure 4.15 NEOS-GR reactor, distillation of *Salvia officinalis L.* (Milestone, Italy).

The internal heating of the *in situ* water within the plant material distends the plant cells and leads to the rupture of glands and oleiferous receptacles. The heating action of microwaves thus frees essential oil and *in situ* water, which are transferred from the inside to the outside of the plant material, and drop out of the microwave reactor by gravity [41]. Figure 4.15 shows the easy application of this technique and the fast isolation of the essential oil from different plants. This extraction method has been scaled up at pilot scale as shown in Figure 4.16 which represents the 75 l reactor MAC-75 [42].

Figure 4.16 Pilot-scale reactor MAC-75 (Milestone, Italy).

4.3.2 Ultrasound-Assisted Extraction (UAE)

Several techniques have been developed to obtain valuable products from plant sources on the industrial scale. Recent trends in extraction techniques are largely focused on finding solutions that minimize or even dispense with the use of solvents. The application of ultrasound (US) to enhance extraction yields began in the fifties at the laboratory level. Nowadays UAE is a well-established technique commonly used to facilitate the extraction of any type of plant matrices. The efficiency of the technique is mainly explained by the phenomenon of cavitation.

Best operative conditions and reactor features are strictly related to the matrix physical properties. While the external glands of plant secretory structures can easily be destroyed by sonication, thus facilitating the release of metabolites into the extraction solvent; harder woody structures need much higher power density to be effective. Ultrasonication produces cavitation phenomena when the acoustic power input is sufficiently high to allow multiple microbubbles to be produced at nucleation sites in the fluid. The bubbles grow during the rarefying phase of the sound wave and then collapse during the compression phase. On collapse, a shock wave passes through the medium. The whole process of gas bubble nucleation, growth, and collapse due to the action of intense sound waves is called *cavitation*. The collapse of the bubbles converts sonic energy into mechanical energy in the form of shock waves equivalent to several thousand atmospheres (300 MPa) pressure. This energy imparts motion to parts of cells which disintegrate when the US energy content exceeds the wall strength. An additional factor which increases cell breakage is microstreaming (very high velocity gradients causing shear stress), which occurs near radially vibrating bubbles of gas generated by the ultrasound. Much of the energy absorbed by cell suspensions is converted to heat, so effective cooling is essential. The amount of protein released by sonication is high; however, a careful monitoring is required to avoid degradation. The main factors that affect cell breakage are the water jets and shock waves generated by US. The mechanical effect of US accelerates the release of organic compounds contained within the plant body by disrupting cell walls, enhancing mass transfer, and facilitating solvent (or any kind of fluid) access to the cell content. This effect is much stronger at low frequencies (18–30 kHz), and practically negligible at 400–1000 kHz, compared for the same wave intensity. UAE has proved to be a versatile technique that has been scaled up with great benefit for industrial production. Often, UAE affords purer extracts in higher yields, in shorter times and working at lower temperatures. The main advantage is the enhanced hydration process which takes place simultaneously with matrix fragmentation without any appreciable chemical degradation. The treatment of pulverized plant material in suspension with US power gives rise to a rapid disruption of cell walls and membranes. Equipment for the large-scale continuous use of US has been available for many years and is widely used by the chemical industry, but has not yet found extensive use in plant cell disruption. The reasons for this may be the conformational lability of some (perhaps most) enzymes to sonication and the damage that they may

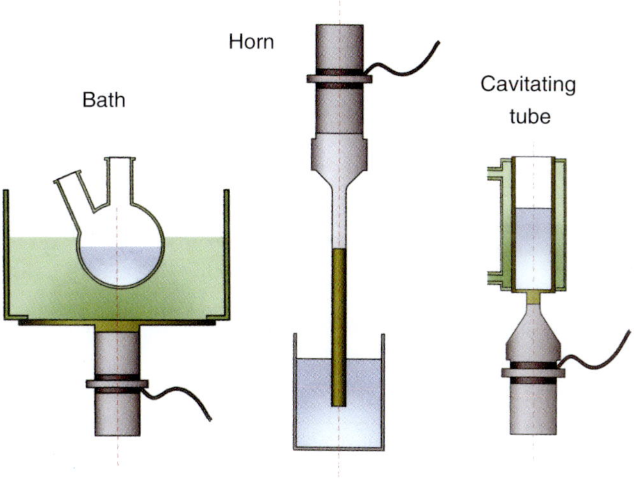

Figure 4.17 Ultrasonic probe system (University of Turin and Danacamerini s.a.s. – Turin, Italy).

experience through oxidation by free radicals, single oxygen, and hydrogen peroxide that is concomitantly produced. The use of radical scavengers (e.g., N_2O) has been shown to reduce this inactivation. As with most cell breakage methods, very fine cell debris particles may be produced which can hinder further processing. Sonication remains, however, a popular, useful, and simple small-scale method for cell disruption. Although acoustic cavitation is known to generate conditions of very high temperature and pressure locally, the activity of the released active compounds and enzymes from the cells remains unaltered. This can be attributed to the fact that the intense conditions only exist for very small intervals of time (typically few microseconds) and hence do not result in any deactivation of the released enzymes.

The US devices can be designed in a batch mode as bath, immersion horn, or cavitating tube (Figure 4.17). However, for a larger scale, processing is conceivable in a continuous mode where multiple units can be combined in a sequential manner, which also increases residence time (Figure 4.18).

Flow UAE is a novel achievement in the field, which exploits mainly multitransducer probe systems either fixed on emitting surfaces or equipped with horns. Normally, the flow-reactor geometries used in UAE can be easily cleaned online in a very efficient and rapid way. In Figures 4.18–4.22, different experimental setups for UAE are shown.

4.3.3 Turbo Extraction

The evolution of high-speed rotating mixers brought up highly efficient turbines suitable for solvent extraction processes, as well as for enhanced steam distillation.

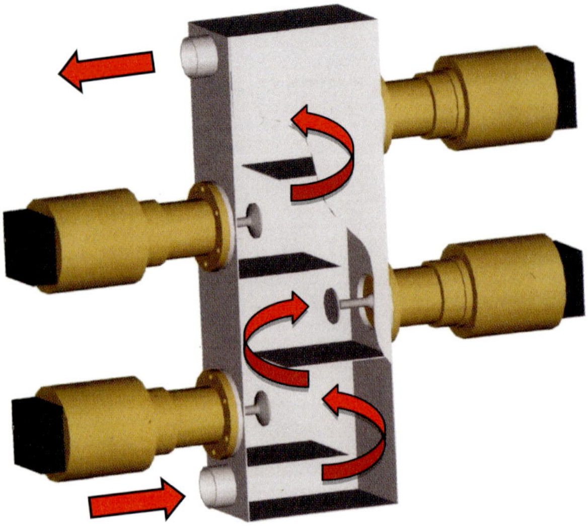

Figure 4.18 Multi-horn flow reactor (University of Turin and Danacamerini s.a.s. – Turin, Italy).

Figure 4.19 Ultrasonic vibrating-plate flow reactor (University of Turin and Danacamerini s.a.s. – Turin, Italy).

Figure 4.20 Transducers fixed on the bottom of the vibrating plate of a multi-horn flow reactor (Figure 4.18).

Both batch devices and flow reactors are commercially available and usually can be used for a wide variety of raw plant materials, in particular for hard dry matrices such as bark, roots, and seeds. Generally, leafy raw materials can be processed as harvested, whereas hard plant materials need to be ground to small sizes in order to optimize the process. In the case of leaves, the too strong mechanical stress may generate microparticles, which are critical in a subsequent filtration step.

In spite of a relatively good mass transfer enabled by this technology, only few examples are reported in the literature. Périno-Issartier [43] describes an apparatus for microwave turbo hydrodistillation (MTHD) and its application for the isolation of the essential oil from dry *Schinus terebinthifolius* berries. The combination of dielectric heating and the mechanical stress from the turbine reduced the hydrodistillation time from 180 to 30 min, thus saving substantial amount of time and energy. An interesting industrial application of turbines for plant extraction

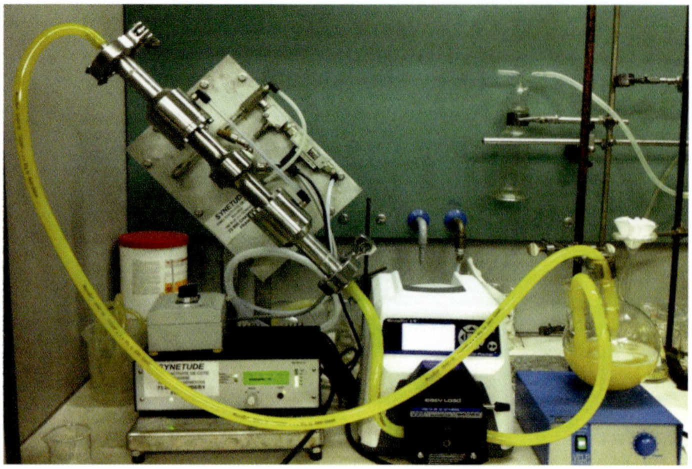

Figure 4.21 Ultrasonic flow reactor Sonitube® (Synetude – Chambery France).

is represented by a robust, compact, and versatile apparatus designed by Lico spa (Arcisate, Italy). A series of dedicated units well suited for grinding, extraction, evaporation, and mixing is presented in Figures 4.23–4.25.

The whole chamber wall is used as an "exchange" surface, ranging from the temperature of liquid nitrogen up to 350 °C. This can also be associated to high vacuum or high pressure with inert gas. The very high speed rotation generates microparticles and microdroplets in close contact, forming a very thin layer of product that covers the wall. In a few seconds, the self-cleaning mixer is empty and ready for the next batch.

4.4 Example Processes for Solid–Liquid Extraction Enhancement

4.4.1 Extraction of Polyphenols from Black Tea – Conventional and Ultrasound-Assisted Extraction

Pretreatment of the raw material and the extraction of valuable components are first steps in processing. Actually, because of lack of physicochemical fundamentals, these unit operations are economically not optimally designed. The first methodical approaches to design optimal processes have been published for the extraction [44–47] as well as for further purification of plant-based extracts [48]. For design and optimization of solid–liquid extraction processes, currently statistical experiment planning and physicochemical modeling are mostly investigated. First and foremost, equilibria as well as mass transport kinetics for target and side components are in focus. Considering technical and economical constraints, these parameters have to be enhanced. Extraction characteristics

Figure 4.22 A multi-horn horizontal reactor (University of Turin and Danacamerini s.a.s. – Turin, Italy).

and hence the equipment depend on the processed raw material and the solvent system [46]. For example, the location of target and side components, the structure of the raw material, and the moisture content can be named as crucial factors on equilibrium and mass transport kinetics. Process intensification and hence enhanced mass transport kinetics and quasi-equilibrium behavior can be achieved by using ultrasound techniques [49]. Using the example of polyphenol extraction from black tea, the influence of ultrasound on mass transport and equilibrium is analyzed and discussed. Differences between conventional and ultrasound-assisted maceration are investigated considering resulting particle size distributions and scanning electron microscope (SEM) images.

4.4.1.1 Material and Methods

For the analysis and discussion of different process concepts, the equilibrium and mass transport kinetics are especially considered. The enhancements are

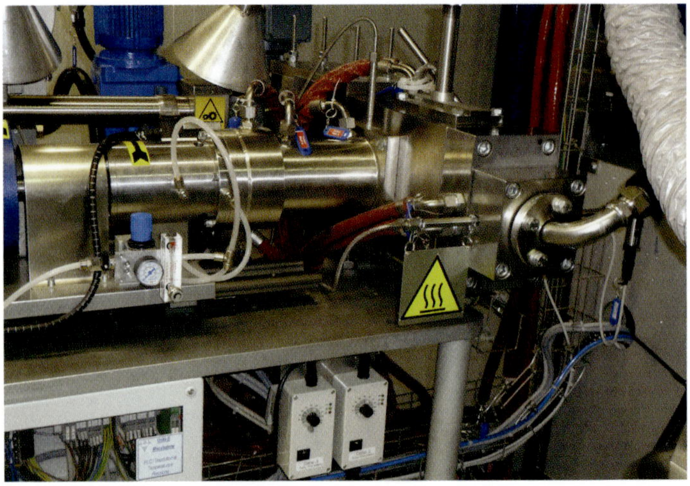

Figure 4.23 Turbo extraction unit. (Kind permission of Lico spa – Arcisate Italy).

Figure 4.24 Turbo extraction: evaporation unit. (Kind permission of Lico spa – Arcisate Italy.)

discussed employing particle size distribution and SEM measurements. The laboratory experiments are carried out with black tea from Kenya. As extracting solvent, pure ethanol as well as a mixture consisting of ethanol/water 90/10 (m/m) are used. The total content of polyphenols is quantified by five -tage maceration (each 24 h), followed by 96 h control percolation with 0.1 BV h^{-1} to ensure total extraction [45]. For the entire amount of polyphenols, that is, a yield of 100%, loading of 211.4 g polyphenols per kilogram of dry substance tea is determined.

UAEs are performed in an ultrasonic extraction reactor PEX1 (R.E.U.S., Contes, France) with 14 × 10 cm internal dimensions and a capacity of 1 l, equipped with a transducer at the base of jug operating at a frequency of 25 kHz with a maximum input power of 150 W. The double jacket allows for the control of extraction temperature by cooling/heating systems. Considering that the actual input power

Figure 4.25 Internal view of the turbo extraction unit. (Kind permission of Lico spa – Arcisate Italy).

from the device is converted to heat which is dissipated in the medium, calorimetric measurements were performed to assess actual ultrasound power, calculated as shown in Equation 4.3.

$$P = m \cdot c_p \frac{dT}{dt} \tag{4.3}$$

Here, c_p is the heat capacity of the solvent at constant pressure ($J\,g^{-1}\,K^{-1}$), m is the mass of solvent (g), and dT/dt is temperature rise per second. Figure 4.26 depicts the equipment used for the comparison of mass transport kinetics and for the determination of the equilibrium line. Mass transport kinetics and phase equilibria are determined at a constant temperature of 40 °C and a solid/liquid ratio of 1/3 (m/m). The solid–liquid equilibrium line is determined in multistage maceration experiments. Each maceration step has to reach equilibrium for this. Owing to additional total extraction and closing of the mass balance for each component in each phase, the equilibrium concentration in solid and liquid phases can be measured.

To close all mass balances, in addition to polyphenol concentration, the amount of water, dry substance, further side components, and the solvent has to be measured in both phases. Table 4.2 shows the necessary methods.

The total amount of polyphenols is determined using the Folin–Ciocalteu test [50] from Seppal. Here, the absorption Abs_i is detected at a specific wave length of 620 nm. By using reference solutions with a defined concentration of $c_{Ref} = 3\,g\,l^{-1}$, the concentrations of polyphenols $c_{PP,i}$ in each fraction can be determined via Equation 4.4.

$$c_{PP,i} = \frac{Abs_i}{Abs_{ref}} \cdot c_{ref} \tag{4.4}$$

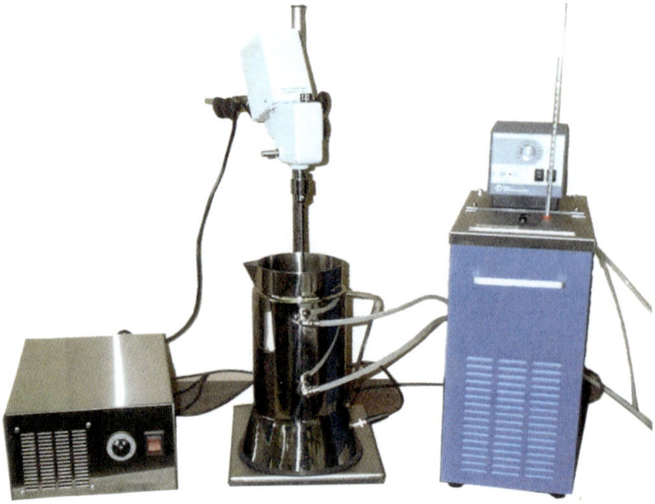

Figure 4.26 R.E.U.S. ultrasound equipment.

Table 4.2 Material and methods.

	In raw material	In liquid phase
Total polyphenol	Multistage maceration	Folin–Ciocalteu test
Dry substance	Dry balance	Dry balance
Water	Dry balance, toluene distillation	Karl–Fischer titration
Solvent mass	Dry balance	Calculation

The dry substance as well as the residual humidity in the raw material is determined using the moisture analyzer Sartorius MA 150. SEM measurements are necessary for analyzing the raw material matrix and hence the quantification of the ultrasonic influence. The measurements are carried out using an FEI/Philips XL30 FEG ESEM. Each sample is compared with a reference sample of fresh tea.

4.4.1.2 Equipment Concepts

The optimal equipment for the extraction of valuable components from plant material depends significantly on the raw material as well as on the processed volume rates [51]. In general, a differentiation between raw material characteristics, physicochemical parameters, and economic factors is made. As raw material characteristics, the

- accessibility of target and side components,
- structure of the raw material matrix,
- moisture content, and
- swelling behavior of the processed raw material

can be named [46]. The physicochemical parameters depending on the characteristics are [51]:

- equilibrium,
- mass transport kinetics and
- fluid dynamics (depending on the equipment).

The design of an economically optimized process further requires taking into account economic factors [48] such as the following:

- Apparatus efforts
- Investment costs
- Operating costs.

Solvent and energy consumption influence the operating conditions significantly. These costs have to be considered with regard to the process design. An optimum between solvent consumption and the target components yield has to be determined. For low-value products, such as sucrose from sugar beets, the solvent consumption, and hence the operating costs have to be kept low. The solvent consumption is subsidiary only for high-value products, such as vanillin from vanilla beans. Yields of nearly 100% compensate the operating costs [51]. Hence, as further constraints in process design, the yield and purity of the extract and, the quality of the product can be named.

Considering the named aspects, different extraction equipment concepts can be investigated. In this study, a first comparison between conventional and ultrasound-assisted maceration is focused on. Characteristics such as the accessibility and mass transport kinetics are analyzed and discussed in detail. Initially, three concepts for investigation (Figure 4.27) are defined. First of all, conventional multistage maceration (Figure 4.27a) is compared to ultrasound-assisted multistage maceration (Figure 4.27b). In addition, the influence of ultrasound is quantified by use of a pretreatment with ultrasound at the first stage and further conventional processing in maceration (Figure 4.27c).

4.4.1.3 Equilibrium Line by Multistage Maceration and Total Extraction

Equilibrium lines can be determined using multistage maceration followed by a control percolation [46, 52]. The equilibrium will shift toward the extract phase, if the slope of the line decreases. This results in high extract-phase concentrations and low solid-phase loadings. In general, three major types of equilibrium lines can arise: the linear type [46, 51], the Langmuir type [46], and the anti-Langmuir type.

The anti-Langmuir type of equilibrium line results from limiting capacities in the liquid phase. The equilibrium relation leads to high solid-phase loadings and low concentrations in the liquid phase. Hence, this is more theoretical in nature [52].

Figure 4.28 depicts the quasi-equilibrium lines for polyphenols from black tea at different solvent compositions (pure ethanol and ethanol/water) and two maceration techniques (with and without ultrasound assistance). Each equilibrium line

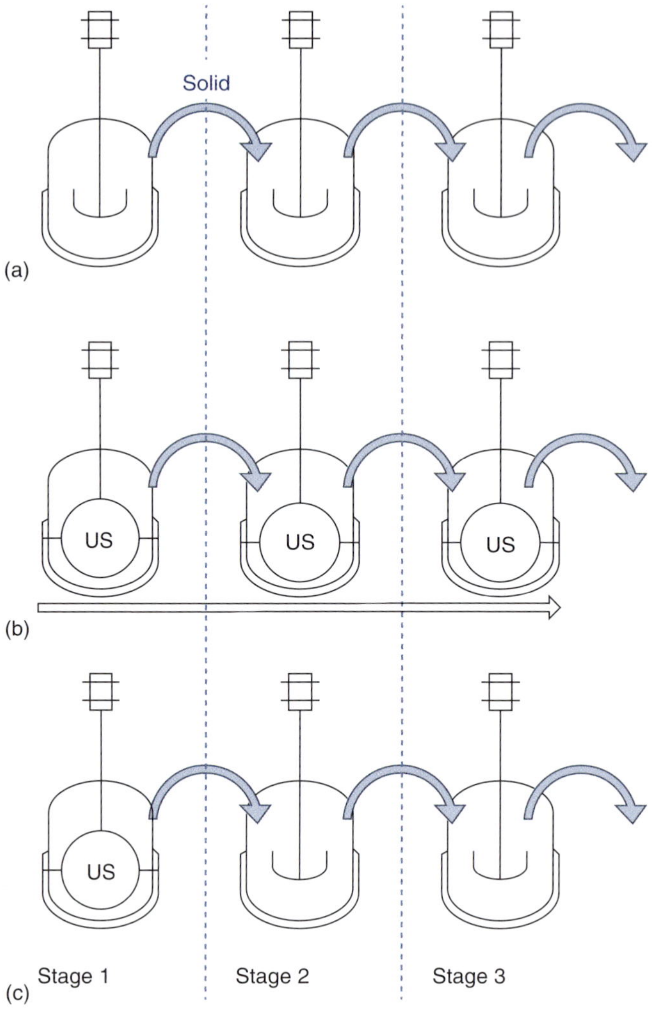

Figure 4.27 (a) Multistage maceration, conventional; (b) multistage maceration, ultrasound-assisted; and (c) multistage maceration, combination of (a) and (b).

show is the Langmuir type. The solvent consisting of ethanol and water shows the best behavior. For a single-stage maceration with a solid/liquid ratio of 1/3 an extract concentration of $c_{\text{Polyphenols}} \sim 21\,\text{g}\,\text{l}^{-1}$ can be reached. A residual loading of about $14\,\text{g}\,\text{l}^{-1}$ results in the solid phase. For pure ethanol, the concentration in the liquid phase ranges between 2 and $4\,\text{g}\,\text{l}^{-1}$ and for the solid phase this is about $22\,\text{g}\,\text{l}^{-1}$. Hence, equilibrium increases toward the solid phase.

Different maximal residual polyphenol loadings q_{max} in the solid phase depend on the technique and solvent used. The maximal residual loading for extraction with pure ethanol lies at about $22\,\text{g/g}_{\text{solid}}$, whereas the loading for the solvent mixture ethanol/water 90/10 (m/m) lies at around $15-17\,\text{g/g}_{\text{solid}}$.

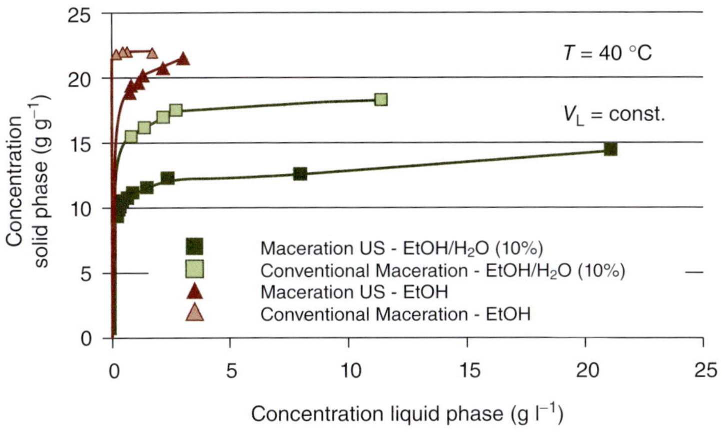

Figure 4.28 Quasi-equilibrium lines for conventional and ultrasound-assisted extraction of tea polyphenols with pure ethanol and ethanol/water 90/10 (m/m).

Lower maximal residual loadings lead to higher concentrations in the liquid phase and hence lower solvent consumption and operating costs, respectively.

Solid-phase loadings below maximal residual loadings are attainable with high solvent consumption. Each of the maceration steps results in a low liquid-phase concentration and therefore in a low decrease in residual loading. A total extraction is economically feasible only with constantly high concentration gradients. This can be realized using percolation equipment. A total extraction is achievable with minimal solvent consumption. Nevertheless, maceration equipment is useful for raw material–solvent systems with a linear equilibrium line having a low slope. A major part of the ingredients is extracted within one equilibrium stage. Low residual concentrations and high extract concentrations are measured. Hence, a lower dilution than for the percolation equipment can be realized with low effort.

As described above, Figure 4.28 depicts the quasi-equilibrium lines, determined via multistage maceration. To gain equilibrium values for higher liquid-phase concentrations, the liquid phase has to be preloaded. Hence, the concentration gradient between the two phases is reduced.

Water can increase the solubility of polar polyphenols in the solvent mixture. In addition, water leads to swelling of the raw material matrix. This results in a higher accessibility for the ingredients.

Furthermore, the ultrasound technique attacks the surface of the raw material particles. Basically, either the whole surface is removed or occasional punctures leading to an increased accessibility can occur [49]. Figure 4.29 depicts the two principles. It has to be kept in mind that the surface structure after treatment depends on the treated raw material and on the exposure time.

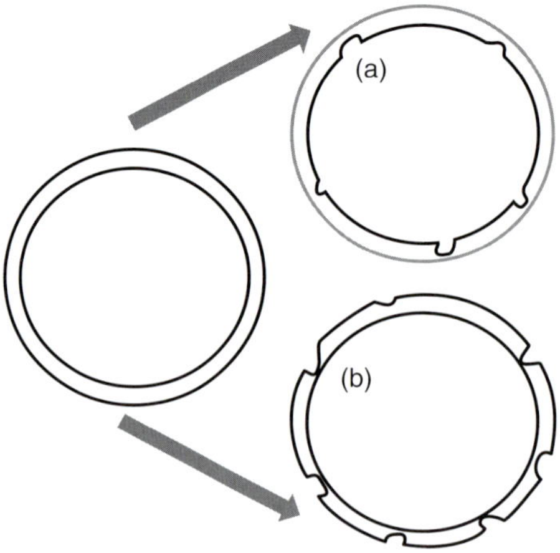

Figure 4.29 Schematic principles of ultrasonic attacks.

4.4.1.4 **Mass Transport Kinetics**

The comparison of mass transfer kinetics for maceration with ethanol and ethanol/water mixture and with regard to both techniques is depicted in Figure 4.30. Owing to higher equilibrium concentration with the use of the ultrasound technique, an absolute higher mass transport occurs. The kinetics is enhanced, whereas the difference of relative mass transport with and without ultrasound lies in the range ±5%, which is within the range of the error of measurement. For both solvents, the same extraction kinetic characteristics can be recognized. Nevertheless, differences occur in equilibrium behavior. Using the mixture of ethanol and water instead of pure ethanol, the equilibrium extract concentration can be increased by about 500%. Differences between the two techniques are around 13% for each solvent. Hence, the choice of solvent is more important than the choice of technique.

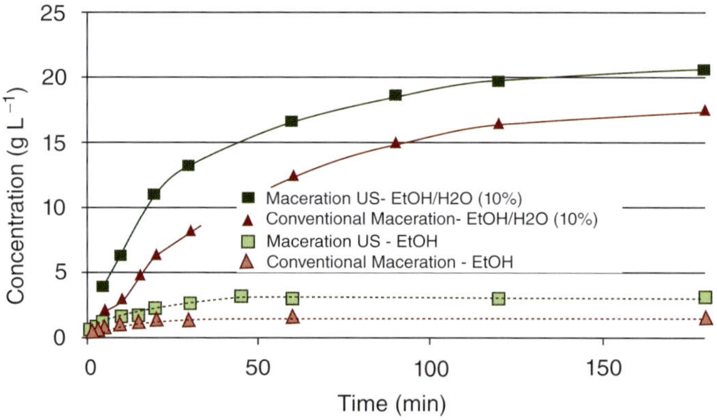

Figure 4.30 Comparison: Conventional and ultrasound-assisted maceration with ethanol and ethanol/water 90/10 (w/w) mixture.

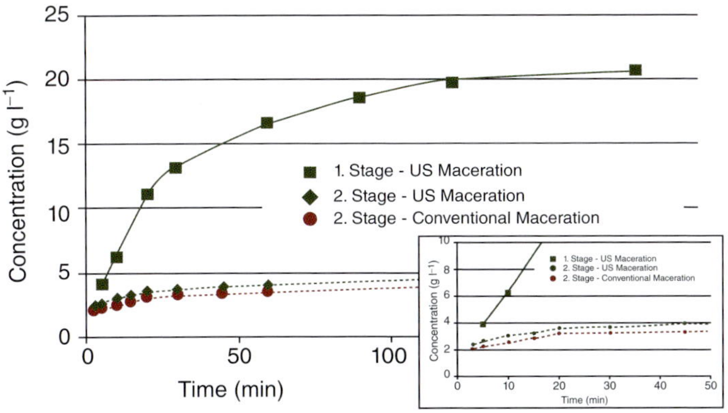

Figure 4.31 Mass transport kinetics in two-stage maceration: first stage ultrasound assisted, second stage with and without employment of ultrasound.

Nevertheless, the use of ultrasound can improve the extraction for a predefined solvent.

For the solvent mixture of ethanol/water, the influence of ultrasound as pretreatment (concept as shown in Figure 4.27c) in comparison to multistage ultrasound-assisted maceration can be investigated using the mass transport kinetics and equilibrium behavior as objectives. Figure 4.31 depicts the extraction kinetics for the first and second maceration stages. Here the first stage is assisted by ultrasound. The second stage is either with or without ultrasound assistance. Even in the second stage, ultrasound leads to higher concentrations in the liquid phase (~13%). The extraction kinetic, as also seen in Figure 4.30, is not enhanced by the use of ultrasound. Quasi-equilibrium is reached between 20 and 30 min. for both cases. This raises the question as to whether the surface is consistently removed by ultrasound or whether an attack occurs more partially and hence whether only the specific surface for mass transport is increased. By employment of particle size and SEM analyses, the principle of attack by ultrasound is investigated. The results are depicted and discussed in the following sections.

The advantages and disadvantages in equilibrium behavior and mass transport enhancement have been discussed. The possibilities for industrial-scale applications have to be determined by use of specific separation costs and hence analysis of operational costs [53–55]. At industrial level, one or two more steps in the extraction process can represent an unacceptable additional cost.

4.4.1.5 Particle Size Distribution

Ultrasound employment leads to higher decrease in particle size than conventional extraction methods (Figure 4.32). The mean particle diameter is reduced from 660 initially to 640 μm using conventional maceration and to 610 μm using ultrasound-assisted maceration. The particle size is about 5% lower with ultrasound employment. In each case, the extraction time is 30 min.

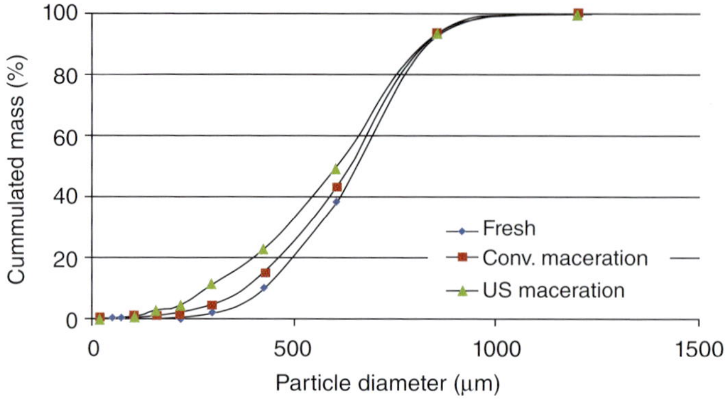

Figure 4.32 Comparison of particle size distribution.

Analyzing the content of dry substance in the extract shows same bias. The content is increased from 2.9 to 4.3% (m/m) with use of ultrasound, which is an increase of about 30–35%. Thereby, the dry substance is defined as a multicomponent mixture containing a high amount of target substances, such as polyphenols. Nevertheless, the raw material matrix is disrupted as well. The extract becomes turbid with the use of ultrasound.

4.4.1.6 **SEM Measurements – Cell Disruption**

Figure 4.33 shows the raw material, the ground black tea leaves. Because of a high degree of grinding (mean particle diameter of about 660 μm), even without further treatment, the different botanical parts can be seen. Exemplarily, the epidermis, stoma, and vascular bundle can be named. An extraction of 30 min leads to a removal of the epidermis, so that further botanical elements become uncovered. Significant differences between the two methods, conventional and ultrasound-assisted maceration, cannot be identified. The figure shows equal parts of the matrix. Owing to the variablility of the leaf particles, no quantification is possible. Each leaf particle is extracted from the overall surface.

4.4.1.7 **Conclusions**

First results for the extraction of polyphenols from black tea are presented. Here, two types of maceration, conventional and ultrasound assisted, as well as the influence of water in the extracting solvent are discussed concerning equilibrium and mass transport kinetics. The differences are analyzed employing the dry substance and polyphenol content in the extract as well as the particle size distribution and SEM images.

The use of ultrasonic extraction leads to higher quasi-equilibrium concentrations in the liquid phase. The content of polyphenols is increased by approximately 15%. Thereby, the solubility of the ingredients, here polyphenols, as well as the accessibility for the solvent are discussed. Ultrasound employment does not compensate the content of water in the solvent. The water leads to higher solubility and swelling of the raw material and hence higher yields.

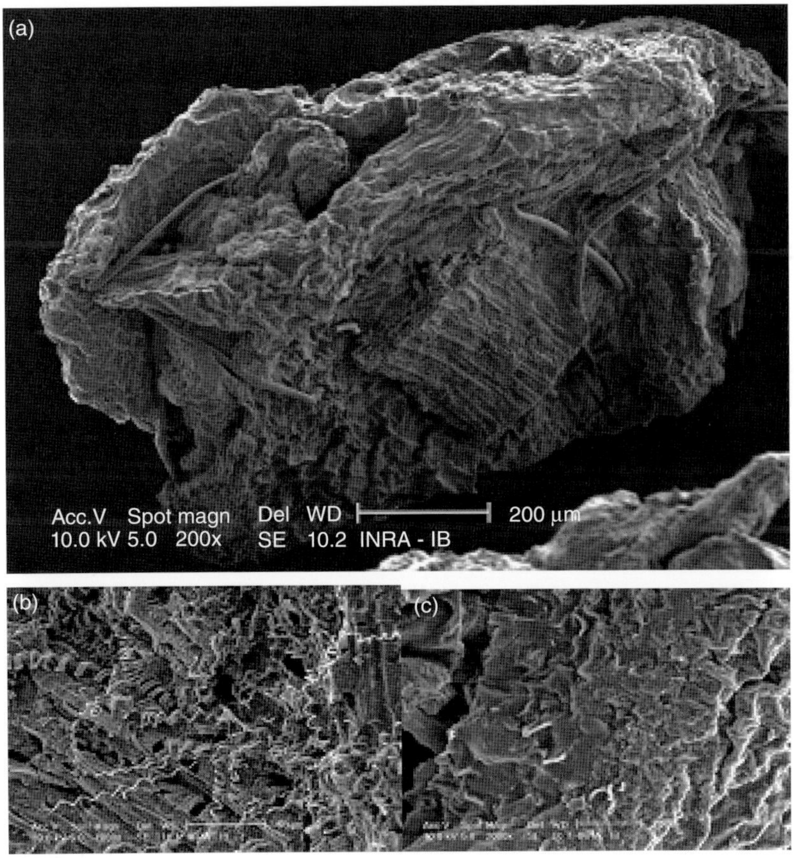

Figure 4.33 Tea leaf analyses via SEM ((a) initially and (b,c) after 30 min maceration with and without ultrasound).

Therefore, ultrasound can assist the extraction for a pre-chosen and optimized solvent.

Via particle size measurements a diminution of the mean particle diameter during extraction is detected. After an extraction time of 30 min, the particle diameter is 5% lower using ultrasound compared to conventional extraction. In addition to polyphenol extraction, an extraction of further leaf elements is detected via analyzing the content of the dry substance in the extract phase. This content is about 30–35% higher when ultrasound is used.

SEM analysis does not show significant differences for tea leaf particles between the two techniques. Because of the high degree of grinding (mean particle diameter of about 600–700 µm), various botanical elements such as epidermis, stoma, and vascular bundle can be identified. During extraction, an increasing number of leaf structures are uncovered. Nevertheless, differences concerning the SEM images between the two maceration types cannot be quantified. The variability of the leaf particles and hence the leaf elements observed via SEM does not allow for any quantification.

Three different equipment concepts are discussed on the basis of the results. Here, the conventional multistage maceration is compared to ultrasound-assisted multistage maceration. In addition, pretreatment with ultrasound as a first maceration step followed by further conventional maceration steps is discussed. In conclusion, ultrasound leads to higher concentrations in liquid phase, even if a pretreatment with ultrasound has been preceding. The yield in the second maceration stage is about 13% higher for the two-stage ultrasound maceration than for single-stage UAE followed by conventional maceration.

4.4.2 Pilot Scale UAE of Clove Buds in Batch and Flow Reactors

Cloves are one of the most ancient and highly prized spices from the Orient and are widely recognized all over the world for their medicinal and culinary qualities. It is an unopened flower bud that grows on a tree belonging to the Myrtaceae family in evergreen Indonesian rain forests (*Eugenia caryophyllus* (Spreng) Bullock and S. G. Harrison; syn. *Syzygium aromaticum* (L.) Merr). Data in the literature is mainly focused on essential oil composition and biological properties. Eugenol is the marker compound in clove essential oil and is of interest as it has been attributed with antioxidant and antimicrobial properties. The other main compounds are eugenol acetate and β-caryophyllene (Figure 4.34).

The plant is widely used as herbal drug in the treatment of dyspepsia, acute/chronic gastritis, and diarrhea as well as in several recipes for food and liqueurs. Eugenol is, in fact, an important flavoring agent also used in cosmetics and body care. Several biological properties have also been attributed to β-caryophyllene; these include anti-inflammatory, antibiotic, antioxidant, anticarcinogenic, and local anesthetic activity. Although hydrodistillation, supercritical carbon dioxide extraction, and superheated water extraction techniques have all been applied to the extraction of clove buds, maceration still remains the most common clove bud extraction method on an industrial scale.

In this case report, three commercial clove batches of different origin (India, China, and Madagascar) are extracted and analyzed by headspace gas chromatography coupled with mass spectrometry (HS-GC/MS). Crude extract yield (dry extract/dry plant w/w %), total phenolic compound content, and antioxidant activity are determined. The aim of the study is to assess the efficiency of both batch and

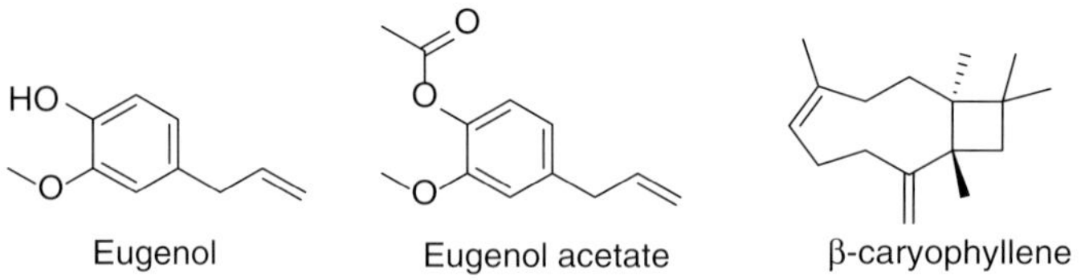

Figure 4.34 Chemical structures of eugenol, eugenol acetate, and ß-caryophyllene.

flow ultrasonic reactors and to compare these more recent methods with classic maceration.

4.4.2.1 Experimental Methods and Reactors

Flow UAE is carried out in a new multi-horn reactor working in continuous-flow mode and at high power density (about 200 W/100 ml) (Figures 4.35 and 4.36). This is a patented sonoreactor designed for process intensification in UAE [56].

Maceration Ground clove samples (1 kg) are dispersed in 20 l of ethanol/water (1 : 1, v/v) in a stainless steel cylindrical vessel under mechanical stirring (15 min). The suspended material is macerated for 15 h at room temperature as determined ideal in a preliminary investigation. The final mixture is filtered and the solvent is removed under vacuum.

Batch UAE Ground clove samples (1 kg) are dispersed in 20 l of ethanol/water (1 : 1, v/v) under mechanical stirring (15 min) and then subjected to UAE for

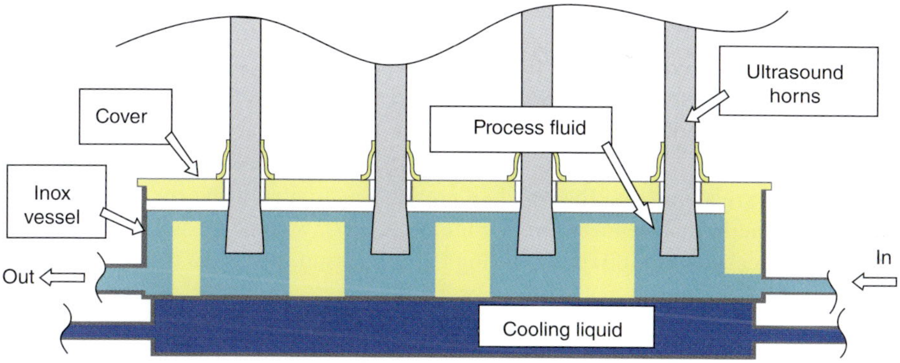

Figure 4.35 Schematic setup of the multi-horn reactor.

Figure 4.36 Setup of the multi-horn reactor in the laboratory.

Figure 4.37 Experimental setup of the 25L-US batch reactor.

45 min, using a 25L-US batch reactor (R.E.U.S. – France) (Figure 4.37). The pilot reactor is equipped with a vibrating plate (frequency 25 kHz, effective power 360 W measured in water, power density 18 W l^{-1}) mounted onto the bottom of the cylindrical reactor. The effective power is estimated using the classic calorimetric method as reported in the literature [57]. Sonication is combined with mechanical stirring (60 rpm). During the extraction, an increase in temperature from 20 to 30 °C within 45 min is observed.

Flow UAE: Ground clove samples (1 kg) are dispersed in 20 l of ethanol/water (1 : 1, v/v) under mechanical stirring (15 min) in a stainless steel cylindrical vessel equipped with a liquid inlet at the top and an outlet valve at the bottom. The suspension is subjected to UAE using a multi-horn flow reactor (four horns, 21.0 kHz) made in collaboration with Danacamerini (Torino, Italy) (Figures 4.35 and 4.36). The total volume of the four PTFE cavities, in which the horns are immersed, is 500 ml. For this extraction, three flow cycles (450, 900, and 1350 ml min^{-1}) are tested by varying the speed of the peristaltic pump (Masterflex L/S Digital Drive, 600 rpm), for an overall time of 45 min. Particles are maintained in suspension because of the turbulence generated by the circulation at all the three flow rates. The highly regular flow of the pump plays a crucial role for the efficiency of the process. The system enables power to be varied up to a maximum of 1 kW, which is monitored by a true reading wattmeter during the experiments. The effective power measured in water is 450 W and the power density 1.12560 W kJ l^{-1}.

The overall energy consumption for one cycle, namely one complete passage of the whole suspension volume through the sonication chamber, is a slightly higher in flow procedures (0.451215 kW hJ^{-1}), which also includes peristaltic pump consumption, than batch UAE (972 kJ, 0.36 kW h^{-1}).

4.4.2.2 Results and Discussion

Yield of Crude Extract Extraction conditions, technique, and reactor type play an important role in process efficiency and the chemical profile of the extract. The highest yields are obtained using flow UAE at high flow rates. In contrast, maceration gives the lowest extraction yields. A simple calculation shows that the three chosen speeds roughly correspond to one, two, and three suspension passages through the sonication chamber. Even though the overall residence time is about the same, a multiple cycle seems to be more effective. Besides the power, the frequency, and the geometry of the reactor, this finding highlights the pivotal role of the flow rate in maximizing extraction yield.

The analyses are performed in triplicate and results are expressed as mean values of the percentage yield ± standard deviation (SD).

Phenolic Content On the basis of our previous studies on the UAE of polyphenols [58], the total phenolic content of the clove bud extracts obtained under different extraction conditions is shown in Figure 4.38. The highest phenolic compound extraction rate is obtained using flow UAE operating at high flow rates. This clearly indicates that, as the sonication flow rate increases, there is a corresponding increase in the total phenolic content of the crude extracts. Total phenolic compound content is expressed as milligrams of gallic acid per liter of extract (GAE/L) and ranges from 190.86 ± 1.25 to 215.02 ± 2.5 mg GAE/L extract.

HS-GC/MS Analysis Volatiles can be released to the headspace at different concentrations depending on their solubility in the matrix, absolute amount,

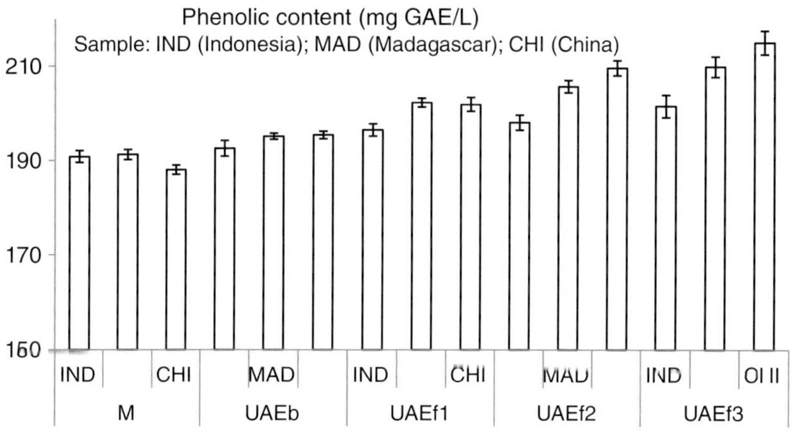

Figure 4.38 Total phenolic compound content is expressed as milligrams of gallic acid per liter of extract.

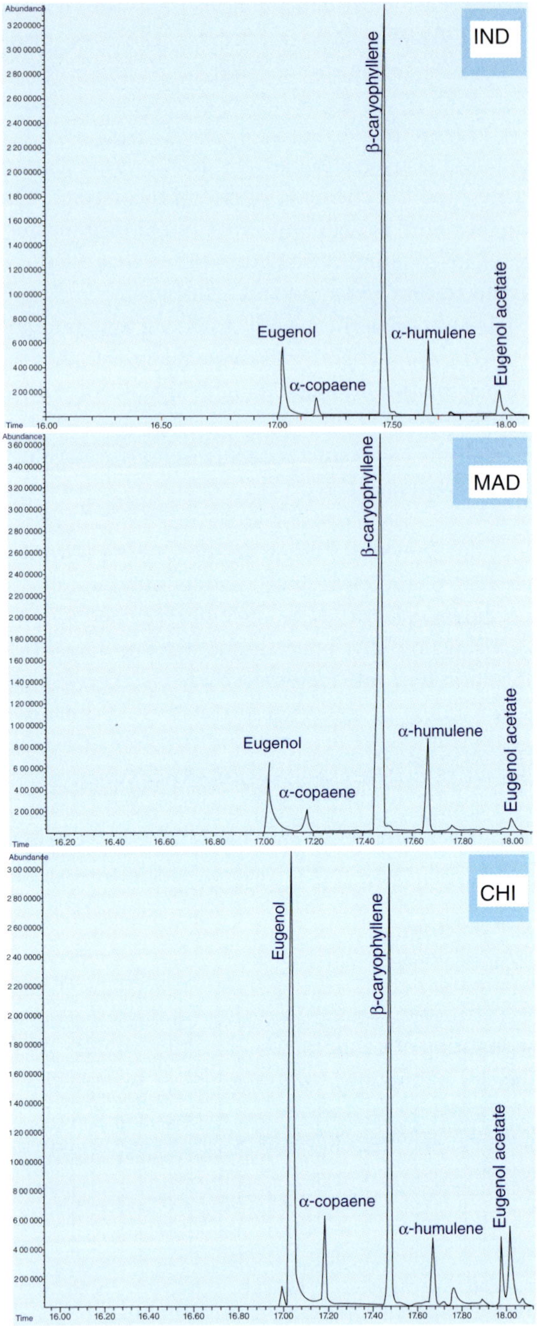

Figure 4.39 Typical HS-GC chromatograms of the three clove buds batches (India, China, and Madagascar).

and volatility. The trapping analysis of the headspace volatiles is closely related to the sensory profile of the final extract. The area percentages of the principal constituents of the clove bud extracts are monitored in the extracts obtained under the various extraction conditions. The highest main constituent area percentages are found in the extracts obtained using UAE at high flow rates (Figure 4.38). Typical chromatograms of the clove bud extract headspace flavor volatiles from the three commercial batches are shown in Figure 4.39.

4.4.2.3 Conclusions

The application of flow UAE to process intensification depends on the development of large-scale multiple transducer reactors operating at high power density. The higher yield and total phenolic content of crude extracts, as well as the excellent antioxidant activity, have all proved that dried clove buds can be easily and efficiently extracted using flow UAE.

4.4.3 UAE and MAE of Lipids from Microalgae

The replacement of classical biofuel crops by microalgae is gaining ever more interest because they can produce up to 10 times more oil per cultivated area than traditional oil plants. Nevertheless, the conversion process is hampered by technical and economic problems that are related to dewatering and extraction. The efficiency of the process can be substantially improved by means of ultrasound and microwaves. Scaling-up feasibility is directly linked to reactor efficiency, energy consumption, environmental impact, and overall cost. In the present case report, the optimization of the extraction process from *Nannochloropsis gaditana* is investigated.

4.4.3.1 Experimental Methods and Equipments

UAE is performed using high-power transducers (probe systems by Danacamerini – Torino, Italy), comprising an immersion horn (19.5 kHz) and a cavitating tube. This is a cup horn-like system consisting of a thin hollow titanium cylinder fixed to a booster (21.5 kHz) [59]. The extraction temperature is kept between 50 and 60 °C by means of a thermostated system (Figure 4.40a,b).

MAE is carried out in a professional multimode oven (2.45 GHz, Microsynth-Milestone, BG Italy) in closed PTFE vessels. The extraction temperature is kept constant at either 60 or 90 °C and monitored by an optical fiber thermometer. The MW device modulates the power input with the aim of keeping the operating temperature constant. The power varies in the range of 25–30 W for the extractions carried out at 60 °C and in the range of 30–35 W for the extraction performed at 90 °C.

A weighed amount of dried microalgae (5 g) is suspended in the solvent (50 ml, ratio of 1 : 10 g ml^{-1}, and separately 250 ml, ratio of 1 : 50 g ml^{-1}). The different

Figure 4.40 (a) US horn, (b) US cavitation tube, and (c) closed MW vessels.

techniques are applied in a time range of 5–60 min and at temperatures from room temperature (rt) up to 90 °C.

A number of solvents are tested; a $H_2O/CHCl_3/MeOH$ 1 : 1 : 2 mixture, a $CHCl_3/MeOH$ 2 : 1 mixture, hexane, acetone, and MeOH. Once the extraction is completed, the mixture is filtered by means of a sintered-glass Buchner funnel and the solvent is evaporated. In the case of the $H_2O/CHCl_3/MeOH$ 1 : 1 : 2 mixture, H_2O and $CHCl_3$ (1 : 1) are added to form a biphasic system after filtration. In the case of the $CHCl_3/MeOH$ 2 : 1 mixture, H_2O is added to form a biphasic system after filtration giving a final ratio of 8 : 4 : 3 for $CHCl_3/MeOH/H_2O$. The organic phase containing the lipid fraction is separated and evaporated under vacuum. When necessary, the aqueous layer is extracted with $CHCl_3$ (1–2 × 20–50 ml).

Generally, the energy consumption in MAE is lower than in UAE. However, fast sonication treatments (5 min) at high power density may be competitive, which is also related to the reactor efficiency. The lowest energy consumption is obtained when the extraction is carried out at 60 °C in 10 min, but the extraction yield is slightly higher at 90 °C in 10 min. In the light of these findings, it is clear that the selection of the best operating conditions needs to be addressed using a wider approach that includes the whole production process, from microalgae cultivation to the final product. To show how far this technology has progressed and to underline the need for further development, the energy consumption of these techniques may be compared with the theoretical maximum energy that can be obtained from microalgae [60].

Conventional extraction procedures with $CHCl_3/MeOH$ mixtures and fast US-assisted extractions and and MAEs with MeOH yield comparable FA content w/w % from dried microalgae (Figure 4.40c). The highest extraction yield and

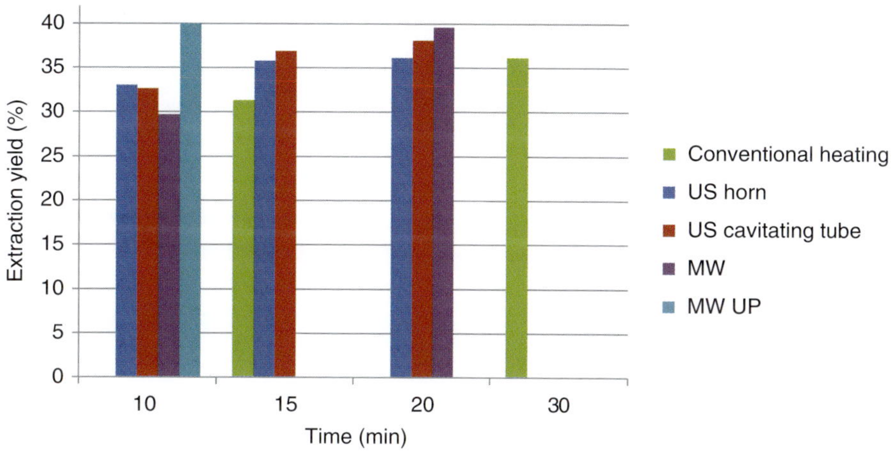

Figure 4.41 Extraction yield (oil %) of *Nannochloropsis gaditana* using MeOH (1 : 10 ratio) and different techniques in 10–30 min.

lowest energy consumption is found to occur under MAE, especially at high temperatures and under pressure (Figure 4.41).

4.4.3.2 Conclusions

This work confirms the advantages of UAE and MAE in the production of bio-oils from microalgae. The best solvent for the classic extraction process is a $CHCl_3$/MeOH mixture and MeOH for UAE and MAE. All of these optimized processes show comparable yields FA/dried microalgae w/w %. Extraction under MW and US requires a lower amount of solvent, avoids chlorinated waste, and proceeds in a shorter extraction time. These techniques also enable one-pot sequential extraction/transesterification for biodiesel production. All these advantages, together with the lower energy consumption in MW reactors, may further reduce the environmental impact of the extraction process. Recent industrial advances in MW-assisted biodiesel production in MW-flow reactors make it easy to expect a fully automated continuous-flow microalgae MAE in the near future.

4.5 Conclusion

The design of more efficient extraction processes that may address the requirements of process intensification and energy consumption reduction has been one of the main research topics in recent years. Safety, sustainability, environmental, and economic factors are all forcing laboratories and industry to turn to nonconventional technologies and greener protocols.

Symbols

Letter	Description	Unit
Abs	Absorption	nm
c	Concentration	$\mathrm{g\,l^{-1}}$
c_p	heat capacity	$\mathrm{J\,K^{-1}}$
m	mass	kg
p	pressure	bar
P	power	J
T	time	s
T	temperature	K
X	mole fraction	—

References

1. Bäcker, W. *et al.* (2003) *Konzeptpapier des Arbeitskreises: Phytoextrakte – Produkte und Prozess,* DECHEMA e.V, Frankfurt.
2. Schönbucher, A. (2002) *Thermische Verfahrenstechnik,* 1st edn, Springer, Berlin, Heidelberg.
3. Voeste, T., Weber, K., Hiskey, B., and Brunner, G. (2006) Liquid-solid extraction, in *Ullmann's Encyclopedia of Industrial Chemistry,* Wiley-VCH Verlag GmbH, Weinheim.
4. Sattler, K. (2001) *Thermische Trennverfahren. Grundlagen, Auslegung, Apparate,* VCH Verlagsgemeinschaft mbH, Weinheim.
5. Jarvis, A.P. and Morgan, E.D. (1997) Isolation of plant products by supercritical-fluid extraction. *Phytochem. Anal.,* **8** (5), 217–222.
6. Hamburger, M., Baumann, D., and Adler, S. (2004) Supercritical carbon dioxide extraction of selected medicinal plants – effects of high pressure and added ethanol on yield of extracted substances. *Phytochem. Anal.,* **15** (1), 46–54.
7. Brunner, G. (2005) Supercritical fluids: technology and application to food processing. *J. Food Eng.,* **67** (1-2), 21–33.
8. Richter, B.E., Jones, B.A., Ezzell, J.L., Porter, N.L., Avdalovic, N., and Pohl, C. (1996) Accelerated solvent extraction: a technique for sample preparation. *Anal. Chem.,* **68** (6), 1033–1039.
9. Benthin, B., Danz, H., and Hamburger, M. (1999) Pressurized liquid extraction of medicinal plants. *J. Chromatogr. A,* **837** (1-2), 211–219.
10. Smith, R.M. (2002) Extractions with superheated water. *J. Chromatogr. A,* **975** (1), 31–46.
11. Kaufmann, B. and Christen, P. (2002) Recent extraction techniques for natural products: microwave-assisted extraction and pressurised solvent extraction. *Phytochem. Anal.,* **13** (2), 105–113.
12. Ondruschka, B. and Asghari, J. (2006) Microwave-assisted extraction – a state-of-the-art overview of varieties. *Chimia,* **60** (6), 321–325.
13. Jacques, R.A., dos Santos Freitas, L., Pérez, V.F., Dariva, C., de Oliveira, A.P., and Caramão, E.B. (2007) The use of ultrasound in the extraction of *Ilex paraguariensis* leaves: a comparison with maceration. *Ultrason. Sonochem.,* **14** (1), 6–12.
14. Vilkhu, K., Mawson, R., Simons, L., and Bates, D. (2008) Applications and opportunities for ultrasound assisted extraction in the food industry – a review. *Innovative Food Sci. Emerg. Technol.,* **9** (2), 161–169.

15. E&E-Verfahrenstechnik *http://www.eunde-verfahrenstechnik.de.*
16. Samtech Extraktionstechnik *http://www.samtech.at/extraktionsverfahren/.*
17. TournaireEquipment *http://www.tournaire.fr.*
18. E&E Verfahrenstechnik *http://www.extraction.de/.*
19. Gea Niro *http://www.niro.de/.*
20. CrownIronWorks (2009) *http://www.crownironworks.com* (accessed 13 August 2014).
21. Schwartzberg, H.G. (1980) Continous counter-current extraction in the food industry. *Chem. Eng. Prog.*, **76** (4), 67–85.
22. Kassing, M., Strube, J., Jenelten, U., and Schenk, J. (2009) Status der Fest-Flüssig-Extraktion von Pflanzenmaterial – Stand der Technik und Modellierung. *Chem. Ing. Tech.*, **81** (8), 1064–1064.
23. Meireles, M.A.A. (2008) *Extracting Bioactive Compounds for Food Products*, CRC Press, Boca Raton, FL.
24. BMA (2009) *http://www.bma-de.com* (accessed 13 August 2014).
25. GEA-Niro (2009) *http://www.gea-niro.com* (accessed 13 August 2014).
26. Poirot, R., Prat, L., Gourdon, C., Diard, C., and Autret, J.M. (2007) Fast batch to continuous solid-liquid extraction from plants in continuous industrial extractor. *Chem. Eng. Technol.*, **30** (1), 46–51.
27. DeSmet (2009) *http://www.desmetballestra.com/.*
28. Harburg-Freudenberger (2009) *http://www.harburg-freudenberger.com* (accessed 13 August 2014).
29. Baerns, M. *et al.* (2006) *Technische Chemie*, Wiley-VCH Verlag GmbH, Weinheim. ISBN: 978-3-527-31000-5.
30. (2011) *Europäisches Arzneibuch*, 7th edn, Deutscher Apotheker Verlag, Stuttgart. ISBN: 376925564X.
31. Sacher, J., García-Llobodanin, L., López, F., Segura, H., and Pérez-Correa, J.R. (2013) Dynamic modeling and simulation of an alembic pear wine distillation. *Food Bioprod. Process.*, **91** (4), 447–456.
32. Glidemeister, E. and Hoffmann, F. (1910) *Die ätherischen Öle*, Miltitz bei Leipzig, Schimmel.
33. Ferhat, M.A., Meklati, B.Y., Smadja, J., and Chemat, F. (2006) An improved microwave clevenger apparatus for distillation of essential oils from orange peel. *J. Chromatogr. A*, **1112** (1-2), 121–126.
34. Johnson, L.A. (2000) Recovery of fats and oils from plant and animal sources, in *Introduction to Fats and Oils Technology*, 2nd edn (eds R.D. O'Brien, W.E. Farr, and P.J. Wan), AOCS Press, Champaign.
35. Sriti, J., Taloub, T., Faye, M., Vilarem, G., and Marzouk, B. (2011) Oil extraction from coriander fruits by extrusion and comparison with solvent extraction processes. *Ind. Crops Prod.*, **33** (3), 659–664.
36. Lampi, A.-M. and Heinonen, M. (2009) Berry seed and grapeseed oils, in *Gourmet and Health-Promoting Specialty Oils* (eds R.A. Moreau and A. Kamal-Eldin), AOCS Press, Urbana.
37. Cravotto, G. (2011) Green extraction techniques for high-quality natural products. *Agrofood Industry High-Tech*, **22** (6), 57–59.
38. Savoire, R., Lanoisellé, J.-L., and Vorobiev, E. (2013) Mechanical continuous oil expression from oilseeds: a review. *Food Bioprocess Technol.*, **6** (1), 1–16.
39. Chemat, F. and Cravotto, G. (2013) *Microwave-assisted Extraction for Bioactive Compounds: Theory and Practice*, Series: Food Engineering Series, vol. 4, Springer Science, New York.
40. Leonelli, C., Veronesi, P., and Cravotto, G. (2013) Microwave-assisted extraction – an introduction to dielectric heating, in *Microwave-Assisted Extraction for Bioactive Compounds: Theory and Practice*, Series: Food Engineering Series, vol. 4 (eds F. Chemat and G. Cravotto), Springer Science, New York.
41. Abert Vian, M., Fernandez, X., Visinoni, F., and Chemat, F. (2008) Microwave hydro-diffusion and gravity: a new device for extraction essential oils. *J. Chromatogr. A*, **1190** (1-2), 11–17.
42. Filly, A., Fernandez, X., Minuti, M., Visinoni, F., Cravotto, G., and Chemat, F. (2014) Solvent-free microwave extraction of essential oil from aromatic herbs:

from laboratory to pilot and industrial scale. *Food Chem.*, **150**, 193–198.

43. Périno-Issartier, S., Abert-Vian, M., Petitcolas, E., and Chemat, F. (2010) Microwave turbo hydrodistillation for rapid extraction of the essential oil from *Schinus terebinthifolius* Raddi Berries. *Chromatographia*, **72** (3-4), 347–350.
44. Bart, H.J. *et al.* (2012) *Positionspapier der Fachgruppe Phytoextrakte – Produkte und Prozesse*, DECHEMA.
45. Kaßing, M., Jenelten, U., Schenk, J., and Strube, J. (2010) A new approach for process development for plant-based extraction processes. *Chem. Eng. Technol.*, **33** (3), 377–387.
46. Kaßing, M., Jenelten, U., Schenk, J., Hänsch, R., and Strube, J. (2012) Combination of rigorous and statistical modeling for process development of plant-based extractions based on mass balances and botanical aspects. *Comput. Chem. Eng.* **35** (1), 109-132.
47. Strube, J., Bäcker, W., and Schulte, M. (2010) Process Engineering and Mini-Plant Technology, in *Industrial Scale Natural Products Extraction* (eds H.J. Bart and S. Pilz), Wiley-VCH Verlag GmbH, Weinheim.
48. Josch, J.P., Both, S., and Strube, J. (2012) Characterization of feed properties for conceptual process design involving complex mixtures such as natural extracts. *Food Nutr. Sci.*, **3** (6), 836–859.
49. Chemat, F. (2011) *Éco-Extraction du Végétal. Procédés innovants et solvants alternatifs*, Dunod, Paris.
50. Obanda, M. and Owuor, P.O. (1997) Flavanol composition and caffeine content of green leaf as quality potential indicators of Kenyan black teas. *J. Sci. Food Agric.*, **74** (2), 209–215.
51. Both, S., Eggersglüß, J., Lehnberger, A., Schulz, T., Schulze, T., and Strube, J. (2013) Optimizing established processes like sugar extraction from sugar beets – design of experiments versus physicochemical modeling. *Chem. Eng. Technol.*, **36** (12), 2125–2136.
52. Both, S., Koudous, I., Jenelten, U., and Strube, J. (2013) Model-based equipment design for plant-based extraction processes – considering botanic and thermodynamic aspects. *C.R. Chim.*, **17** (3), 187-196.
53. Blass, E. (1997) *Entwicklung verfahrenstechnischer Prozesse. Methoden, Zielsuche, Lösungssuche, Lösungsauswahl*, Springer, Berlin.
54. Hurme, M. and Järveläinen, M. (1995) Combined process synthesis and simulation system for feasibility studies. *Comput. Chem. Eng.*, **19** (Supp. 1), 663-668.
55. Both, S., Helling, C., Namyslo, J., Kaufmann, D., Rother, B., and Strube, J. (2013) Resource efficient process technology for energy plants. *Chem. Ing. Tech.*, **85** (8), 1282–1289.
56. Daghero, P. and Cravotto, G. (2012) Plant composition comprising the phytocomplex of a plant species and process for preparing same. EP Patent 2520182 A1, Priority Application IT 2011-TO390.
57. Kimura, T., Sakamoto, T., Leveque, J.-M., Sohmiya, H., Fujita, M., Ikeda, S., and Ando, T. (1996) Standardization of ultrasonic power for sonochemical reaction. *Ultrason. Sonochem.*, **3** (3), S157–S161.
58. Alexandru, L., Cravotto, G., Giordana, L., Binello, A., and Chemat, F. (2013) Ultrasound-assisted extraction of clove buds with batch- and flow-reactors: a comparative study on a pilot scale. *Innovative Food Sci. Emerg. Technol.*, **20**, 167–172.
59. Cravotto, G., Boffa, L., Mantegna, S., Perego, P., Avogadro, M., and Cintas, P. (2008) Improved extraction of vegetable oils under high-intensity ultrasound and/or microwaves. *Ultrason. Sonochem.*, **15** (5), 898–902.
60. Bermúdez Menéndez, J.M., Arenillas, A., Menéndez Díaz, J.Á., Boffa, L., Mantegna, S., Binello, A., and Cravotto, G. (2014) Optimization of microalgae oil extraction under ultrasound and microwave irradiation. *J. Chem. Technol. Biotechnol.*, **89**, 1779–1784.

5
Fundamentals of Process-Intensification Strategy for Green Extraction Operations

Tamara Allaf and Karim Allaf

5.1
Process-Intensification Strategy PI-S from High Capacity to High Controlled Quality Industrial Manufacturing

It is well known that over a long period of history, the main difference between artisanal and industrial production was mainly in *production capacity*. Industrialization at those times meant exclusively large production capacity. Industrial engineers enthusiastically recognized scaling up (normally associated with a lower cost) as the only factor ensuring control of the market. They designed, built, and used increasingly large production units. Interest was only focused on spreading the equipment efficiency and size, consequently reducing production cost. High availability and low cost of energy sources (coal, petroleum oil, etc.) allowed accepting *energy-intensive* and *poorly controlled* industrial *processes and reactors*. Seeking above all to reduce costs, industrial engineers were hardly motivated by environmental impacts, and the future of by-products. Even the quality of finished products was not the principal objective and *industrial production* was for a long time synonymous with poor quality, especially when compared to homemade products. Thus, for a long period, industry manufacturing was mono-dimensional, only animated by greater production capacities, increasingly higher amounts of human needs, and so on. However, recently, these large production capacities have included awareness and requirements of consumers, taking into account the risks to humans and the environment.

5.2
What Does "Intensified Industrial Manufacturing" Mean?

The need to reduce the use of hazardous materials has become more widespread. Knowledge about the reduction in the natural resources is now widespread. The apparently irreversible changes in atmospheric characteristics are an essential element in the choice of an industrial process. Real economic risks systematically arise from the overcapacity of both agriculture production and industrial plant

Green Extraction of Natural Products: Theory and Practice, First Edition.
Edited by Farid Chemat and Jochen Strube.

manufacturing, without any real policy allowing adequate preservation processes. Consequently, various complementary constraints have invariably defined the sustainable technologies for industry. This has resulted in defining an innovative charter having as its basis the *pluridimensional process intensification strategy* (**PI-S**) revealed by the numerous criteria for *unit operation performance, final product quality*, and *equipment reliability.*

5.2.1
Unit Operation Performance

The PI-S in terms of unit operation performance mainly aims at the following:

- reducing specific energy consumption, while greatly increasing yields;
- improving environmental and impacts by using natural raw materials, recovering residue and "wastes";
- availing of advantages and benefits of better process kinetics.

Indeed, PI-S is vital in defining innovative technology approaches, resulting in the following:

- High-quality final products;
- more competitiveness in terms of manufacturing as well as transport cost.

Thus, some technologies leading to high product quality have either not been used or have been very rarely used so far because they involve high manufacturing cost (e.g., freeze-drying) in terms of energy, consumables and/or equipment, and so on or high transport cost (freezing, canning, etc.).

The environmental impact of the new approach will be one of the key factors of new processes that are being introduced. Inevitably, environment considerations increasingly should weigh in the new intensified processes in the context of global climate change. To ensure that the intensification is quite relevant in terms of environment impact, researchers and engineers have to lower the total energy consumption in order to reduce CO_2 emission and for considerable recovery of valuable "solid wastes," transforming them into value-added coproducts. Progressively, industrial processes have to adequately reduce the emission of "greenhouse gases" as well as contaminating/contaminated water and effluents.

5.2.2
Final Product Quality

The One of the objectives of PI-S is to achieve high quality control, which requires progressively more attributes and constraints that depend on the nature of the product. Thus, in the case of food, cosmetics, or pharmaceutical products, policy of sustainable development with high quality appears to be stressed owing to opposing and conflicting requirements. Sensorial/organoleptic contents such as flavor (taste and aroma), color, and texture must mainly be defined and coupled with functional and nutritional features:

- contain vitamins, flavonoids, polyphenols, antioxidants, unsaturated fatty acids, minerals, and so on;
- follow safety and health preservation – this type of products must satisfy legally acceptable safety standards, in terms of microbiology contents (absence of bacteria, spores, etc.);
- have reduced level, or even be completely free, of natural or synthetic allergens (allergen-free products);
- have reduced, or even be completely free of natural or induced antinutritional compounds, for example, leguminous substances such as lupine, soya, acrylamide in fried snacks, and volatile risky compounds.

The following are other attributes of the final product quality requirements:

- high convenience revealed through fast and easy consumption as consumer ease (fast rehydration of numerous dried vegetables; easy grinding for obtaining dried powders, etc.);
- high functional behavior;
- environmentally friendly packaging (e.g., edible coating). Final products have to be easily stored and transported, with low storing energy.

5.2.3 Equipment Reliability

PI-S in terms of equipment in the main part aims at obtaining highly reliable reactors. It results in the following:

- reduction in the maintenance effort;
- increase in the possibility to control the process well;
- easy cleaning of the machine.

Thanks to the development of new electronic and automation means, it has become much easier to design equipment well adapted with a process once the process is properly defined. For safety reasons, equipment must be capable of being easily and effectively cleaned. This will provide the opportunity of extending the treatment domain to different products. Some new automated maintenance methods can also be easily designed, which allows higher equipment reliability.

The main research work concerned with intensification of equipment reliability is in terms of kinetics: *the faster the kinetics, the shorter the residence time, the smaller the reactor, the more homogeneous the treatment*. This steadily allows for the following:

- higher quality control;
- design of a machine that will be small in size and have a large capacity.

For some very short time operations, the machines can be equally efficient whether they are operated by continuous or by batch mode. Industrial equipment is not anymore closely linked to the highest scale. Since the new high-performance process is well defined, equipment

- can be flexible;
- does not necessarily need scaling up (even if it can be scaled down);
- can be located as close as possible to the raw material zone.

These aspects can easily increase safety and reduce environmental risks resulting from precarious and hazardous materials.

5.3 Intensification Strategy as a Pluridimensional Approach

5.3.1 Objectives of Intensification Strategy

The innovative charter of intensification is a pluridimensional strategy for final *product quality* and *equipment reliability*, both strictly depending on an *effective unit operation*. It results in defining a new industrial era based on innovative processes to obtain the highest quality of final products, with the least impacts in terms of global cost and environment for a sustainable development.

5.3.2 Specific Case of Food Industry

In the specific case of the food industry, bridges can be established between the quality assurance methods QAMs (e.g., quality risk management, hazard analysis and critical control points, HACCPs; hazard operability analysis, HAZOP, etc.), and the properly defined PI-S. The control of the processes can be validated and even their effectiveness quantified, thus combining safety and commercial value. This leads to multicriteria optimizations of manufacturing conditions. It should increase plant reliability to be perfectly appropriate to the new innovative intensified process. Cost and environmental impacts increasingly become very important key factors.

5.3.3 PI-S as a Continual Progressing-Development Strategy

Although PI-S is a continual progressing-development strategy, significant improvement of the final product quality, unit operation performance, and equipment reliability should be obtained through research work.

In fact, as pointed out by Allaf and collaborators in their many papers on intensification of drying, solvent extraction, essential oil extraction, and so on, the central feature of PI-S is to improve the process kinetics, which can result in

- small-size plants;
- high yields and capacities;
- lower energy consumption, and so on.

5.4
Fundamentals for Starting Basis Analyses

P-IS cannot be well defined or established by simply carrying out experimental studies and comparatively determining the impacts of various new technologies. The gap between characteristics of usual unit operations and required objectives should be studied via fundamental understanding of the various coupled phenomena.

5.4.1
Intensification Procedure

5.4.1.1 Intensification Cycle

Systematically, the definition of process, equipment, and final product should imply

- the fundamental analyses of the process;
- the online and final assessments of the products;
- the appropriate plant design.

Therefore, intensification strategy for more sustainable industrial manufacturing always follows recurrent approaches (Figure 5.1) using innovative

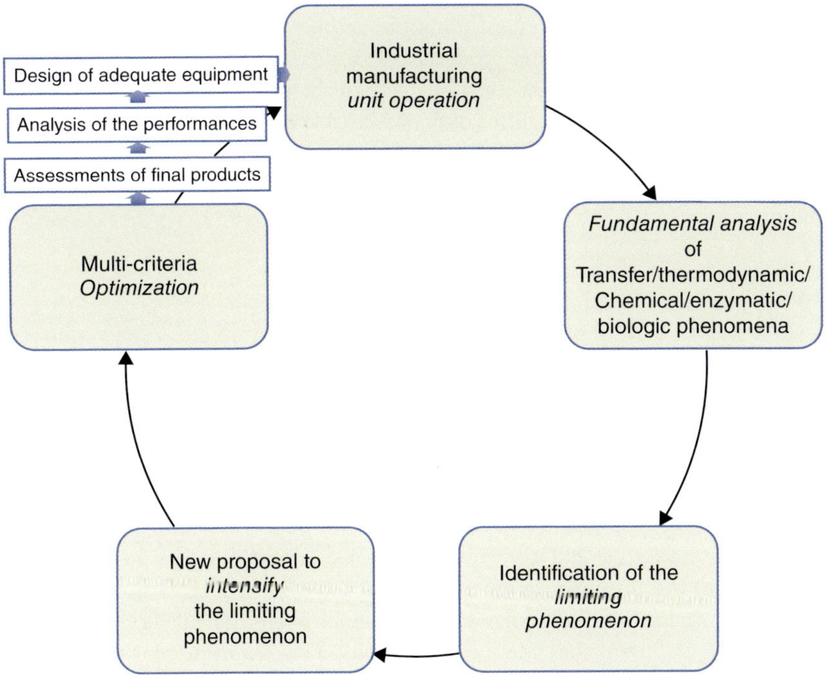

Figure 5.1 Scheme of multilevel successive process-intensification strategy PI-S.

fundamental concepts to analyze the phenomena, solve the limiting process, adopt a new strategy of green industrial production, optimize the operation in order to couple high quality, great kinetics, low cost, and effectively environmentally friendly process:

Since 1988, Allaf and collaborators have founded a new integrated intensification strategy by analyzing the phenomena and, mainly, their kinetics. In this intensification procedure, fundamental studies are the starting basis stage. Its main objective should be to recognize the limiting phenomenon and possibly identify the ways to increase kinetics, improve quality, and make available process performance. This task is completely distinctive from empirical modeling, which does not allow explaining the physical phenomena but is usually limited to being used for controlling and/or automating unit operations, normally in very limited processing conditions.

From the fundamental modeling, we have to understand the processes, analyze the phenomena, and establish correlations between structure and functional behavior without necessarily controlling the industrial process. Some pretreatment operations can be stated from their direct impacts on kinetics. They are grinding, swelling, and, in many vegetables, rupturing highly resistant cell walls.

5.4.1.2 Multi-cycle Intensification Procedure

Considering the main part of the objectives, it is evident that the PI-S needs recurrent cycles each beginning by identifying the gap between current and required situations, followed by achieving fundamental approaches to propose new intensified methods. The new process must be characterized through the identification of structural and functional qualities of the products and the quantification of process performance together with equipment engineering. This results in implementation of a new specific intensified manufacturing operation from fundamental modeling.

5.4.1.3 Intensification Charter

The whole intensification procedure obviously defines a strategic policy requiring synergy by combining the efforts of various scientific researchers from academic communities, engineers from manufacturing and production industries, and experts on markets and international cultures. It is the challenging way to contribute to the new sustainable development of "human industrial civilization." It has been also necessary to continue achieving pluridimensional and multilevel analyses involving academia researchers, industry engineers, nutritionists, and specialists in marketing and commerce. This was done in order to accelerate the adoption of the proposed innovative sustainable technologies and to largely extend their application domains. Progressive PI-S is vital in allowing active sustainable development that adequately involves different industrial sectors and various controlled attributes.

5.4.2 Specificities of Instant Controlled Pressure DIC Drop in Process Intensification Strategy PI-S

As intensification strategy is based on studying transfer laws, and as the functional aptitudes greatly depend on structural characteristics, it is highly important to identify the specific impacts of instantaneous transformations and to compare them to conventional thermodynamics. Indeed, instant controlled pressure drop DIC (détente instantanée contrôlée) frequently plays a specific role allowing natural vegetables to

- completely be textured;
- get adjusted technological aptitudes for extraction;
- have increased extraction kinetics;
- provide higher yields and better quality of extracts.

In research works and various industrial applications of Allaf and collaborators, the use of instantaneous thermodynamics-based processes has contributed to increase the mass transfers with mass diffusion as the limiting process [33]. Moreover, in several cases, DIC has the advantage adopting at industrial manufacturing scale the same processing conditions optimized at laboratory scale.

5.4.2.1 Introduction

Before studying the DIC (French for "instant controlled pressure drop"), Allaf and Allaf [1] exposed the true thermodynamic temperature θ. The temperature θ of a system is the average density per mole of the kinetic energy of translational fluctuation of the system particles (Table 5.1). It correlates with T (in K) through:

$$\theta = \frac{3}{2}RT = \frac{1}{2}M\langle V_t^2\rangle \tag{5.1}$$

where

θ is the true thermodynamic temperature of the system (J mol^{-1})
R is the universal ideal gas constant ($R = 8.314$ J mol^{-1} K^{-1})
T is the temperature expressed in Kelvin K
M is the molar mass of the particles forming the system being studied
$\langle V_t\rangle$ is the mean translational velocity of the particles; it is a random three-dimensional isotropic velocity with a statistically homogeneous value $\langle V_t^2\rangle = \langle V_x^2\rangle + \langle V_y^2\rangle + \langle V_z^2\rangle = 3\langle V_x^2\rangle$.

Decompression is a transformation bringing the system from high-temperature/high-pressure state to a lower pressure. It normally implies an autovaporization of an amount of volatiles in order to bring residual water and moisture to reach the new equilibrium (boiling) temperature level. If the operation of pressure drop is achieved instantaneously, it induces, during a very short time

Table 5.1 The true thermodynamic temperature θ (J mol^{-1}); cases of H_2O vapor and air.

T (°C)	T (K)	θ (J mol^{-1})	Average of translational fluctuation velocity $<V_t>$	
			$\langle V_t \rangle_v$ of vapor (m s^{-1})	$\langle V_t \rangle_a$ of air (m s^{-1})
−273.15	0	0	0	0
−18	255.15	3182.00	595	468
0	273.15	3406.40	615	485
20.01	293.16	3656.00	637	502
37.01	310.16	3868.00	656	516
47.59	320.74	4000.00	667	525
127.78	400.93	5000.00	745	587
207.97	481.12	6000.00	817	643
288.15	561.3	7000.00	882	695

(some dozens of milliseconds), a non-isotropic velocity with a statistically homogeneous value limited to only two dimensions $\langle V_t^2 \rangle = \langle V_y^2 \rangle + \langle V_z^2 \rangle = 2\langle V_x^2 \rangle$. The vapor phase system should have, during this very short time of dozens of milliseconds, an absolute temperature (T in K) or a true thermodynamic temperature (θ in J mol^{-1}) limited to 2/3 of the initial level. This asymptotic value can be much lower than the equilibrium temperature level (boiling temperature) at the pressure-drop vacuum pressure level. The autovaporization is then much more effective and the amount of vapor much higher.

5.4.2.2 Transfer Phenomena in Instantaneous Controlled Pressure Drop DIC Treatment

DIC treatment mainly consists on baring a material to an instantaneous pressure drop usually toward a vacuum. It takes place after a short period of time at high temperature/high pressure, carried out through one of various ways of obtaining saturated steam pressure, high-temperature compressed air, high-pressure microwave, and so on.

Pressure usually is higher than the equilibrium level of the temperature higher than 60 °C, usually up to 180 °C (Figure 5.2 relates the case of saturated steam at 151 °C and 0.6 MPa). Treatment conditions strictly depend on the product and the objectives of decontamination, texturing, autovaporization, extraction, and/or drying. In volatile extraction, the temperature of DIC treatment is typically higher than the boiling temperature of water at atmospheric pressure usually used with the standard steam distillation, to let the vapor pressure of each volatile compound be higher. This high-temperature stage is a short-time treatment followed by the abrupt pressure drop and the final vacuum is about 5 kPa. DIC pressure-drop rate ($\Delta P/\Delta t$) is higher than 0.5 MPa s^{-1}, which simultaneously triggers the following:

- the autovaporization of volatile compounds;
- the instant cooling of the product: this stops thermal degradation;
- the interstitial structure swelling or even the rupture of the cell and secretion element walls.

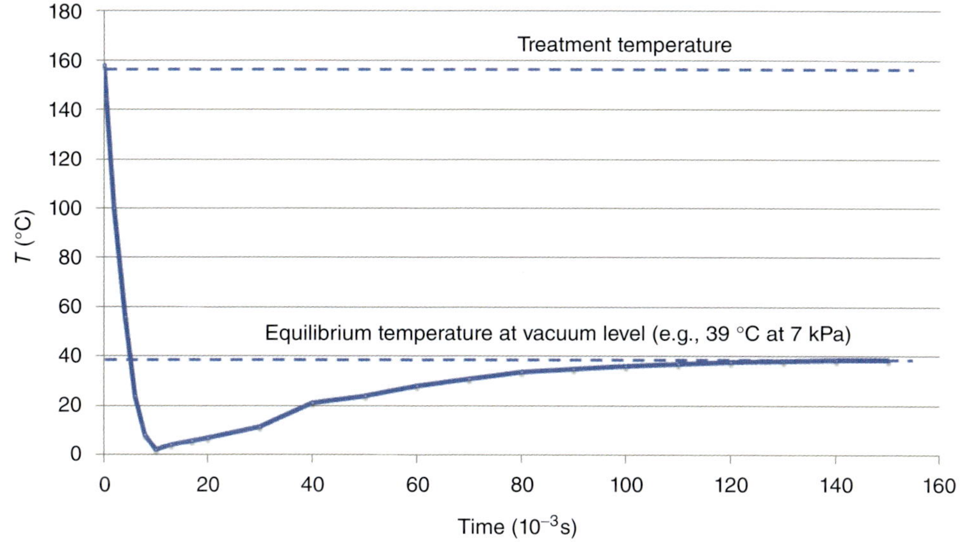

Figure 5.2 History of temperature during the pressure-drop stage toward a vacuum (treatment temperature: $T_t = 158\,°C$; vacuum level equilibrium temperature $T_e = 28\,°C$).

Heating stage The heating stage allowed the product initially at T_a to reach the saturated steam pressure temperature T_t, first at the surface by convection/condensation, and then followed by a conduction process within the product. Since the latent heat of vaporization $L = 2107.42\,\text{kJ kg}^{-1}$, the specific sensible heat of dry matter $c_{p,d} = 1.13\,\text{kJ kg}^{-1}\,\text{K}^{-1}$ and of water $c_{p,W} = 4.182\,\text{kJ kg}^{-1}\,\text{K}^{-1}$; the amount of condensed steam

$$\mathrm{m}_v = \frac{(c_{p,d} + W c_{p,W})(T_t - T_a)}{L} = 0.111\,\text{kg steam/kg dry matter dm} \tag{5.2}$$

The water content turned out to be

$$W_t = W_i + \Delta W = W_i + \frac{(c_{p,d} + W c_{p,W})(T_t - T_a)}{L} = 26.01\%\ \text{db} \tag{5.3}$$

The specific heat absorbed by the material becomes

$$Q = (c_{p,d} + W c_{p,W})(T_t - T_a) = 232\,\text{kJ kg}^{-1}\,\text{of dry matter dm} = 64.5\,\text{Wh kg}^{-1}\,\text{of dm} \tag{5.4}$$

By assuming the part of convection is negligible in the heating process, the heat flow from the saturated steam toward the surface is supposed to be

$$\text{Heating of surface kinetics: } \frac{dQ}{dt} = k_{\text{condensation}} S_{\text{eff}} L\,(T_t - \mathrm{T}) \tag{5.5}$$

In terms of kinetics, because the initial vacuum step allowed the steam injected to almost completely fill the open superficial and dead-end pores, it led to a close contact between both mediums, and a higher effective exchange surface S_{eff}. Heating was then a very short-time process.

After this very short-time washing step, heat and water were transferred and diffused from the surface within the material toward the core, through conduction according to a Fourier-type law, and Fick-type law, respectively:

$$\vec{\varphi} = -\lambda_{\text{eff}}.\vec{\nabla} T \tag{5.6}$$

$$\frac{\rho_W}{\rho_d}(\vec{v}_W - \vec{v}_d) = -D_{\text{eff}}\vec{\nabla}\left(\frac{\rho_W}{\rho_d}\right) \tag{5.7}$$

which converted, by assuming $\vec{v}_d = 0$ and ρ_d = constant, and by adding the continuity of both heat and water,

$$\frac{\partial \rho_W}{\partial t} = -\vec{\nabla}.\,(D_{\text{eff}}\vec{\nabla}\rho_W) \tag{5.8}$$

$$\frac{\partial T}{\partial t} = -\vec{\nabla}.\,(\alpha_{\text{eff}}\,\vec{\nabla} T) \tag{5.9}$$

The duration (t_c) of the high-temperature stage of DIC must generally be defined so that both temperature (T) and moisture content (W) are uniform within the product. It is worth noting that, whatever the water content W (dry basis), temperature T, porosity ratio, and average size of pores, the effective diffusivities α_{eff} and D_{eff} usually are about 10^{-7} and about 10^{-10} $m^2\,s^{-1}$, respectively. Since deep heating is achieved much faster, t_c was defined in order to reach water content homogeneity. In the present case, t_c had to be higher than 6.5 s.

Pressure-Drop Stage Numerous instantaneous phenomena simultaneously occur when the pressure is reduced abruptly from the high temperature–high pressure state to a vacuum. They are autovaporization/cooling and expansion/glass transition processes.

Autovaporization/Cooling Since the autovaporization of overheated water and volatile compounds is an adiabatic process, it inexorably induces an adequate cooling of the residual material. When the pressure abruptly and instantaneously drops, the temperature instantly falls to the lowest level T_d before increasing till equilibrium temperature T_e is reached. This results in a mass of vapor much higher than in quasi-static autovaporization.

The mass of DIC autovaporized water produced per kilogram of dry matter dm is

$$m_v = \frac{(c_{p,d} + W_t\, c_{p,W})(T_t - T_d)}{L} > 0.129\,\text{kg steam/kg dry matter dm} \tag{5.10}$$

where T_d is the temperature reached by the product just after the pressure drops.

Thus, just after the pressure instantaneously drops, the residual water content W_r of the product is

$$W_r = W_t + \frac{(c_{p,d} + W' c_{p,W})(T_d - T_t)}{L} < 13.1\%\ \text{db} \tag{5.11}$$

5.4.2.3 DIC – Texturing

The strictly correlated instantaneous processes of autovaporization of volatile molecules present in the porous matrix and cooling occur because of the abrupt decompression toward a vacuum. The mass of vapor thus supplied is sufficient to develop great mechanical constraints within the porous material without reaching too high a processing temperature [1], with a greater likelihood of achieving glass transition.

While structure evolvement largely depends on the amount of vapor generated and the matter's viscoelastic behavior, which is a function of temperature and water content, the preservation of the swelled new structure requires touching the low temperature and water-content glass transition. This feature is very important in explaining why the pressure has to drop toward a vacuum. This occurs in order to let the amount of vapor to increase considerably and equilibrium temperature reaches the glass transition temperature level needed to obtain a number of starch-free products.

Texturing Model The specific volume of this generated vapor once it reaches the equilibrium temperature T_e is assumed to be

$$V_v = \frac{(c_{p,d} + W_t\, c_{p,W})(T_t - T_d)}{L} \frac{R\, T_e}{M_W P_v} \tag{5.12}$$

This, in the present considered conditions, is

$$V_v > 3.652\ \text{m}^3\,\text{kg}^{-1}\,\text{dm} \tag{5.13}$$

while the volume V_a of initial pores, which is assumed to be the difference between the specific and the intrinsic volumes of the material, $V_a = \left(\frac{1}{\rho_{\text{specific}}} - \frac{1}{\rho_{\text{intrinsic}}}\right)$, did not exceed

$$V_a = \left(\frac{1}{1023} - \frac{1}{1318}\right)\ \text{m}^3\,\text{kg}^{-1}\,\text{dm} = 0.000219\,\text{m}^3\,\text{kg}^{-1}\,\text{dm} \tag{5.14}$$

In the asymptotic conditions of instantaneous pressure drop, the highest value of generated constraining pressure within this confined volume can be

$$\Delta P_{\max} = \frac{m_v}{M_W}\frac{R\,T_e}{V_a} - P_v = \frac{(c_{p,d} + W_t c_{p,W})(T_t - T_d) R\,T_e}{M_W L\left(\frac{1}{\rho_{\text{specific}}} - \frac{1}{\rho_{\text{intrinsic}}}\right)} - P_v \tag{5.15}$$

The asymptotic highest value of generated constraining pressure within this confined volume in the present case of DIC is then

$$\Delta P_{\max} > 83.5\ \text{MPa} \tag{5.16}$$

The amount of autovaporized water, m_v, and the asymptotic highest value of the constraining pressure difference $\Delta P_{P\max}$ between the pore and the surrounding atmosphere are very important parameters whatever the texturing operation is. However, when pressure drops toward the atmospheric pressure as it is carried out for extrusion-cooking, steam explosion, puffing, and popping, their values m_v and $\Delta P_{P\max}$ were, both, less than one-half of that in DIC:

$$m_v = 0.056\,\mathrm{kg}^{-1}\,\mathrm{kg\,dm} \quad \text{and} \quad \Delta P_{\max} = 44.3\,\mathrm{MPa} \tag{5.17}$$

It is worth noting that the constraining pressure difference ΔP is the driving force in the temporal evolution of the pore size versus time:

$$\mu\, 2\pi R \frac{dR}{dt} = \frac{1}{2}\pi R^2 \Delta P \tag{5.18}$$

The change rate of the pore is thus proportional to ΔP, the difference between the pressure inside the pore and the external medium, and inversely proportional to viscosity. However, as this pressure difference depends on the initial value and evolves according to the pore size, we can write a relation similar to that used by Arhaliass *et al.* [2].

Glass Transition High levels of humidity and temperature during the DIC heating step and, more precisely, just before the pressure instantaneously falls, often leads to greater likelihood of the material showing viscoelastic behavior. Preservation of the new honeycomb expanded structure is closely related to the glass transition of the material. The instantaneous pressure drop, more especially in the case of DIC treatment, implies decrease in both humidity and temperature of the material. This can often increase the possibility of glass transition of the material.

Indeed, experimental measurements of the glass transition temperature obtained at each level of moisture content can be correlated with the Gordon–Taylor relationship [3], formulated according to Allaf [4]:

$$T_g(W) = \frac{T_{g,m} + kWT_{g,w}}{1 + kW} \tag{5.19}$$

$$(T_{g,m} - T_g) = (T_g - T_{g,w})kW \tag{5.20}$$

W is the moisture content (dry basis), $T_{g,m}$ and $T_{g,w}$ are the glass transition temperatures of the dry material and pure water, respectively. Note that $T_{g,w}$ is suggested by Orford *et al.* [5] to be: $T_{g,w} = -139\,°\mathrm{C}$. Finally, k is the Gordon–Taylor correlation parameter, which depends on the material. For $k = 1$, the glass transition can be obtained as shown in Figure 5.3.

5.4.3 Mass Transfer by Permeability

Allaf and Allaf [1] assumed that the vapor once generated is transported by permeability from the material core toward the surrounding atmosphere. This process is carried out immediately after the pressure drops instantaneously. It continues during the vacuum step, as long as the total internal pressure is higher than the surrounding external pressure. Under DIC conditions, the total pressure of the vapor generated in the porous material just after the pressure drop is so high that

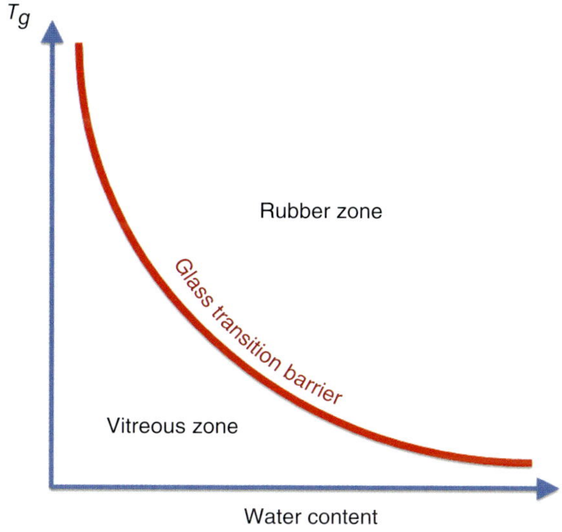

Figure 5.3 Evolution of glass transition versus water content dry basis W.

the transport operation is achieved in very short time [6]. In various cases of membrane, porous material, and so on, permeability mass transfer of the general Darcy type occurs with the total pressure gradient as the "driving force:"

$$\rho_v(\vec{v}_v - \vec{v}_d) = -\frac{K}{\nu_v}\vec{\nabla}P \tag{5.21}$$

The permeability K (in m^2) depends on the solid matrix structure, porosity, and tortuosity, while the kinematic viscosity ν_v (in m^2 s^{-1}) depends on the temperature and nature of transferred fluid (density, molecular size, etc.). The time frame of instantaneous autovaporization was measured through a 1000-image-per-second high-speed camera [6]. The expansion-time was estimated between 20 and 200 ms, depending on the material and the DIC conditions.

Vapor flow can be calculated by integrating a Darcy-type law between the pore radius R_o and the external radius R_s of a supposed spherical shape granule:

$$m_v = \frac{4\pi K}{\nu_v}\frac{(P_{\text{globule}} - P_{\text{ext}})}{\left(\frac{1}{R_o} - \frac{1}{R_s}\right)} \tag{5.22}$$

There is a decrease in the total pressure (P_{globule}) in the pore versus the transfer time (t). The pore and globule radius (R_o and R_s, repectively), are assumed to be constant (absence of expansion) as is the temperature Equation (5.22) becomes

$$\frac{dP_{\text{globule}}}{dt}\frac{MV_{\text{pore}}}{RT} = -\frac{4\pi K}{\nu_v}\frac{(P_{\text{globule}} - P_{\text{ext}})}{\left(\frac{1}{R_o} - \frac{1}{R_s}\right)} \tag{5.23}$$

$$\Delta P = \Delta P_{\max} \exp\left(-\frac{3KRT}{\nu_v}\frac{t}{MR_o^3\left(\frac{1}{R_o}-\frac{1}{R_s}\right)}\right) \tag{5.24}$$

The characteristic Darcy mass transfer time is

$$\tau = \frac{MR_o^3\left(\frac{1}{R_o}-\frac{1}{R_s}\right)\nu_v}{3KRT} \tag{5.25}$$

Usually, the time t_v of the vacuum stage (f) is defined to allow an effective vapor transfer toward the surrounding medium.

5.5 Processes of Extraction

Of late, numerous industrial sectors such as the biochemical, food, nutraceutical, cosmetic, pharmaceutical, perfume, flavor, and aroma sectors have increasingly started opting for granting the natural plant-based active molecules to the detriment of synthetic compounds. Nevertheless, the main issues and disadvantages of processes of extraction are

- the low yields and weak kinetics, which results in a low production capacity;
- the high energy consumption;
- the low quality of extract;
- the weak capability of use of residual solids, with normally negative environment impact.

Researchers and industry engineers are attempting to solve these issues by defining and adapting innovative ways to meet the requirements related to a relevant PI-S.

5.5.1 Extraction of Volatile Compounds

Different processes of extraction of plant-based volatiles such as essential oil are carried out using the vapor pressure of each compound. They usually concern the well-known hydrodistillation (solid–liquid followed by a liquid vapor interaction) and steam distillation (solid–gas interaction), where heat and mass are delivered and transferred from boiling water or steam using both convection and condensation processes.

Issues connected with these conventional operations have implied weak performances and low quality of essential oil such as that extracted with thermal degradation of residual solid. These issues are mainly linked to the low technology aptitudes owing the plant structure and its ability vis-à-vis the heat and mass (liquid and vapor) transfers, the kinetics of different processes. This allows researchers to identify the limiting process and, thus, to propose a better definition of an appropriate PI-S, possibly leading to lower energy consumption and better environment impact.

The production at high temperature of vapor from volatile molecules such as water, essential oils, and solvents is followed by several stages of condensation usually achieved by adjusting low-temperature decantation assuring the separation of immiscible liquids of different specific gravity. The kinetics of hydrodistillation and steam distillation is known to be slow with a long processing time (up to 24 h and sometimes longer, thus inducing degradation of sensitive molecules). They both involve a large consumption of energy [7–13].

5.5.1.1 Kinetics

Heat and Fluid Surface Interaction Heat and fluid first interact at the surface, thus quickly extracting the superficial part of volatiles. The starting accessibility δX_s reveals this diffusion-free short-time stage. It strictly depends on the difference of vapor pressure between the exchange-surface and the surrounding ambient. Allaf and Allaf [1] proposed to define this stage through the *washing process* model:

$$\text{Essential oil EO vapor flow: } \dot{m}_{v,\text{EO}} = \frac{dm_{v,\text{EO}}}{dt} = k_{\text{EO}} \langle A \rangle [p_{\text{EO},X-s} - p_{\text{EO,cond}}] \tag{5.26}$$

$$\text{Water vapor flow generation: } \dot{m}_{v,W} = \frac{dm_{v,W}}{dt} = k_W \langle A \rangle [p_{W,X-s} - p_{W,\text{cond}}] \tag{5.27}$$

$$\text{Convection/condensation heat flow generation: } \dot{Q} = h\,A\,[T_{\text{steam}} - T_{X-s}] \tag{5.28}$$

$$\text{Heat flow balance: } \dot{Q} = \dot{m}_{v,\text{EO}}\,L_{v,\text{EO}} + \dot{m}_{v,W}\,L_{v,W} \tag{5.29}$$

From Equations (5.28) and (5.29) Allaf and Allaf [1] claimed that the main part of the absorbed heat Q is used for evaporating water, while the part required for the essential oils is normally very restricted, owing to the low values of both volatility ($p_{\text{EO}} \ll p_W$) and latent evaporation heats ($L_{v,\text{EO}} \ll L_{v,W}$).

Internal Transfer Processes

Internal Heat Transfer After this short-time starting stage, because usually plants are assumed to be made up of porous material, coupled heat and mass transfer processes are assumed to occur and take place successively within the matrix [6, 14, 15].

The heat source temperature can define the temperature repartition within the product. In steam extraction, the saturated steam ensures that heat transfer can be done by convection/condensation at the external surface of the solid matrix. Subsequently, once the highest temperature level (very close to the saturated steam temperature) is reached, a gradual spread of heat within the solid occurs through a similar conduction phenomenon. The presence of water and other volatile molecules within the porous structure implies specific complex occurrence combining *convection-vaporization/condensation inside the pores.* This induces an intensified heat transfer phenomenon similar to the Fourier conduction model where the driving force is the temperature gradient with usually very high effective conductivity.

Mass Transfer of Internal Liquid and Vapor The natural structure of plants is normally limited process by mass diffusivity because it is much lower than effective heat diffusivity.

It is normally assumed that essential oil transfer within the porous material is a gas diffusion phenomenon with the gradient of vapor pressure of each volatile compound as the driving force. A Fick-type law with an effective diffusivity D_{eff} should be coupled with the internal heat transfer through conduction:

$$\vec{\nabla}.(-\lambda_{\mathrm{eff}}\vec{\nabla}T) + (\rho_d c_{pd} + \rho_{\mathrm{EO}} c_{p\mathrm{EO}} + \rho_w c_{pw})\frac{\partial T}{\partial t} + \frac{\partial}{\partial t}\left[\frac{\psi}{R_{GP}T}\left(p_{\mathrm{EO}} M_{\mathrm{EO}} L_{\mathrm{EO}} + p_w M_w L_w\right)\right] = 0 \quad (5.30)$$

Allaf and Allaf [1] assumed that the main part of the transferred heat is used to evaporate essential oils and water. However, as the external vapor pressure is saturated, the internal water vapor pressure p_w in the porous material should approximately be constant:

$$-\lambda_{\mathrm{eff}}\vec{\nabla}.(\vec{\nabla}T) + \psi M_{\mathrm{EO}} L_{\mathrm{EO}}\frac{\partial}{\partial t}\left[\frac{p_{\mathrm{EO}}}{R_{\mathrm{GP}}T}\right] = 0 \quad (5.31)$$

They also assumed that the mass transfer of essential oil vapor is directed by a Fick-type law:

$$\frac{(p_{\mathrm{EO}}/T)}{\rho_d}(\vec{v}_{\mathrm{EO}} - \vec{v}_d) = -D_{\mathrm{eff_{EO}}}\vec{\nabla}\left(\frac{p_{\mathrm{EO}}/T}{\rho_d}\right) \quad (5.32)$$

By assuming $\vec{v}_d = 0$ and ρ_d = constant, Equation 5.32 may be transformed into

$$(p_{\mathrm{EO}}/T)\vec{v}_{\mathrm{EO}} = -D_{\mathrm{eff_{EO}}}\vec{\nabla}(p_{\mathrm{EO}}/T) \quad (5.33)$$

For one dimension (r) and a homogeneous and isotropic medium, it becomes

$$(p_{\mathrm{EO}}/T)v_{\mathrm{EO}} = -D_{\mathrm{eff_{EO}}}\left(\frac{\partial(p_{\mathrm{EO}}/T)}{\partial r}\right) \quad (5.34)$$

$$-\lambda\frac{\partial^2 T}{\partial r^2} + \psi M_{\mathrm{EO}} L_{\mathrm{EO}}\frac{\partial}{\partial t}\left[\frac{p_{\mathrm{EO}}}{R_{\mathrm{GP}}T}\right] = 0 \quad (5.35)$$

To achieve the diffusion operation, the essential oil vapor should flow to be directed from $r = 0$ toward $r = R$:

$$v_{\mathrm{EO}} > 0 \quad (5.36)$$

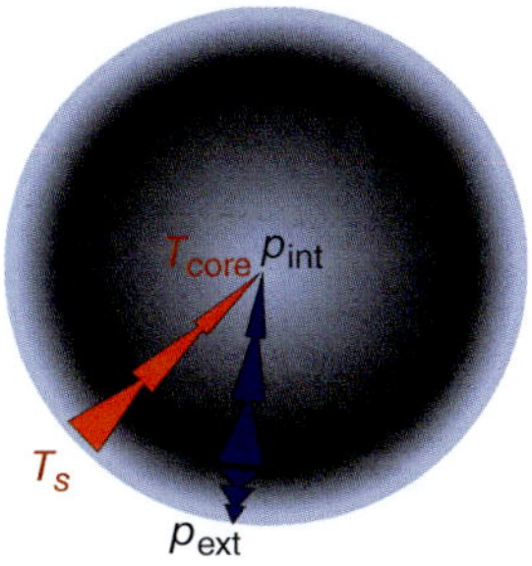

Figure 5.4 The paradoxical situation of the extraction of volatile compounds during steam distillation: impact of coupled essential oil, and heat transfers in terms of "front progression" kinetics.

$$\frac{\partial(p_{EO}/T)}{\partial r} < 0 \tag{5.37}$$

However, as the exchange surface has the highest values of both temperature as well as p_{EO}/T; this decreases near the particle core, implying a paradoxical situation:

$$\frac{\partial(p_{EO}/T)}{\partial r} > 0 \tag{5.38}$$

It is completely opposite to what is required for an extraction operation [16, 17]. Then, in the standard steam extraction of essential oils, the operation is achieved using a long-consuming time "front progression" process (Figure 5.4).

5.5.1.2 **Intensification of Essential Oil Extraction**

The PI-S in the specific case of volatile extraction versus standard steam distillation should aim at solving the main paradoxical issues. Thus, this involves complete disengagement of diffusion mass transfer from conduction heat transfer. A first intensification was carried out through microwaves, which homogeneously generate heat inside the product independently of the conduction phenomenon. Mass transfer with vapor pressure gradient as driving force for a similar Fick diffusion can then be achieved according to increasing r (from the core toward the exchange surface), and the paradoxical situation can be reduced or even avoided. Another very effective key was to involve Darcy-type vapor transfer from inside the material toward the surrounding medium; the driving force is then the gradient of the total pressure. Indeed, by creating instant autovaporization through instantaneous pressure-drop technology DIC [4] from high temperature, while establishing vacuum in the surrounding medium, the paradoxical situation is completely removed and the kinetics is quite considerable. The DIC extraction of volatile compounds is then a highly relevant short-time (2–3 min) operation.

Thus, DIC instantaneous autovaporization can be an appropriate solution to the prejudicial paradoxical phenomenon. It provides great intensification of volatile extraction. Here again, the role of water is very important in the driving force because of its higher vapor pressure, normally 10 times greater than that of any other natural volatile molecules.

5.5.2 Case of Solvent extraction

5.5.2.1 Introduction

Solvent extraction is usually used to recover compounds from either a solid or a liquid. It is a very important operation in numerous industries such as the chemical and biochemical industries, as also the food, cosmetics, and pharmaceutical industries [18].

Solvent extraction has embraced numerous standard techniques including several methods that all consist of a solid/liquid interaction in order to dissolve the desired components [19]. Allaf and Allaf [1] itemized them as follows:

- *Percolation*: Usually a very hot solvent is flushed onto a bed of finely divided solids. This method is used in the preparation of coffee.
- *Decoction*: The solid is immersed in a boiling liquid solvent. This is a brutal operation that should be reserved for the extraction of non-thermolabile active ingredients. However, this process is very fast and is sometimes necessary.
- *Infusion*: The solid is immersed in a heated solvent without boiling, followed by cooling the mixture. The preparation of tea is a typical example.
- *Maceration*: The solid is immersed in a cold solvent. The operation is usually long and with poor results; however, it is the only method used in the case of a set of fragile molecules. To be effective, maceration can last 4–10 days, which may present problems in terms of fermentation or bacterial contamination, particularly if the solvent used is water. These phenomena can cause rapid degradation of active molecules. In order to avoid or limit these drawbacks, maceration can be carried out in a covered container and, in some cases, in a refrigerator.
- *Digestion*: Especially used in perfumery and pharmacy, digestion is a hot maceration. As it is faster, it generally does not involve any degradation or bacterial contamination problems.

5.5.2.2 Extraction Process Issues

In all these methods, mass transfer is hindered and limited by the interstitial structure and the cell walls of the matrix. Therefore, the solvent extraction process involves solid/liquid or liquid/liquid interactions in order to dissolve the desired components. A solvent is a liquid capable of "dissolving" desired components from a solid or a liquid thus generating an extract solution, solvent + solute. A second stage of separation normally allows recovery of a major part of the solvent to obtain the pure extract preferentially. Other harmful weaknesses of the conventional solvent extraction processes are their risky impact, more particularly when petroleum-based organic solvents are used. The difficulty in removing their traces from both extract and residual cake is also very deleterious. The use of a special solvent such as super- or subcritical fluids was recently considered as an innovative and relevant way to solve this aspect. However, as it is a very costly operation, it remains attractive only in the manufacture of high value-added ingredients and natural pharmaceutics. Since these operations can involve non-dangerous compounds such as CO_2 or water, they would be a relevant answer to possibly use the residual solids, and thus have a positive environmental impact. However, in terms of kinetics, all these solvent extraction processes are similar because they are limited during the washing stage by exchange surface value and during the fluid diffusion stage by low diffusivity within the porous solid. The disadvantages are significant in terms of equipment because of the time-consuming processes. Here again, a PI-S should be used in order to define a significant way for higher process performance, better product quality, and more reliable and adequate equipment. Other routes to reduce conventional solvent extraction issues and to improve some important aspects have been proposed by using ultrasound, microwaves, accelerated solvent extraction, pulsed electric field PEF, and so on. These options have not definitively resolved the main issues. Thus, an adequate PI-S should be further discussed. Process performance should aim at growing the kinetics, reducing energy consumption and simultaneously getting greater starting accessibility, δX_s, effective diffusivity, D_{eff}, and availability, X_∞, of different natural cases, with the best preservation of quality of the final product.

Identification of Limiting Process and Intensification of Unit Operation During solvent extraction, the transfer of active molecules to the surroundings may occur through diffusion whose driving force results from a concentration gradient of the solute within the solution from the most concentrated solid surface. The operation starts with a superficial solute dissolution in the solvent. This occurs at the surface (external process of washing) and is immediately transported within the surrounding solvent medium. The solid/liquid interaction is achieved by close contact between the solid and the solvent. During the extraction, the concentration of solute in the solid varies continuously, with a nonstationary mass transfer.

At the exchange surface, the system quickly tends to quasi-equilibrium state, the diffusion is near zero, and the extraction process practically stops. In contrast, if the liquid phase is continuously renewed through adequate stirring, the diffusion at the exchange surface continues until complete exhaustion of the solid phase, at least from what is available.

Following this "washing" stage, a series of successive processes occurs, reflecting the interaction between the solid initially containing the solute and the extraction solvent. These successive processes involve the following:

1) the solvent diffusing within the porous solid matrix;
2) the internal solute dissolving in the solvent already filling the pores of the solid matrix;
3) the solute diffusing in the filled-in solvent pores;
4) external transfer of the solute from the exchange surface to the surrounding solute/solvent solution.

The extraction kinetics is usually expressed as the solute concentration in the solid versus time, $X = f(x)$. Since these four processes are successive in terms of kinetics, the operation rate follows the slowest process, which is considered as the limiting process.

PI-S first involves the analysis of these various processes the operation and identification of the limiting process as the starting basis to intensify the whole operation.

Dissolution of Solute in the Liquid Solvent The operation of dissolving a solute in a solvent, is defined from the thermodynamic equilibrium in saturation concentration, ϖ_e, which is a function of the solvent, the compound to extract, and the temperature [20]. Usually, process of dissolution of the desired solute is not the limiting process unless the solvent and/or temperature are inappropriate. The adequate extraction solvent and temperature are normally selected according to the desired solutes. However, the nature and the variety of other compounds can also be considered in order to facilitate subsequent separation.

Thus, it is normally easy to avoid having dissolution as the limiting process. In practice, usually it is a quick and even instantaneous process. Normally, the dissolution process of the solute in the solvent does not drive the extraction process.

Exchange Surface and Transport of Solute in the Surrounding Solvent Since the solute dissolution close to the exchange surface is assumed to be fast, its transfer to the surrounding medium is achieved, via either diffusion and/or convection.

The transfer of solute from the exchange surface toward the external solvent should not be the limiting process of the extraction operation, unless the operation is carried out without adequate stirring. Intensification of this external transfer is easy. It only requires adopting a convective instead of a diffusional transfer. This can be accomplished through adequate agitation of the outside "solvent medium." The external transfer resistance becomes negligible. This is why agitation is often included as a primordial part of the extraction process [20].

At the washing step, the pure interaction between the exchange surface and the convective medium of external solvent is revealed through the "starting accessibility" parameter, δX_s:

$$\delta X_s = k_e \mathrm{SESA}_{\mathrm{eff}}(\varpi_e - \varpi_{\mathrm{solvent}})\delta t \tag{5.39}$$

Usually, in kinetic study, the starting accessibility, δX_s, does not imply any diffusion source or effect; it is the density of solute per unit of dry matter extracted by the initial interaction between the solvent and the exchange surface and is expressed as kilograms of solute per kilogram of dry matter or %db.

It is obviously easy to carry out the "external" process intensification. An appropriate stirring can possibly be used to increase the value of k_e. The use of crushed material leads to higher specific exchange surface area, $\mathrm{SESA}_{\mathrm{eff}}$. Finally, regeneration and renewal of the solvent in contact with the material at the exchange surface can improve the external transfer phenomena ($\varpi_{\mathrm{solvent}} \rightarrow 0$). This clearly explains the benefits generated by the use of countercurrent extractors.

Once the external transfer phenomena are perfectly intensified, solute extraction kinetics from the exchange surface reaches its highest level [18]. Only the internal transfer should then be studied as possible source of limiting the extraction process.

Internal Diffusion Transfer Once the external transfer is adequately intensified the internal solute/solvent transfer becomes the possible limiting process. The effective diffusion process can play a specific role here and be considered as the limiting process. Indeed, whatever the usual pore size, internal convection in the pore is negligible. Two-level diffusion process is achieved through solvent transfer within the plant structure and solute transfer within the solvent solution. The first is a solid/liquid interaction with the gradient of apparent density ratio of solvent to solid as driving force. The second is a liquid/liquid interaction through the gradient of the apparent density ratio of solute to solvent.

> As the first usually is much quicker, the main part of the kinetics appears as solute-in-solvent transfer in the pores of the solid matrix, closely depending on the porosity of the material.

Thus, during this operation, the Fick-type diffusion of solute emerges as the major transport phenomenon occurring within the solid matrix [21]. Therefore, the operation cannot be intensified through any external mechanical or thermal changes [1]. Only modifications of shape (by grinding), expansion processes improving the porosity, and ultrasonic treatments increasing the apparent agitation of molecules in the pores can improve this operation.

The structure of the natural porous material is normally prejudicial for effective extraction kinetics. This reveals low technological aptitudes at various levels of process performance, extract quality, and equipment reliability. The conventional manner to increase the operation was done by modifying the shape through grinding. However, crushing of raw material is not easy and, moreover it only increases the specific surface area, keeping the effective diffusivity constant.

Besides, a bibliographic analysis has shown that the effective diffusivity (D_{eff}) of the entire extraction operation of liquids in plant matrices has a magnitude that is generally close to 10^{-11} to 10^{-10} m^2 s^{-1}, while heat diffusivity is usually between 10^{-8} and 10^{-6} m^2 s^{-1}, depending on water content and porosity.

This confirms, in a way, that the solvent diffusion into a solid matrix must often be the slowest process and appears to be the "main" limiting process throughout the operation. The intensification operation lies in the expansion of this natural structure as a real method for improving the processing ability of the plant regarding solvent extraction.

As the natural structure is the principal cause of the inconveniences in the whole extraction process, adequate texturing would be the crucial pretreatment that overcomes these problems. Accordingly, instant controlled pressure drop (DIC) technology was defined as an appropriate way to improve the internal structure of matter while preserving its chemical composition, even for heat-sensitive molecules. The resulting porous structure enhances mass transfer and intensifies both effective diffusivity and starting accessibility, thereby increasing the overall kinetics. Additional effects of DIC treatment on the overall performance of the extraction are to facilitate the grinding and obtain expanded granule powders. This leads to significantly higher yields, lower energy consumption, and better environment impact is. Various compounds and different ingredients issued from DIC-assisted solvent extraction of botanical sources have been studied in terms of composition and functional properties. Academic research has been converted into manageable solutions for industrial use in a short time. Thus, manufacturing new equipment for the unit operation could dovetail with an approach combining fundamental analysis with rigorous process modeling, and empirical studies using experimental designs. Moreover, close collaborations between scientists from botany, thermodynamic, analytical chemistry, and mechanic academic disciplines, and process and mechanic industrial engineering have been highly appropriate for a relevant intensification strategy PI-S to adopt.

Instant controlled pressure drop DIC, has also increasingly become a highly effective industrial reality. The new concept of DIC-expanded granule powder is a relevant industrial method for increasing the kinetic parameters of starting accessibility, δX_s, and effective diffusivity, D_{eff}, while possibly reaching greater availability X_∞ of numerous natural compounds. The preservation and even the improvement of functional quality coupled with a highly reliability of equipment perfectly explains how relevant such PI-S is.

5.5.2.3 **Kinetic Modeling**

As already shown, a stirred external solvent can immediately dissolve the solute that is initially accessible on the exchange surface, to remove it far from the plant, usually by convection [22–24]. Therefore, in any solvent extraction operation performed on a porous solid material, the first interaction between the solvent and the exchange surface, so-called "washing," takes place during a short period of time [25].

Subsequently, the main part of the operation is controlled through various penetration/diffusion processes of the solute/solvent within the material (capillarity, molecular diffusivity, etc.). There, the driving force of the solvent (solute) transfer is the gradient of apparent density ratio of solvent to solid (solute to solvent). In both approaches, a Fick-type law can be used.

In such conditions, formulation of [4] can be used in both cases:

$$\frac{\rho_{\text{solvent}}}{\rho_d}(\vec{v}_{\text{solvent}} - \vec{v}_d) = -D_{\text{eff, solvent}-d}\,\vec{\nabla}\left(\frac{\rho_{\text{solvent}}}{\rho_d}\right) \tag{5.40}$$

and

$$\frac{\rho_{\text{solute}}}{\rho_{\text{solvent}}}(\vec{v}_{\text{solute}} - \vec{v}_{\text{solvent}}) = -D_{\text{eff, solute}-\text{solvent}}\,\vec{\nabla}\left(\frac{\rho_{\text{solute}}}{\rho_{\text{solvent}}}\right) \tag{5.41}$$

To model the global diffusion process, a Fick-type law can be considered with the gradient of the ratio solute/solid densities using an effective diffusivity D_{eff} ($m^2\ s^{-1}$) as the main process coefficient [18, 26]:

$$\frac{\rho_{solute}}{\rho_d}(\vec{v}_{solute} - \vec{v}_d) = -D_{eff}\,\vec{\nabla}\left(\frac{\rho_{solute}}{\rho_d}\right) \tag{5.42}$$

Since the operation does not involve any swelling or shrinkage of the porous material, it is possible to simplify this phenomenological transfer law by adopting a presumably motionless solid matrix frame. This also suggests that the change with time of solid porous medium density is negligible ($\vec{v}_d = 0$ and $\rho_d =$ constant):

$$\rho_{solute}\vec{v}_{solute} = -D_{eff}\,\vec{\nabla}\rho_{solute} \tag{5.43}$$

By including continuity, it becomes

$$\frac{\partial\rho_{solute}}{\partial t} = -\vec{\nabla}.\,(D_{eff}\,\vec{\nabla}\rho_{solute}) \tag{5.44}$$

The effective diffusivity D_{eff} can be considered constant if both structural and thermal distributions are assumed uniform within the solid. This allows the equation to become [27]

$$\frac{\partial\rho_{solute}}{\partial t} = -\,D_{eff}\nabla^2\rho_{solute} \tag{5.45}$$

Further, for one-direction radial flow, it becomes

$$\frac{\partial\rho_{solute}}{\partial t} = -\,D_{eff}\frac{\partial^2\rho_{solute}}{\partial r^2} \tag{5.46}$$

The time t of this diffusion stage should be considered after the washing stage time, t_o. The solutions required for this diffusion equation closely depend on the initial and boundary solvent conditions. Allaf and Allaf [1] adopted the classical Crank's solutions according to the granule geometry [28] while inserting the time, t_o, which corresponds to a time when the extraction process is in the pure diffusion area. They are

- for an infinite plate shape with $r_d =$ thickness/2:

$$\frac{X_\infty - X}{X_\infty - X_{t_o}} = \sum_{i=1}^{\infty}\frac{8}{(2i-1)^2\pi^2}\exp\left(-\frac{(2i-1)^2\pi^2 D_{eff}}{4{r_d}^2}(t-t_o)\right) \tag{5.47}$$

- and for a spherical shape with $r_d =$ radius:

$$\frac{X_\infty - X}{X_\infty - X_{t_o}} = \sum_{i=1}^{\infty}\frac{6}{i^2\pi^2}\exp\left(-\frac{i^2\pi^2 D_{eff}}{{r_d}^2}(t-t_o)\right) \tag{5.48}$$

To study the kinetics and use the experimental data to make a model of the diffusion process, Mounir and Allaf [29] proposed to exclude X values between $t = 0$ and $t = t_o$ from the experimental data used for studying the diffusion part. The extrapolation of the diffusion model was thus obtained till $t = 0$. The calculated value X_o should be different from that at $X_i = 0$, and the starting accessibility δX_s becomes

$$X_o - X_i = X_o = \delta X_s \tag{5.49}$$

Solvent extraction kinetics is then defined through the yield values, X_∞, the starting accessibility δX_s and the effective diffusivity D_{eff}. The yield X_∞ and the starting accessibility, δX_s, which is the amount of solute directly extracted by solvent washing the exchange surface, are usually both expressed in grams of solute/gram dm (dry material).

5.5.3 Conclusion: Process Intensification Strategy: How to Use PI-S Solvent Extraction Processes?

Sequential fundamental studies of solvent extraction process done after each procedure highlight four operations followed by an adequate intensification of the solution [27]:

1) Agitation of the surrounding solvent allows the transfer of the solute from the exchange surface to the outside by convection instead of diffusion. This way should increase the starting accessibility (δX_s).
2) Grinding normally reduces the depth of diffusion of a liquid within the granule. It also increases the exchange surface area and makes the solute on the surface more accessible. This operation should also increase the starting accessibility (δX_s).
3) Texturing aims at reducing mass transfer resistance. Indeed, the natural plant structure and more specifically the cytoplasmic membrane and cell walls dramatically restrict the liquid transfer processes. Hence, by breaking these membranes whether via enzymatic treatments, DIC swelling, and so on, mass transfer kinetics can significantly increase. This operation can facilitate the grinding, increase the effective diffusivity (D_{eff}), and improve the starting accessibility (δX_s). It may also increase the yields, X_∞.
4) Ultrasonic assistance takes place at
   a. external exchange surface (increasing the starting accessibility δX_s) where it may imply a higher value of effective coefficient of mass convection;
   b. internal pore solvent (increasing the effective diffusivity D_{eff}) where it may induce mass convection instead of solute-in-solvent diffusion. Here, it worth noting that the pore size distribution plays a significant role.

   It is important to highlight the strategic place of instant controlled pressure drop DIC technology in all these possibilities. Indeed, in all the solvent extraction operations and whatever the part of the plant, it is possible to intensify the operation through expansion, which increases the porosity, the specific surface area, and the availability of the desired compounds.

   Besides an adequate expansion should facilitate the grinding and reach expanded granule powders. This increases the porosity and the volume of the pores, improves solute transfer, and completely modifies the internal ability of the material versus the extraction.

The physical effects of ultrasound and cavitation give rise to strong convection in the medium through various physical phenomena such as microstreaming, microturbulence, shock waves, and microjets. In other words, ultrasonic treatment implies an internal motion/agitation of the liquid within the pores, leading to solute transfer by convection instead of or coupled to diffusion [30]. Both effects can increase mass transfers of solvent within the solid as well as solute within the solvent [30, 31].

5) Microwave treatment

The main specificity of microwaves as a heat source is its penetrating power in materials normally independently of the conductivity of the processed materials. A number of factors may be taken into account to adopt the use of microwaves:

a. geometry such as sizes of products to be processed, and the difficulty of using other heating methods;
b. Thermal sensitivity;
c. Cost: products have to be high-added value to tolerate treatment with microwaves.

An important element in the development of microwave and design systems is the ability to model the interactions between matter and electromagnetic waves.

The correlations between dielectric properties (usually linked with humidity and/or solvent presence) and temperature are important for simulation and modeling processes.

The most important parameter in the microwave treatment is $P \tan \delta$. It indicates the product's ability to transform the absorbed energy into heat. So it must be a combination of a moderate ε' to good penetration and high loss (maximum of ε'').

Products to be processed by microwave must be dipolar, which, fortunately, is the case of many natural plant products. However, as the thermal conductivity of materials is low, the heat is not dissipated quickly in the surrounding zone.

In a microwave-assisted unit operations, heat generation and transfer within a plant as a porous medium is described by the following equation:

$$\dot{q} - \vec{\nabla}.\vec{\varphi} = \rho_s(c_{ps} + Wc_{pw})\frac{\partial T}{\partial t} + \psi\frac{\partial\left(\frac{M_w L_w p_w}{R_{\mathrm{GP}}T}\right)}{\partial t} \tag{5.50}$$

where the second term in the LHS stands for heat transfer, which normally follows the conduction law. The first term in the RHS stands for the sensitive heat accumulation and the second term for the heat dissipated during evaporation.

The first term in the LHS, $\dot{q}$, stands for the microwave volume absorbed power or, in other words, the heat produced per unit volume per unit time; it is also called *absorption rate density (ARD)* and can be expressed by the following relation:

$$\dot{q}(\vec{r}) = \pi f \varepsilon_o \varepsilon''(\vec{r}) |E(\vec{r})|^2 \tag{5.51}$$

where $|E(r)|$ is the amplitude of electric field intensity, which varies depending on the point defined by the vector $(\vec{r})$. If the thickness l_p of the heated material is sufficiently large and the loss factor is constant, power distribution can follow Lambert's law:

$$\dot{q}(x) = \pi f \varepsilon_o \varepsilon'' |E_o|^2 \exp\left(-\frac{x}{d_p}\right) \tag{5.52}$$

where $|E_0|$ stands for the amplitude of electric field intensity at the material surface $(x = 0)$, and d_p for the penetration depth [32]:

$$d_p = \frac{c}{2\pi f \sqrt{2\varepsilon''}} \left(\sqrt{1 + \left(\frac{\varepsilon''}{\varepsilon'}\right)^2} - 1 \right)^{-1/2} \tag{5.53}$$

At the penetration depth d_p from the surface $(x = 0)$, the power density is equal to 36.788% of the power density on the surface.

5.6 Conclusion

A relevant PI-S in the specific case of the unit operations of extarction was studied through the well-known procedure. Therefore, we

1) analyzed successive transfer processes at the fundamental level;
2) identified the slowest phenomenon;
3) proposed an adequate solution referred to increasing kinetics, reducing energy consumption, improving environment impacts, and improving final product and coproduct quality;
4) design adequate equipment.

After achieving the fundamental approaches and studying different specific cases of various specific plants, different kinetic parameters and functional properties were identified. We finally carried out new modeling studies to sequentially apply this PI-S. This chapter highlights the impact of the parameters of instant controlled pressure drop thus allowing sequential optimization processes and concludes with the design of high-reliability processing equipment.

References

1. Allaf, T. and Allaf, K. (2014) *Instant Controlled Pressure Drop (D.I.C.) in Food Processing*, Springer, New York.
2. Arhaliass, A., Legrand, J., Vauchel, P., Fodil-Pacha, F., Lamer, T., and Bouvier, J.M. (2009) The effect of wheat and

maize flours properties on the expansion mechanism during extrusion cooking. *Food Bioprocess Technol.*, **2**, 186–193.
3. Gordon, M. and Taylor, J.S. (1952) Ideal copolymers and the second-order transitions of synthetic rubbers. I. non-crystalline copolymers. *J. Appl. Chem.*, **2**, 493–500.
4. Allaf, K. (1982) *Transfer Phenomena and Industrial Applications*, Lebanese University, Faculty of Science, Beirut.
5. Orford, P.D., Parker, R., and Ring, S.G. (1990) Aspects of the glass transition behaviour of mixtures of carbohydrates of low molecular weight. *Carbohydr. Res.*, **196**, 11–18.
6. Allaf, K. (2009) *Essential Oils and Aromas: Green Extraction and Application*, Chemat F., New Delhi, pp. 85–121.
7. Bendahou, M., Muselli, A., Grignon-Dubois, M., Benyoucef, M., Desjobert, J.-M., Bernardini, A.-F., and Costa, J. (2008) Antimicrobial activity and chemical composition of Origanum glandulosum Desf. essential oil and extract obtained by microwave extraction: Comparison with hydrodistillation. *Food Chem.*, **106**, 132–139.
8. Bocchio, E. (1985) *Hydrodistillation of Essential Oils: Theory and Applications*, vol. 63, Société d'expansion technique et économique, Paris.
9. Cassel, E., Vargas, R.M.F., Martinez, N., Lorenzo, D., and Dellacassa, E. (2009) Steam distillation modeling for essential oil extraction process. *Ind. Crops Prod.*, **29**, 171–176.
10. Chemat, F. and Lucchesi, M.E. (2006) Microwave assisted extraction of essential oils, in *Microwaves in Organic Synthesis*, Chapter 22 (ed. A. Loupy), Wiley-VCH Verlag GmbH, Weinheim.
11. Ferhat, M.A., Meklati, B.Y., Smadja, J., and Chemat, F. (2006) An improved microwave Clevenger apparatus for distillation of essential oils from orange peel. *J. Chromatogr. A*, **1112**, 121–126.
12. Gámiz-Gracia, L. and Luque de Castro, M.D. (2000) Continuous subcritical water extraction of medicinal plant essential oil: comparison with conventional techniques. *Talanta*, **51**, 1179–1185.
13. Lucchesi, M.E., Chemat, F., and Smadja, J. (2004) Solvent-free microwave extraction of essential oil from aromatic herbs: comparison with conventional hydro-distillation. *J. Chromatogr. A*, **1043**, 323–327.
14. Allaf, T., Besombes, C., Mih, I., Lefevre, L., and Allaf, K. (2011) Decontamination of solid and powder foodstuffs using DIC technology, in *Advances in Computer Science and Engineering* (ed. M. Schmidt), InTech, Croatia.
15. Besombes, C., Berka-Zougali, B., and Allaf, K. (2010) Instant controlled pressure drop extraction of lavandin essential oils: fundamentals and experimental studies. *J. Chromatogr. A*, **1217**, 6807–6815.
16. Al Haddad, M. (2007) *Contribution théorique et modélisation des phénomènes instantanée dans les opérations d'autovaporisation et de déshydratation*, Université de La Rochelle, La Rochelle.
17. Al Haddad, M., Mounir, S., Sobolik, V., and Allaf, K. (2008) Fruits and vegetables drying combining hot air, DIC technology and microwaves. *Int. J. Food Eng.*, **4**, 1556–3758.
18. Allaf, K., Besombes, C., Berka-Zougali, B., Kristiawan, M., Sobolik, V., and Allaf, T. (2011) in *Enhancing Extraction Processes in the Food Industry* (eds N. Lebovka, E. Vorobiev, and F. Chemat), CRC Press, Taylor & Francis Group, Dublin, pp. 255–302.
19. Leybros, J. and Frémeaux, P. (1990) Extraction solide-liquide. Aspects théoriques. Techniques de l'Ingénieur J1 077 06.
20. Schwartzberg, H.G. and Chao, R.Y. (1982) Solute diffusivities in leaching processes. *Food Technol.*, **2**, 73–86.
21. Aguilera, J.M. and Stanley, D.W. (1999) *Microstructural Principles of Food Processing and Engineering*, Aspen Publication.
22. Amor, B.B., Lamy, C., Andre, P., and Allaf, K. (2008) Effect of instant controlled pressure drop treatments on the oligosaccharides extractability and microstructure of Tephrosia purpurea seeds. *J. Chromatogr. A*, **1213**, 118–124.

23. Ben Amor, B. (2008) *Maîtrise de l'aptiture technologique de la matière végétale dans les opérations d'extraction de principes actifs: texturation par Détente Instantanée Contrôlée DIC*, Université de La Rochelle, France.
24. Ben Amor, B. and Allaf, K. (2009) Impact of texturing using instant pressure drop treatment prior to solvent extraction of anthocyanins from Malaysian Roselle (*Hibiscus sabdariffa*). *Food Chem.*, **115**, 820–825.
25. Fernández, M.B., Perez, E.E., Crapiste, G.H., and Nolasco, S.M. (2012) Kinetic study of canola oil and tocopherol extraction: parameter comparison of nonlinear models. *J. Food Eng.*, **111**, 682–689.
26. Allaf, T., Mounir, S., Tomao, V., and Chemat, F. (2012) Instant controlled pressure drop combined to ultrasounds as innovative extraction process combination: fundamental aspects. *Proc. Eng.*, **42**, 1164–1181.
27. Mounir, S. and Allaf, K. (2008) Three-stage spray drying: new process involving instant controlled pressure drop. *Drying Technol.*, **26**, 452–463.
28. Crank, J. (1975) The mathematics of diffusion. (2nd ed.). Oxford: Clarendon Press.
29. Mounir, S. and Allaf, K. (2009) Study and modelling of dehydration and rehydration kinetics within porous medium. Paper Presented at AFSIA, Lyon, France.
30. Toma, M., Vinatoru, M., Paniwnyk, L., and Mason, T.J. (2001) Investigation of the effects of ultrasound on vegetal tissues during solvent extraction. *Ultrason. Sonochem.*, **8**, 137–142.
31. Mason, T.J. (2000) Large scale sonochemical processing: aspiration and actuality. *Ultrason. Sonochem.*, **7**, 145–149.
32. Kitchen, R. (2001) *Radio Frequency RF and Microwave Radiation Security*, Reed Elsevier Group, Oxford.
33. Allaf, K. (2002) Analysis of instantaneity in thermodynamic processes: fundamental laws. 1st Franco- Lebanese Symposium on technologies and studies on Process Engineering and Biochemistry, Lebanese University, Beyrouth, Lebanese.

6
Panorama of Sustainable Solvents for Green Extraction Processes

Iraj Koudous, Werner Kunz, and Jochen Strube

6.1
Introduction

A proper choice of solvents is of utmost importance to conceive extraction processes and subsequent purification steps. On the one hand, the solvent must ensure a sufficient solubility of valuable substances, while on the other, the economic aspects and HSE (health, safety, environment) profile of the solvent must be taken into account. The extraction agent must further exhibit a high selectivity and it must be chemically inert, nonflammable, temperature stable, and of low or negligible corrosiveness.

In the field of solid-state extraction, the following classes of solvents are possible:

- Aqueous media
- Organic solvents
- Aqueous two-phase systems
- Supercritical fluids
- Ionic liquids or deep eutectic solvents.

Obviously, there is a strong tendency worldwide toward the use of "green" extraction agents and solvents. But what is a "green" or "sustainable" solvent? According to Capello *et al.*, it minimizes the environmental impact [1]. Estévez [2] suggests a more precise definition, saying that a "green solvent has to bear a number of environment, health and safety characteristics that differentiate it from existing hazardous solvents." But even this definition remains quite vague. There are 12 principles of green chemistry and 12 further ones for engineering; the question is which and how many of them must be fulfilled for the solvent to be considered as "green" or "sustainable."

It should be also mentioned that often there is confusion between "green" solvents and "bio-sourced" ones. Bio-sourced or bio-based solvents are made mostly of biomass feedstock, with the help of chemical or biochemical transformations. However, these transformations are not necessarily "green" and even the solvents, although "bio-sourced," may not have a favorable HSE profile. For example, the

Green Extraction of Natural Products: Theory and Practice, First Edition.
Edited by Farid Chemat and Jochen Strube.

bio-based furfuraldehyde is toxic and carcinogenic and as such is certainly not a green solvent.

By contrast, even petrochemical-based solvents can have a good HSE profile and can be considered as green and "eco-friendly." An example is MMB (3-methoxy-3-methyl-butan-1ol), which is synthesized from iso-butylene (C4 fraction of cracked naphtha). It is of very low toxicity and has a low vapor-pressure and therefore can be reasonably called a *green solvent*.

Once solvents are considered to be green, the next step is to compare their physical properties to those of classical solvents to see what type of solvent can be replaced by greener ones and for which processes. To do this, modeling tools are quite useful, but also simple classifications are helpful. In 2011, Jessop [3] published a paper about green solvents in which he critically reviewed the state of art in the field of green solvents in general. Figure 6.1 is taken from this paper and it presents green solvents in so-called Kamlet–Taft plots. The corresponding parameters are a measure of solvent polarity or polarizability (π^*) and of their basicity or hydrogen-bond accepting ability (β). It is easy to see that large domains are not yet covered by green solvent alternatives, for example, high basicity/low polarity aprotic solvents, meaning that amines are still difficult to replace.

Jessop's plots show another dilemma. The Kamlet–Taft parameters must be determined experimentally, and because they have not yet been measured for all relevant green solvents, the plots are necessarily incomplete. Therefore, there is a strong need to identify tools that predict almost quantitatively the solubility of a product to be extracted in this solvent and allow comparison to the solubilization power of a classical solvent. Further, the tool should be able to estimate partition coefficients of a chemical product between two different solvents, to predict vapor pressures in mixtures, and other properties such as viscosity and surface tension. While the two last properties are still almost impossible to predict from first principles, there are indeed *in silico* tools available now that permit one to evaluate phase equilibria.

The thermodynamic key parameter is the chemical potential of the components or, more precisely, the excess potential of a product in a mixture expressed in terms of activity coefficients. They can be approximated, for example, via group contribution methods such as the UNIFAC model. This is explained in Section 6.2.1.

As a first approximation, the compatibility of two substances follows the principle "like mixes with (or dissolves in) like." A very popular approach quantifies this principle. This is the Hansen solubility parameter (HSP) approach, as it is briefly introduced in Section 6.2.2. However, as the UNIFAC model, it requires a considerable experimental data set to determine the polarity, dispersion interaction, and hydrogen bonding power of a given solvent and extraction agent. By far the most powerful approach today is the so-called COSMO-RS (conductor-like screening model for real solvents) model, which is based on quantum-mechanical calculations and requires no or very little experimental input to calculate compatibilities of molecules, mixing and dissolution power, vapor pressures in mixtures, partition coefficients, and so on. This approach will be discussed in more detail in Section 6.2.3. Much better predictions of extraction yields and selectivity, and

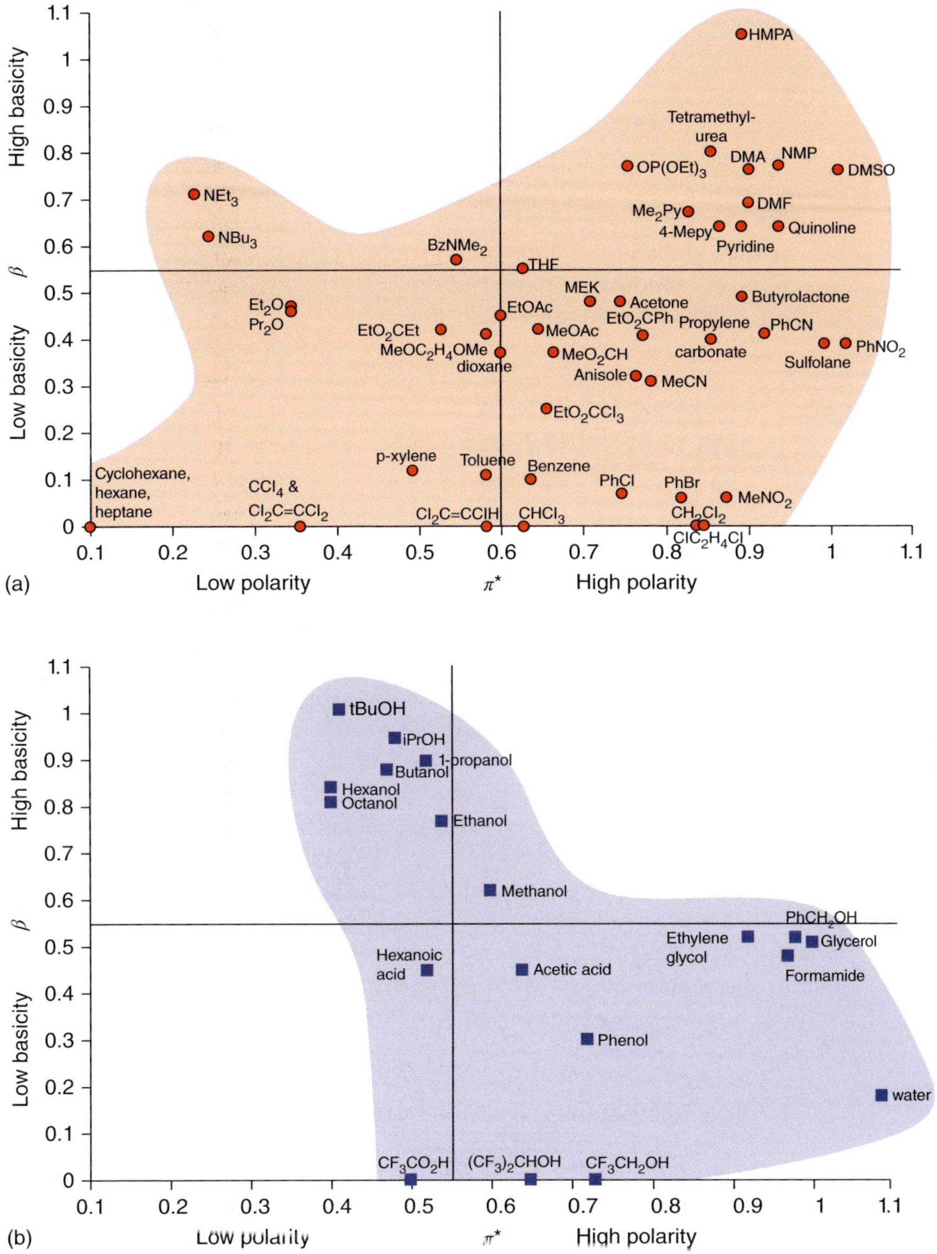

Figure 6.1 Plots showing (a) common aprotic solvents and (b) common protic solvents as a function of their π^*- and β-values according to [3].

consequently, cost-effectiveness, are expected with COSMO-RS. The combination of such prescreening tools with a proper preselection of eco-friendly solvents is one of the keys for future extraction processes. The resulting solvent selection for green solid–liquid extraction is discussed in Section 6.3. Finally, it is worth having a look at some particular solvents, namely, ionic liquids and deep eutectic solvents (DESs) that are supposed to play an increasing role in extraction science. A critical look at these solvents is given in Section 6.4.

However, the "best" extraction process is insufficient without the consideration of the subsequent purification steps and a careful process design including the modeling of distillation processes, membranes, crystallization, liquid–liquid extraction, and chromatography. Ways to predict optimized extraction process design are discussed in Section 6.5.

6.2 Thermodynamic Models of Mixing and Dissolving

6.2.1 UNIFAC and Modified UNIFAC

So far, systems for the assessment of thermodynamic properties based on the group contribution method, such as the UNIFAC model, have proven themselves [4]. The UNIFAC model is based on the fact that the number of all chemical compounds is much larger than the number of different molecule groups which they are composed of. Therefore, a large number of chemical compounds can be developed from a relatively small number of functional groups [5].

Figure 6.2 shows the breakdown of ethanol molecule into a CH_3, a CH_2, and an OH-group. In the UNIFAC-model, a compound mixture is not considered as a mixture of molecules, but a mixture of molecule groups. Interactions between molecule groups are treated in the same way. The type of interaction only depends on the functional group and not on the compound in which it is contained. Analogous to the UNIQUAC-model, the activity coefficient, which can be used to describe thermodynamic properties, consists of an entropic size correction and a residual term resulting from the group interactions and is defined as follows:

$$\ln \gamma_i = \ln \gamma_i^{\text{entropic}} + \ln \gamma_i^{\text{enthalpic}} = \ln \gamma_i^C + \ln \gamma_i^R \tag{6.1}$$

The entropic term sums up all effects which relate to molecular size and shape and is described as the combinatorial contribution $\ln \gamma_i^C$. This combinatorial contribution can be calculated as follows:

$$\ln \gamma_i^C = \ln \frac{\Phi_i}{x_i} + \frac{z}{2} q_i \ln \frac{\Theta_i}{\Phi_i} + \Phi_j \left(l_i - \frac{r_l l_j}{r_j} \right) \tag{6.2}$$

$$\Phi_i = \frac{x_i \cdot r_i}{\sum_j x_j \cdot r_j} \tag{6.3}$$

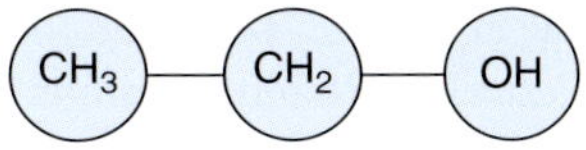

Figure 6.2 Breakdown of the ethanol molecule into its functional groups.

$$\Theta_i = \frac{x_i \cdot q_i}{\sum_j x_j \cdot q_j} \tag{6.4}$$

$$r_i = \sum_k v_k^{(i)} \cdot R_k \tag{6.5}$$

$$q_i = \sum_k v_k^{(i)} \cdot Q_k \tag{6.6}$$

Φ_i are the volume contributions and Θ_i the surface contributions, r_i the volume parameters and q_i the surface parameters, by which the volume and surface contributions can be calculated using a suitable bias. Volume and surface parameters can be generated from X-ray structural data or from simple geometric correlations and can be taken from the literature [6].

The enthalpic term sums up all effects resulting from energetic interactions between the functional groups and is called the *residual contribution*. For its calculation, the following equation is needed:

$$\ln \gamma_i^R = \sum_k v_k^{(i)} \cdot (\ln \Gamma_k - \ln \Gamma_k^{(i)}) \tag{6.7}$$

By introducing the group activity coefficient $\ln \Gamma$, it is possible to separate molecular interactions in the mixture from the interactions in the respective pure substances. It is defined as

$$\ln \Gamma = Q_k \cdot \left[1 - \ln \left(\sum_n \Theta_m \cdot \Psi_{mk} \right) - \sum_m \frac{\Theta_m \cdot \Psi_{km}}{\sum_m \Theta_n \cdot \Psi_{nm}} \right] \tag{6.8}$$

$$\Theta_m = \frac{X_m \cdot Q_m}{\sum_n X_n \cdot Q_n} \tag{6.9}$$

$$X_m = \frac{\sum_j v_m^{(j)} \cdot x_j}{\sum_j \sum_n v_n^{(j)} \cdot x_j} \tag{6.10}$$

Θ_m are the surface and X_m the molar contributions. By the definition of the interaction parameter (Ψ_{nm}) of the main groups,

$$\Psi_{nm} = \exp\left(-\frac{a_{nm}}{T}\right) \tag{6.11}$$

the residual contribution can be calculated.

The modified UNIFAC model builds on the same basic equations; however, the volume part of the entropic term is modified to better describe strongly asymmetric and complex molecules.

$$\ln \gamma_i^C = 1 - V_i + \ln V_i - 5q_i l \left(1 - \frac{V_i}{F_i} + \ln \frac{V_i}{F_i}\right) \tag{6.12}$$

$$V_i = \frac{r_i^{\frac{3}{4}}}{\sum_j x_j r_j^{\frac{3}{4}}} \tag{6.13}$$

The exponent 3/4 was found empirically. To better represent temperature dependencies, the interaction parameter is expanded by a polynomial extension:

$$\Psi_{nm} = \exp\left(-\frac{a_{nm} + b_{nm}T + c_{nm}T^2}{T}\right) \tag{6.14}$$

A further improvement of the modified UNIFAC model over the conventional approach is the extension of the main- and side-group parameters. More than 1000 parameters are published in literature [7]. This large database is another advantage of the UNIFAC model. In particular, the extended group contribution methods provide good results for many terpenes and flavonoids. Nevertheless, in case of very complex molecules, they fail also [8].

6.2.2
The Hansen Solubility Parameters

"Like dissolves like" is a good concept for mixing and solubilizing, but are there some simple rules to define what is "like"? Joel Henry Hildebrand, professor of chemistry in Berkeley and pioneer in solution chemistry in the first half of the twentieth century proposed a first suitable concept in this direction [9]. He introduced the so-called Hildebrand solubility parameter δ, which is related to the cohesion energy density E_{coh} and the molar volume V_m in the following way: $\delta = (E_{\text{coh}}/V_m)^{\frac{1}{2}}$. E_{coh} is the difference between the enthalpy of evaporation and the energy of the ideal gas: $E_{\text{coh}} = \Delta H_{\text{vap}} - RT$ (R: gas constant, T: temperature). As a consequence, the Hildebrand parameter is easy to determine, provided that the substance (liquid) has a sufficient vapor pressure. If not, as for salts that are liquid at room temperature (i.e., ionic liquids), δ can be inferred from inverse gas chromatography.

According to Hildebrand, two substances are miscible if the δ values of both components are similar, because the Gibbs energy of mixing ΔG_M can be written as

$$\Delta G_M = \Phi_1 \Phi_2 V_M (\delta_1 \delta_2)^2 - \Delta(TS)^{\text{ideal}} \tag{6.15}$$

where Φ_i is the volume fraction of component i, V_M is the mean molar volume of the two components and S is the ideal entropy of mixing.

Polystyrene has a solubility parameter of 18.6 $MPa^{1/2}$, and for ethyl acetate δ (in SI units) is 18.2 $MPa^{1/2}$, which is "quite" similar. Indeed, polystyrene is soluble in ethyl acetate, whereas it is insoluble in ethanol ($\delta = 26.2$ $MPa^{1/2}$).

The concept is very simple and easily extendable to multicomponent systems. Mean values can be calculated from the individual δ parameters, the mean volume, and the composition of the mixtures. For example, a mixture of one-third (in volume) of acetone with two-thirds of toluene has a mean value of $(19.7 \times 1/3 + 18.3 \times 2/3) = 18.7$ $MPa^{1/2}$, which is the same value as that of chloroform. Indeed, several resins show a similar solution behavior in both media.

Of course, it cannot be expected that one parameter is sufficient to predict miscibilities and solubilities. In practice, this concept works often only when the components are supposed to form so-called regular mixtures (for which Hildebrand initially developed his model) and when the mean molar volume is small. For strong interactions involving polar interactions and/or hydrogen bonds, it usually fails. Water is characterized by a Hildebrand parameter $\delta = 48.0$ $MPa^{1/2}$, whereas $\delta^{\text{nitromethane}} = 25.1$, which is close to the value of ethanol. However, ethanol is very soluble in water, whereas nitromethane is not.

Among numerous attempts to improve Hildebrand's approach, HSPs [10] are the most prominent ones and still widely used in industry. Hansen divided Hildebrand's δ parameter into three contributions: $\delta^2 = \delta^2{}_{\text{D}} + \delta^2{}_{\text{P}} + \delta^2{}_{\text{H}}$, corresponding to $E = E_{\text{D}} + E_{\text{P}} + E_{\text{H}}$, where the different terms denote dispersion (D) and permanent dipole (P) interactions and hydrogen bonding (H), respectively. Two components are compatible (miscible or soluble) if the three parameters are similar. To calculate the distance (R_a) between the HSPs in the so-called Hansen space, the following formula is used:

$$R_a{}^2 = 4(\delta_{\text{D2}} - \delta_{\text{D1}})^2 + (\delta_{\text{P2}} - \delta_{\text{P1}})^2 + (\delta_{\text{H2}} - \delta_{\text{H1}})^2 \tag{6.16}$$

For a given compound, a certain interaction radius R_0 is defined (usually inferred from experiments). Finally, the relative energy difference (RED) $\text{RED} = R_a/R_0$ determines the concept of "like dissolves like." For $\text{RED} < 1$ the compound is like and will dissolve, for $\text{RED} = 1$, a partial solubility is predicted, and for $\text{RED} > 1$ no solubility should exist.

As an example, let us come back to the case of very different water solubilities of ethanol and nitromethane. The polarity parameter δ_{P} (nitromethane) = 15.8 is even closer to the corresponding water parameter (δ_{P}, (water) = 16.0) than δ_{P}, (ethanol) = 8.8. If there were no hydrogen bonding, nitromethane would be even more water soluble than ethanol. However, δ_{H} (H_2O) = 43.3, δ_{H} (nitromethane) is only 5.1, and δ_{H} (ethanol) = 19.4, which explains while ethanol is water soluble, nitromethane not.

Hansen's concept is often successfully applied in the case of polymer solutions, liquid mixtures, and also gas solubilities. It was even extended to some less usual cases such as the skin permeability of solvents. It was found that liquids with this property have HSP values around $\delta_{\text{D}} = 17.6$, $\delta_{\text{P}} = 12.5$, and $\delta_{\text{H}} = 11.0$, such as

DMSO ($\delta_D = 18.4$, $\delta_P = 16.4$, and $\delta_H = 10.2$). Numerous examples can be found in the Web and in several textbooks.

Of course, the HSPs require considerable experimental work, from which they are deduced. Impurities can have a significant influence, especially when they are charged (salts, polyelectrolytes, proteins). Further, effects such as those of temperature or pressure are usually not taken into account. Further, the RED allows only a rough classification into "soluble," "insoluble," and "partially soluble." In addition, binding between molecules is more complex than the mentioned Hansen parameters. The molecular shape is also relevant, but is not considered. Another limitation is the usage of the Hansen model for solids, because there is no enthalpy of evaporation for solids and this parameter is necessary for the determination of HSP.

Therefore, the whole concept cannot be more than a first-order approximation to solubilization (after the zeroth-order approximation of Hildebrand). Nevertheless, it is a valuable concept, provided that a sufficient parameter set is available.

6.2.3
COSMO and COSMO-RS

The regular solution theories of Hildebrand and Hansen, as briefly presented in the previous section, are based on the concept of maximum similarity of solute and solvent with respect to the intermolecular interactions. As stated in the section, they reflect the old alchemist's concept of "*similia similibus solvuntur.*" This rule has proven to be very useful, but it stays on the level of a relatively crude, qualitative understanding of solvation phenomena. Nowadays, we can gain more quantitative and theoretically more thorough inside knowledge of solvation and solubilization based on a molecular description of the intermolecular interactions. The currently most useful framework in this regard is provided by the COSMO-RS theory [11] which is briefly described in the following. More comprehensive descriptions of COSMO-RS can be found elsewhere [12–14].

The starting point of COSMO-RS are quantum chemical calculations of the solute and solvent molecules. In contrast to standard quantum chemistry which considers molecules in vacuum, for COSMO-RS these calculations are done in the presence of a virtual conductor embedding the molecules. This is achieved by employing the conductor-like screening model (COSMO). Apart from the energies and geometries of the molecules, these DFT/COSMO calculations provide the conductor-polarization charge density σ on the surface of the molecules, which turns out to be a very detailed descriptor of the local surface polarity. As shown in Figure 6.3, the σ-surfaces, that is, the molecular surfaces color-coded by σ, and the σ-profiles, that is, the histograms $p^i(\sigma)$ of the molecular surface with respect to the surface polarity σ, already give a vivid description of the similarities and differences of molecules.

The σ-profiles are real fingerprints of the molecules with respect to surface polarity. For example, hexane has a rather neutral (green) surface which results in a narrow double peak in the σ-profile (curve). The two peaks result from

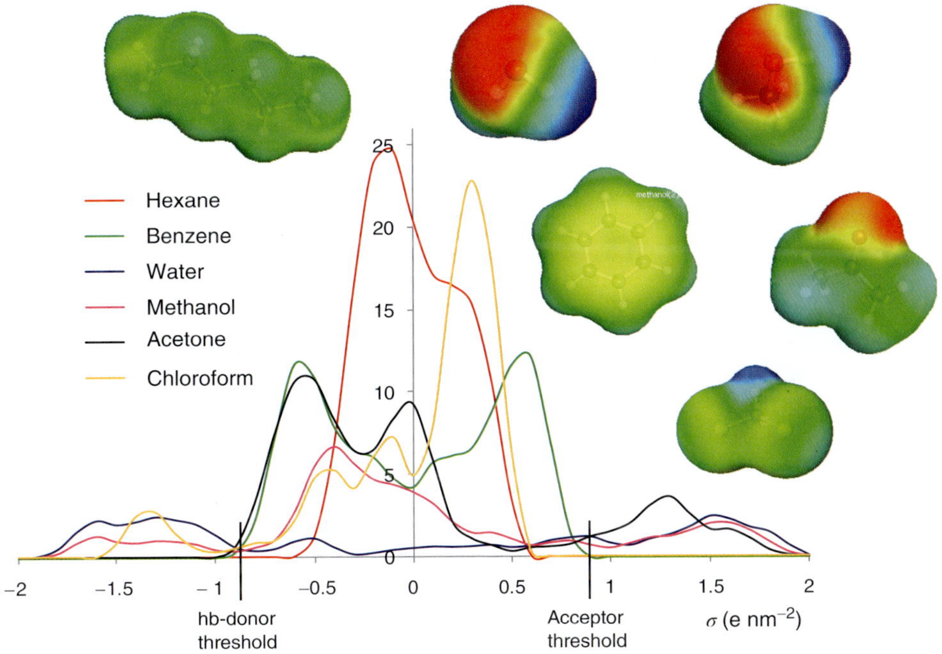

Figure 6.3 σ-surfaces and σ-profiles of six representative molecules.

the slightly different polarities of carbon and hydrogen surface. Owing to its quadrupole moment, the σ-surface of benzene is already slightly yellow in the π-face region, and slightly blue at the hydrogens, which translates into the broader double peak structure (green curve) in the σ-profile. The σ-surface of water is dominated by blue and red regions, resulting from the very polar hydrogen atoms and the lone-pair regions of the oxygen atom. As a result, the σ-profile of water (blue line) is dominated by two pronounced peaks in the hydrogen bond (hb) donor region ($\sigma < -1\,\mathrm{e\,nm^{-2}}$) and in the hb acceptor region ($\sigma > 1\,\mathrm{e\,nm^{-2}}$), with very few neutral surface in between. Methanol (pink σ-profile) has almost the same, but slightly less polar oxygen feature in the σ-profile. On the other hand, it has only one hb donor and thus half the intensity in the donor region, and a slightly polarized, that is, slightly shifted, alkane feature in the nonpolar σ-region. Owing to its sp^2 oxygen, acetone shows a different σ-profile in the hb-acceptor region than methanol and water, and it has no polar surface in the hb-donor region. The σ-profile of chloroform is dominated by a large peak of weakly polar surface resulting from the three chlorine atoms, while the hydrogen surface is sufficiently polar to act as an hb donor.

The fundamental concept of COSMO-RS is the quantification of the molecular interactions based on the surface polarities σ and σ' which form a molecular contact. Charge and dipole interactions of molecules in liquids are quantified by the so-called electrostatic misfit energy:

$$e_{\text{misfit}}(\sigma, \sigma') = \frac{1}{2}\alpha'_{\text{misfit}}(\sigma + \sigma')^2 \tag{6.17}$$

The misfit energy is zero if both surfaces have the same but opposite polarity and it increases squared with the misfit of σ and σ'. In addition to electrostatic interactions, hydrogen bonding leads to strong interactions between molecules. Hydrogen bonds (hb) are contacts of very polar surfaces of opposite polarity. Obviously, such contacts include a lot of electrostatic interaction energy. The additional hb interaction energy can be quite accurately quantified by an expression as

$$e_{\text{hb}}(\sigma, \sigma') = c_{\text{hb}}(T)\min(0, \sigma\sigma' - \sigma_{\text{hb}}^2) \tag{6.18}$$

This simple equation expresses the increase of hydrogen bond strength with the polarity of donor and acceptor polarity as soon as the absolute values of the polarities of donor and acceptor overcome a threshold value σ_{hb}. The temperature dependence of e_{hb} captures the loss of entropy going along with the formation of a hydrogen bond.

Using these expressions for the surface interactions, we can now calculate the chemical potentials $\mu_S(\sigma)$ of surface segments in a liquid system S:

$$\mu_S(\sigma) = -\frac{kT}{a_{\text{eff}}}\ln \int p_S(\sigma)\exp\left\{-\frac{a_{\text{eff}}}{kT}\left(e_{int}\left(\sigma, \sigma'\right) - \mu_S(\sigma')\right)\right\}d\sigma \tag{6.19}$$

where $e_{int}(\sigma, \sigma')$ is the sum of misfit and hb interactions, a_{eff} is the average size of a contact surface segment, and

$$p_S(\sigma) = \frac{\sum_i x_i p^i(\sigma)}{\sum_i x_i A^i} \tag{6.20}$$

is the relative composition of the liquid system S with respect to surface polarity σ, which can be easily evaluated from the σ-profiles $p^i(\sigma)$, surface areas A^i, and mole fractions x_i of the components of a pure or mixed liquid system S. The function $\mu_S(\sigma)$, which is called *σ-potential,* is a characteristic function of the liquid system S, describing its affinity for molecular surface of varying polarity σ. The σ-potentials are excellently suited to compare and classify solvents and to quantify solvent properties.

Finally, the standard chemical potential μ_S^X of a molecule X in system S can be calculated by integration of the segment chemical potential

$$\mu_S^X = \int p^X(\sigma)\mu_S(\sigma)d\sigma + kT\ln\gamma_{\text{comb}}(X, S) \tag{6.21}$$

where the combinatorial activity coefficient $\gamma_{\text{comb}}(X, S)$ provides a small, but significant size correction, which is well known in chemical engineering thermodynamics and which can be evaluated on the basis of COSMO surface areas and COSMO volumes of solute and solvent molecules.

By Equation 6.21, the COSMO-RS can provide the chemical potential of almost every chemical compound as a function of composition and temperature in almost every liquid system, including pure solvents, mixtures, micellar systems, ionic liquids, and many more. By this, it gives access to activity and partition coefficients, vapor pressures, solubilities, enthalpies, and entropies of solution, and complete phase behavior of liquid systems. As a result, COSMO-RS can be applied to a very wide range of solvation-related problems, especially for screening and selecting solvents and solubility-enhancing additives. A general overview of the current application areas is given in Ref. [13]. In the following, we will consider examples of three applications elucidating the power of COSMO-RS in its COSMOtherm software implementation [15].

6.2.3.1 Example 1: Mutual Solubility of Acetone with Benzene, Chloroform, and Carbon disulfide

The DFT/COSMO files of more than 6000 common compounds, with a focus on common solvents, are stored in the COSMObase database. As a result, COSMO-RS calculations for these compounds can be done within seconds with the COSMOtherm software. For our example, we chose the four compounds propanone (acetone), benzene, $CHCl_3$ (chloroform), and CS_2 (carbon disulfide) from the COSMObase and loaded them into COSMOtherm. Next we just chose the VLE/LLE panel, selected the three binaries of interest, set the temperature to 25 °C, and submitted these calculations. After 10 s the programs returned a results table, providing all calculated information about the three binary vapor–liquid equilibria systems, including excess enthalpies and free energies, activity coefficients, partial pressures, and more, which could be graphically visualized in a graphics panel. In Figure 6.4 the total pressure curves are shown.

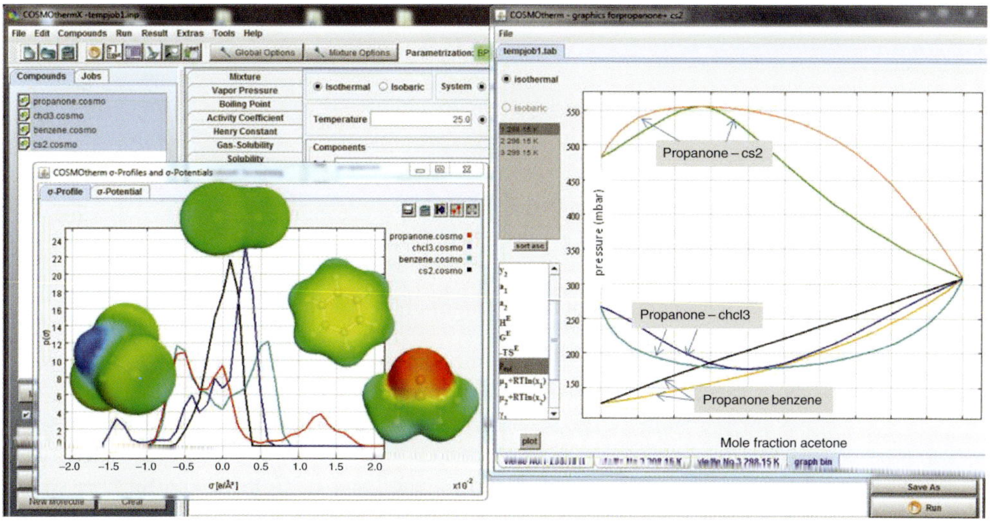

Figure 6.4 Graphical visualization of the COSMOtherm workflow and results for three binary systems.

Despite correctly predicting complete miscibility for all three systems, in agreement with experiment, COSMO-RS predicted quite different VLE behaviors for the three binary mixtures of acetone. Although the σ-profiles of benzene and acetone are substantially different on the right side, the VLE with benzene is nearly ideal (straight line of the vaporization curve). This is due to the fact that on the left side the σ-profiles are almost identical, leading to equally bad interactions of the polar oxygen parts of acetone in acetone–benzene mixtures as in pure acetone. The situation is very different for acetone–chloroform mixtures. Here, the very polar hydrogen surface of chloroform, which was lacking suitable partners in pure chloroform, is a very good contact partner for the acetone oxygen, because their polarities are nearly perfectly opposite. As a result, the free energy of acetone–chloroform mixtures is lower than that of the pure solvents, leading to a reduction of the vapor pressure and even to a maximum boiling point azeotrope at about 40 mol% of acetone. The situation is the opposite for the mixture of acetone with carbon disulfide (CS_2). As can be seen from the σ-surface and the σ-profile, the surface of carbon disulfide is even less polar than that of acetone, and hence it is an even worse partner for acetone, leading to a positive curvature of the vapor pressure curve and a minimum boiling point azeotrope at about 30 mol% of acetone.

The HSP method would yield Hansen distances of 12.7, 8.7, and 11.8 (15.4) for the three binaries of acetone with benzene, chloroform, and carbon disulfide, respectively, where the number in brackets is an alternative value for the polar interaction parameter of carbon disulfide tabulated in the HSP tables [16]. Typically, a distance of 8 is considered as limit of good solubility. Hence, in this sense, chloroform would be considered as a fair solvent of acetone, and the other two as bad solvents for acetone. It is apparent that COSMOtherm yields a much more detailed and accurate description of the physicochemical behavior of these mixtures.

6.2.3.2 Example 2: Solubility Screening for Indigo

Indigo is one of the oldest known dyes and it is still of industrial importance. The largest problem in the use of indigo is its low solubility. Therefore, it is interesting to perform a solubility screening for indigo and to discuss the results. The solubility of indigo in water at 25 °C is referenced in the OECD SIDS report [17] as $0.99 \times 10^{-6}\,g\,g^{-1}$, corresponding to 0.0001 wt%. For convenience, all other solubilities will also be referenced in weight % in this example. Taking this value as reliable reference, we have performed a solvent screening for 20 solvents, including some standard solvents and a collection of liquids listed as solvents for indigo in various literature sources. The results of the screening are listed in Table 6.1. As a check for reliability of the solubility screening, we may take the relative solubility of indigo in water and octanol. The COSMO-RS prediction for the solubility in octanol is 0.074%, and thus 2.87 log-units higher than the solubility in water. This value is in excellent agreement with the octanol/water partition coefficient of indigo, which is listed in the same OECD SIDS report as $\log P_{OW} = 2.7$. From the reference solubility COSMO-RS estimates a fusion free energy of ΔG_{fus} of

Table 6.1 Results of COSMO-RS solubility screening for indigo.

Rank	Solvent	Solubility (wt%)
1	H_2SO_4	100.0000
2	Dimethylsulfoxide	11.9093
3	Phenol	10.4013
4	Propanone	6.5225
5	Pyridine	4.3329
6	Pyrrole	4.1934
7	Chloral hydrate	2.9799
8	n-Cyclohexyl-2-pyrrolidone	2.1431
9	Aniline	1.7467
10	Chloroform	1.6337
11	Diethylether	1.4813
12	Acetic acid	1.2209
13	Nitrobenzene	0.6988
14	Benzene	0.5339
15	Camphor	0.4953
16	Ethanol	0.4694
17	Liquid indigo	0.3500
18	1-Pentanol	0.1523
19	1-Octanol	0.0742
20	Hexane	0.0140
21	*n*-Hexadecane	0.0037
22	Water	0.0001

3.34 kcal mol^{-1}. From this, the solubility of solid indigo in a virtually liquid indigo, that is, the ideal solubility, can be estimated to be 0.35%. This implies that according to the HSP method, no other solvent should be better than that, because in the HSP method, a solvent with identical solubility parameters to those of the solute should be optimal.

As can be seen in Table 6.1, COSMO-RS indeed predicts higher solubility than 0.35% in 16 of the 22 solvents. The best solvent found is sulfuric acid, in which COSMO-RS predicts complete miscibility, and which is also technically used as solvent for indigo. The reason for the high solubility of indigo in sulfuric acid can be explained on the basis of the σ-profiles of the two compounds which are displayed in Figure 6.5. The σ-profile of indigo is dominated by a huge double peak in the less polar region mainly resulting from the two phenyl rings. In addition, there is a strong one in the range of 1.5 e nm^{-2} arising from the lone pair regions of the two carbonyl oxygens, and a donor peak at about the opposite position arising from the two N-H donors, but the height of this donor peak, that is, the amount of hb-donor surface area, is only about one third of the height of the hb-acceptor peak, because the donor is sterically not well accessible in indigo. As a result, indigo can form about four hydrogen bonds per molecule in pure indigo, two as a donor, and two as an acceptor, but more than half of the hb-acceptor capacity of indigo stays unused.

The σ-profile of sulfuric acid shows a very polar hb-donor peak reaching out as far as to $-2.5\,e\,nm^{-2}$, resulting from the very acidic and thus very polar hydrogen atoms, see also figure 6.5. On the other side the acceptor peak resulting from the oxygen lone pairs is centered at only $0.9\,e\,nm^{-2}$, identifying the sulfuric acid oxygen atoms as very weak hb-acceptors. Hence, in pure sulfuric acid, the very strong donors only find very weak acceptors as partners. As a result, if we mix sulfuric acid with indigo, the sulfuric acid donors can instead form hydrogen bonds with the excessively available, moderately polar hydrogen acceptors of indigo, leading to a large gain in energy and thus to high solubility. A similar indigo solubilization mechanism is given for phenol, pyrrole, chloral hydrate, and chloroform. Phenol also has a quite polar hb-donor and less polar hb-acceptors than indigo, leading to favorable phenol–indigo hydrogen bonds, and pyrrole has a moderately strong N–H donor, but no acceptors, thus leading to additional hydrogen bonds in a mixture with indigo. Similar arguments count for chloral hydrate and chloroform.

Interestingly, the indigo solubilization mechanism is quite opposite for the other solvents that were in the top 10 of our indigo solvent screening. Dimethylsulfoxide, propanone, pyridine, *N*-cyclohexyl-2-pyrrolidone, and aniline, all have stronger hb-acceptors than indigo, while having no, or only very weak, hb-donors on their own. Hence, in mixtures of these solvents with indigo, the hydrogen bonds between indigo donors and acceptors can be replaced by stronger hydrogen bonds between the indigo donors and the solvent acceptors, leading to an energy gain and thus to solubilization (Figure 6.5).

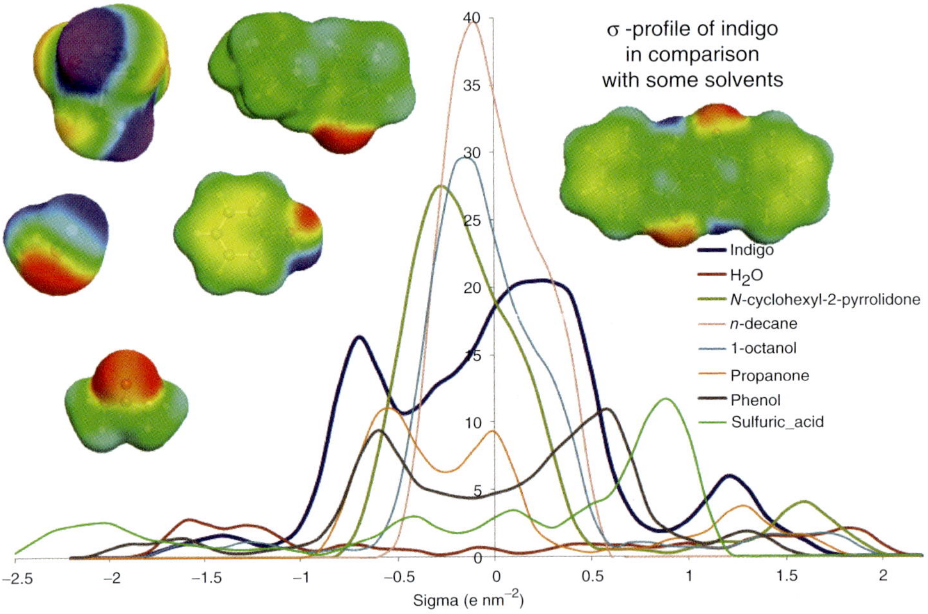

Figure 6.5 Some σ-profiles for the indigo solvent screening example.

Even the solubility behavior in mixtures can be predicted and rationalized with COSMO-RS. It is found that water acts as a strong antisolvent for indigo in *N*-cyclohexyl-2-pyrrolidone. Even small amounts of water cause massive fallout of indigo in this solvent. In agreement with this observation, COSMO-RS predicts a reduction of indigo solubility by 1 order of magnitude when 10% water is added to *N*-cyclohexyl-2-pyrrolidone. The reason of the reduction is also obvious: While in pure *N*-cyclohexyl-2-pyrrolidone, the stronger hydrogen bonds between indigo and *N*-cyclohexyl-2-pyrrolidone are the driving force for the solubilization, upon the addition of water, the more polar donors of water start to compete and cast the less polar donors of indigo out of these strong hydrogen bonds, thus distorting the driving force for solubilization of indigo in *N*-cyclohexyl-2-pyrrolidone.

Summarizing, this indigo example vividly demonstrates the power of the COSMO-RS for the rationalization and prediction of solubility.

6.3 Solvent Selection for Green Solid–Liquid Extraction

Solvent selection is usually based on solubility parameters and the polarity of the target component and solvent. Using a clustering of similar substance groups, relevant solvents may initially be selected, see Figure 6.6. For industrial applications, often the 1-parameter model from Hildebrand is used, because the required parameters of each component for multiparameter models are typically missing.

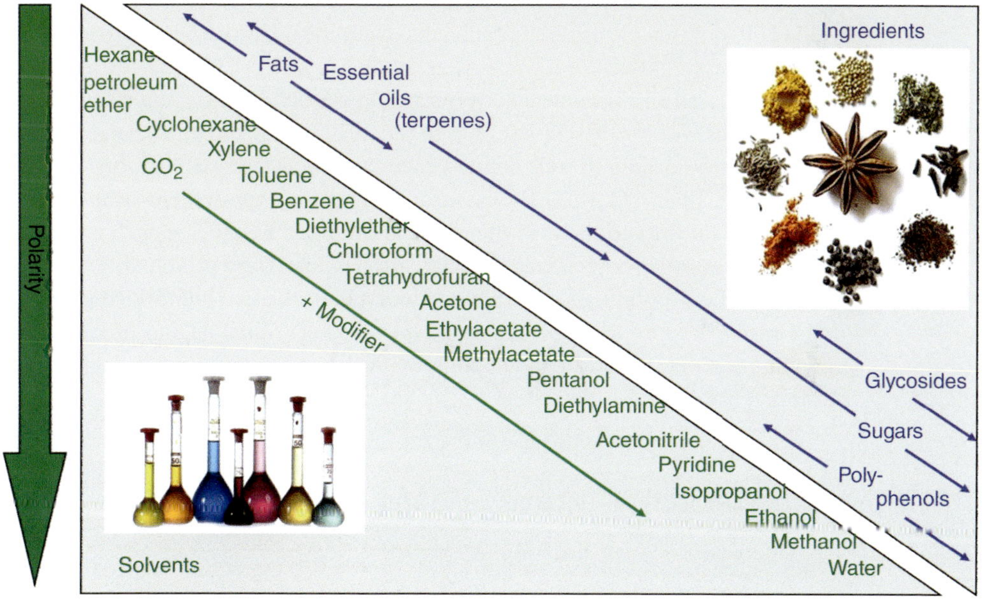

Figure 6.6 Solvent selection [18].

Another option is to assess the solubility via the activity coefficient, which depends on the composition of the solvent and is often very difficult to determine experimentally, because of the complexity of the solution. Besides the extraction yield, the selectivity of a solvent for the target compared to side components is an important factor, as this will have a paramount influence on the further purification steps. Clearly, regulatory constraints and consumer demands especially for pharmaceutical and food applications have to be integrated into the selection process as well, which limits further the number of applicable solvents drastically.

Nevertheless, there are modern thermodynamic models available now that can take into account molecular interactions in complex liquids. As explained in Section 6.2.3, the probably most powerful one of these *in silico* approaches is COSMO-RS. In the next section, it is applied to the classification of green solvents.

But even having such excellent tools, it should always be kept in mind that especially in plant extraction, the substrate is a complex mixture that never can be modeled in full detail. Numerous side products are usually present, and they may influence the activity coefficients of the extract and extraction solvent decisively. In this case, proper extraction predictions can be expected neither from COSMO-RS nor from any other method.

6.3.1
General Green Solvent Ranking with COSMO-RS

In 2012, Aubry and coworkers published a landmark paper on green chemistry, entitled "Panorama of sustainable solvents using the COSMO-RS approach" [19]. This work is an extension of an earlier study, in which COSMO-RS had been applied to classify 153 classical solvents and to subdivide them in a rigorous and unambiguous way into 10 classes [20]. In the 2012 paper, 138 green solvents have been chosen according to well-defined criteria such as flammability, toxicity, and HSE profile. Then COSMO-RS was used to compare the solvent properties of these green solvents to those of the classical ones and to allocate each green solvent to 1 of the 10 solvent classes. The allocation is based on a principal component analysis of the σ-potentials, as explained in detail in the two mentioned papers.

The result is given in Table 6.2. This table is a very valuable and unique starting point for any industrial application, for which more sustainable solvents are desired.

6.3.2
Concrete Example: Solid–Liquid Extraction of Carnesol and Carnosic Acid from Sage

As described in the previous sections, the choice of an extraction method is essential both for the design of solid–liquid extraction and subsequent purification of the product. In this section, the focus is on solid–liquid extraction.

Table 6.2 Classification of 138 green solvents into 10 classes [19].

NAME	I_{tox}	VOC	Fire	CAS
Cluster II: Weak electron pair donor bases				
Acetone	5	+	*R11*	67-54-1
N,N-Dimethyloctanamide	4	–	–	1118-92-9
Methyl 5-(dimethylamino)-2-methyl-oxopentanoate	**4**	**–**	**–**	**1174627-68-9**
2-Pyrrolidone	**5**	**–**	**–**	**616-45-5**
Cluster III: Aprotic dipolar				
Acetyltributyl citrate	5	–	–	77-90-4
Benzyl benzoate	**4**	**–**	**–**	**120-51-4**
Butylacetate	4	+	R10	123-86-4
Butyl laurate	5	–	–	106-18-3
1,4-Cineol	4	+	R10	470-67-7
1,8-Cineol	4	+	R10	470-82-6
Cyclopentyl methyl ether	**4**	**+**	**R11**	**5614-37-9**
Dibutyl sebacate	**5**	**–**	**–**	**109-43-3**
Diethyl adipate	**5**	**–**	**–**	**141-28-6**
Diethyl glutarate	**4**	**–**	**–**	**818-38-2**
Diethyl phthalate	**5**	**–**	**–**	**84-66-2**
Diethyl succinate	**5**	**+**	**–**	**123-25-1**
Diisoamylsuccinate	n.f	–	–	818-04-2
Diisobutyl adipate	**5**	**–**	**–**	**141-04-8**
Diisobutyl glutarate	**5**	**–**	**–**	**71195-64-7**
Diisobutyl succinate	**5**	**–**	**–**	**925-06-4**
Diisooctylsuccinate	n.f	–	–	2915-57-3
Dimethyl adipate	**4**	**–**	**–**	**627-93-0**
Dimethyl glutarate	**5**	**+**	**–**	**1119-40-0**
Dimethyl phthalate	**5**	**–**	–	**131-11-3**
Dimethyl succinate	**5**	**+**	**–**	**106-65-0**
N,N-Dimethyldecanamide	4	–	–	14433-76-2
Dimethylisosorbide	5	–	–	5306-85-4
Dioctylsuccinate	n.f	–	–	14491-66-8
1,3-Dioxolane	**4**	**+**	**R11**	**646-06-0**
Ethyl acetate	**5**	**+**	***R11***	**141-78-0**
Ethyl laurate	5	–	–	106-33-2
Ethyl linoleate	5	–	–	544-35-4
Ethyl linoleate	6	–	–	1191-41-9
Ethyl myristate	5	–	–	124-06-9
Geranyl acetate	5	–	–	105-87-3
Glycerol triacetate	4	–	–	102-76-1
Glycerol-1,2,3-tributyl ether	n.f	–	–	131570-26-1
Glycerol-1,2,3-triethyl ether	5	+	R10	162614-45-1
Glycerol-1,2,3-trimethyl ether	4	+	R10	20637-49-4
Glycerol-1,3-dibutyl ether	4	–	–	2216-77-5
Isoamyl acetate	6	+	R10	123-92-2
Isobutyl acetate	**5**	**+**	**R11**	**110-19-0**
Isopropyl acetate	5	–	R11	108-21-4
Isopropyl myristate	6	–	–	110-27-0
Isosorbide dioctanoate	5	–	–	64893-70-4
Methyl abietate	5	–	–	127-25-3
Methyl acetate	**5**	**+**	**R11**	**79-20-9**
Methyl laurate	4	–	–	111-82-0
Cluster III: Aprotic dipolar (continued)				
Methyl linoleate	5	–	–	112-63-0
Methyl linolenate	5	–	–	301-00-8
Methyl myristate	5	–	–	124-10-7
Methyl oleate	4	–	–	112-62-9
Methyl palmitate	4	–	–	112-39-0
Dimethyl2-methyl glutarate	**4**	**–**	**–**	**14035-94-0**
2-Methyltetrahydrofuran	5	+	R11	96-47-9
Menthanyl acetate	5	+	–	58985-18-5
n-Propyl acetate	**5**	**+**	**R11**	**109-60-4**
Terpineol acetate	5	n.f	–	8007-35-0
Tributyl citrate	5	–	–	77-94-1
Triethyl citrate	5	–	–	77-93-0
Cluster IV: Aprotic highly dipolar				
Dimethyl sulfoxide	**5**	**+**	**–**	**67-68-5**
2-Furfuraldehyde*	3	+	–	98-01-1
Propylene carbonate	**4**	**–**	**–**	**108-32-7**
γ-Valerolactone	5	+	–	108-29-2
Cluster V: Apolar				
Butyl myristate	5	–	–	110-36-1
Butyl palmitate	5	–	–	111-06-8
Butyl stearate	6	–	–	123-95-5
Cyclohexane	**5**	**+**	**R11**	**110-82-7**
p-Cymene	4	+	R10	99-87-6
β-Myrcene	5	+	R10	123-35-3
Decamethylcyclo-pentasiloxane	**4**	**+**	**–**	**541-02-6**
Dipropyleneglycol	**5**	**–**	**–**	**110-98-5**
Ethyl oleate	5	–	–	111-62-6
Ethyl palmitate	5	–	–	628-97-7
Isopropyl palmitate	5	–	–	142-91-6
d-Limonene	4	+	R10	5989-27-5
Methyl stearate	5	–	–	112-61-8
Isododecane	**5**	**–**	**–**	**31807-55-3**
Perfluorooctane	**3**	**+**	**–**	**307-34-6**
α-Pinene	4	+	R10	80-56-8
β-Pinene	4	+	R10	127-91-3
Terpinolene	4	+	–	586-62-9
Cluster VII: Amphiprotic				
Benzyl alcohol	**4**	**+**	**–**	**100-51-6**
1-Butanol	4	+	R10	71-36-3
Cyclademol	5	–	–	25225-09-6
1-Decanol	4	–	–	112-30-1
Dihydromyrcenol	4	+	–	18479-58-8
1,3-Dioxolane-4-methanol	5	–	–	5660-53-7
Ethanol	5	+	R11	64-47-5
Ethylhexyllactate	n.f			6283-86-9
Ethyl lactate	5	+	R10	97-64-3
Geraniol	4	–	–	106-24-1
Glycerol-1,3-diethyl ether	4	–	–	4043-59-8
Cluster VII: Amphiprotic (continued)				
Glycerol-1,2-dibutyl ether	n.f	–	–	91337-36-9
Glycerol-1,2-dietyl ether	n.f	–	R10	1756-20-1
Glycerol-1,2-dimethyl ether	n.f	+	R10	40453-77-8
Glycerol-1,3-dimethyl ether	5	+	R10	623-69-8
Glycerol-1-butyl monoether	5	–	–	624-52-2
Glycerol-1-ethyl monoether	4	–	–	1874-62-0
Glycerol-2-butyl monoether	n.f	–	–	100078-36-2
Glycerol-2-ethyl monoether	n.f		R10	22598-16-9
Glycofurol (n=2)	**5**	**–**	**–**	**52814-38-7**
N,N-Diethylolcapramide	5	–	–	136-25-5
Caprylic acid diethanolamide	5	–	–	3077-30-3
Isoamyl alcohol	5	+	R10	123-51-3
Isopropyl alcohol	5	+	R11	67-63-0
Methyl ricinoleate	5	–	–	141-24-2
Menthanol	5	–	–	498-81-7
Nopol	4	–	–	128-50-7
1-Octanol	4	+	–	111-87-5
Oleic acid	6	–	–	112-80-1
Oleyl alcohol	5	–	–	143-28-2
PolyEthyleneGlycol 600	**6**	**–**	**–**	**25322-68-3**
Solketal	5	+	–	100-79-8
Ricinoleic acid	5	–	–	141-22-0
α-Terpineol	4	–	–	98-55-5
β-Terpineol	5	–	–	138-87-4
Tetrahydrofurfurylic alcohol	4	+	–	97-99-4
Cluster VIII: Polar protic				
1,3-Dioxan-5-ol	5	+	–	4740-78-7
1,3-*Dioxolane-4-methanol*	5	+	–	5464-28-8
Ethylene glycol	4	+	–	107-21-1
β-Farnesene	5	–	–	18794-84-8
Furfurylic alcohol*	3	+	–	98-00-0
Glycerol	5	–	–	56-81-5
Glycerol carbonate	5	–	–	931-40-8
Glycerol-1methyl monoether	4	–	–	923-39-2
Glycerol-2-methyl monoether	5	–	R10	761-06-8
5-(Hydroxymethyl)furfural	4	–	–	67-47-0
3-Hydroxypropionic acid	5	–	–	503-66-2
3-Methoxy-3-methyl-1-butanol	**4**	**–**	**–**	**56539-66-3**
PolyEthyleneGlycol 200	**6**	**–**	**–**	**112-60-7**
1,3-Propanediol	4	–	–	504-63-2
Propylene glycol	6	+	–	57-55-6
Cluster IX: Organic acidic compounds				
Acetic acid	4	+	R10	64-19-7
Propionic acid	4	+	–	79-09-4
Cluster X: Polar structured				
Water	6	–	–	7732-18-5
Ionic liquids				
Choline acetate	n.f	–	–	14586-35-7
3Butyl-1-methylinidazolium tetrafluoroborate	**4**	**–**	**–**	**174501-65-6**

To reduce the experimental effort, solvents may be classified in terms of solubility using thermodynamic models such COSMO-RS. Here, the general applicability of COSMO-RS for solvent screening is shown. The knowledge gained can be transferred to green solvents and extraction methods. As an example system, the solid–liquid extraction of carnosol and carnosic acid from sage is selected.

Sage (*Salvia officinalis*) belongs to the family of Lamiaceae [21]. The plant acts as a strong antioxidant because of the radical-scavenging properties of the target compounds carnosic acid, carnosol, and rosmarinic acid. In addition, carnosic acid possesses antimicrobial properties. Therefore, sage extracts are often used to treat infections of oral and pharynx mucosa or gingivitis. Furthermore, it contains a number of side products, such as thujone, camphor, cineol, and oleanolic acid, with thujone and camphor being mainly responsible for the typical aromatic smell of sage leaves [22].

The valuable substances occurring in sage can be allocated to different chemical classes such as di- and triterpenes, or flavonoids. Carnosic acid and carnosol are diterpenes, whereas thujone, camphor, and cineol represent monoterpenes.

The calculation of solubility using the COSMO-RS model has been explained in detail in Section 6.2.3. Only the application to a complex sample system will be shown in this section. Owing to the complexity of the molecules, it is not possible to calculate absolute values for the solubility. These differ in other physicochemical property data significantly from reality, as shown in Chapter 5. Nevertheless thermodynamic models such as COSMO-RS can be used for performing a classification of the solvents. By means of clustering of similar substance groups, relevant solvents may initially be selected (Figure 6.6). The clustering of substance groups can also be performed using COSMO-RS (Figure 6.7).

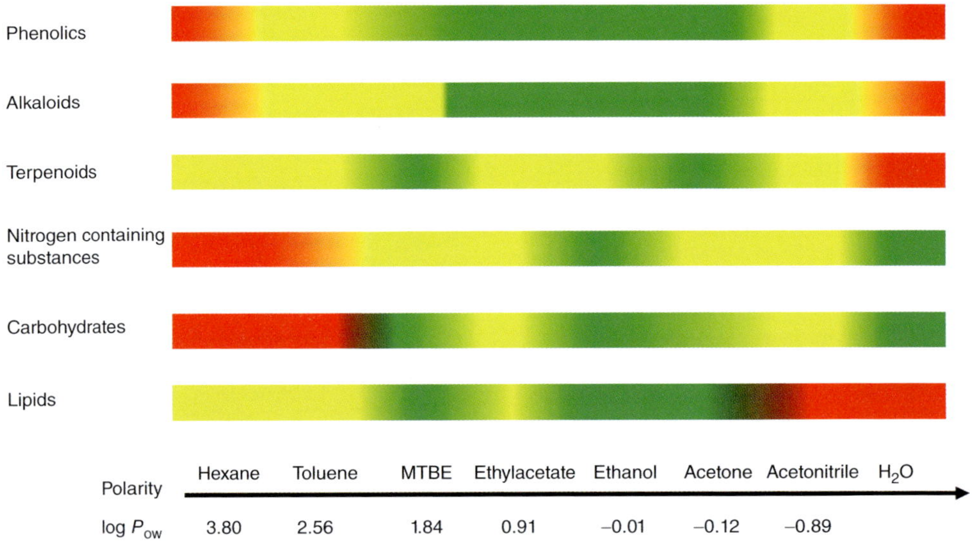

Figure 6.7 Clustering of substance groups with COSMO-RS: green = best solvent, yellow = moderate solvent, and red = inadequate solvent.

Table 6.3 COSMO-RS ranking of solvents for the example system salvia.

	Main components		Side components		
Solvent ranking	**Carnosol**	**Carnosic acid**	**Thujone**	**Camphor**	**Eucalyptol**
1	Propanone	Propanone	Propanone	Propanone	Hexane
2	Ethylacetate	Ethylacetate	Ethylacetate	Ethylacetate	Propanone
3	Ethanol	Ethanol	Hexane	Ethanol	Ethylacetate
4	Hexane	Hexane	Ethanol	Hexane	Ethanol
5	Water	Water	Water	Water	Water

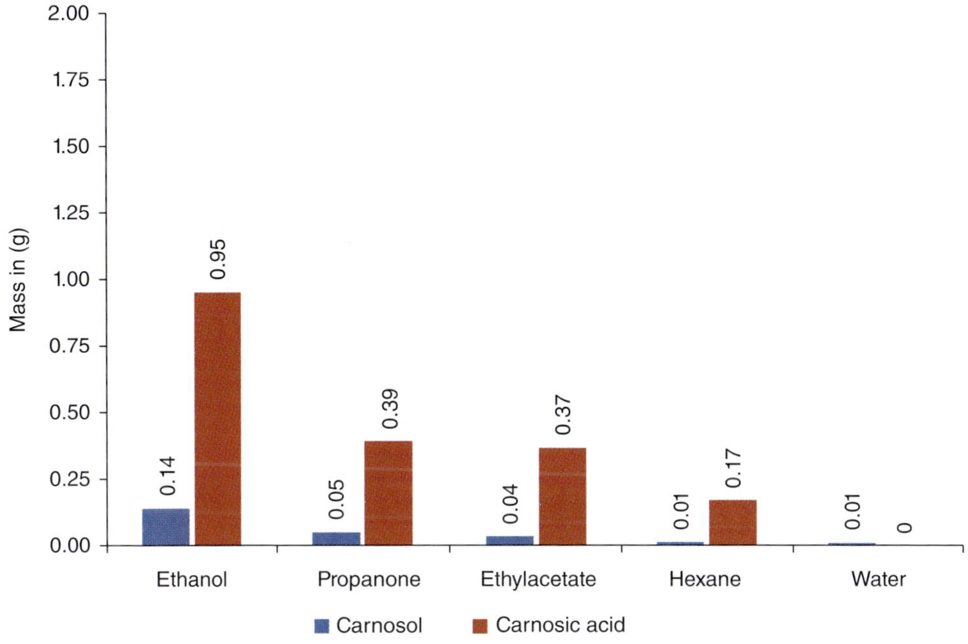

Figure 6.8 Amount of carnosol and carnosic acid after extraction.

With the aid of these clustering methods, industrially important solvents, such as ethanol, ethyl acetate, acetone, hexane, and water are selected for the solid–liquid extraction of carnosol and carnosic acid. It is known that water is not a suitable solvent for the extraction of terpenoids. Nevertheless, it is screened here in order to illustrate the applicability of COSMO-RS.

Table 6.3 shows the results of the solvent rankings. Acetone is well suited to solve the target components carnosol and carnosic acid and the side components thujone and camphor compared to other solvents. For eucalyptol, hexane is the best solvent.

6.3.3 Experimental Validation of COSMO-RS Solvent Ranking

In order to validate the solvent ranking using COSMO-RS, extraction experiments were carried out with sage. In each case, 10 g sage is extracted for 24 h with the respective solvents. Figure 6.8 presents the results of these extraction experiments.

The experimental data concerning the worst solvents hexane and water for the extraction of carnosol and carnosic acid agree with the COSMO-RS ranking (Figure 6.8).

Carnosol and carnosic acid are more soluble in ethanol than in acetone and ethyl acetate according to the COSMO-RS calculations. The experimental data do not reflect these results. More carnosol and carnosic acid could be extracted with ethanol than with acetone and ethyl acetate. The reason for this could be that in solid–liquid extraction processes not only the solubility of one substance but also the presence of many other substances and many other factors play a role.

As described in Chapter 2, the accessibility of the product, which is a function of grinding degree and hence the particle size and porosity, is an important factor in solid–liquid extraction processes. Other important factors are the localization of the substances in plant material, the structure of the raw material, and the humidity as key factors influencing the equilibrium state. If the solid matrix is polar but the components nonpolar, then the accessibility of the target components must first be improved with a suitable polar solvent. The use of a nonpolar solvent without previous swelling of the matrix would lead to poor yields. These factors cannot be taken into account in the COSMO-RS calculations. Therefore, the experimental results show that carnosol and carnosic acid have a much better accessibility for ethanol than for acetone and ethyl acetate.

The influence of accessibility can also be illustrated using the example of the extraction of side components (Figure 6.9). Here, the COSMO-RS ranking differs from the experimental values. Only for water as the least suitable solvent, the experimental data agree with the calculated values.

The side components thujone, camphor, and eucalyptol are volatile components of the essential oil of sage and have a very good solubility in solvents such as hexane or acetone. Here, the COSMO-RS calculations are consistent with the reality. However, owing to the swelling behavior, the availability of these substances for ethanol is better than that of hexane or acetone. Thus, higher yields are obtained with ethanol (Figure 6.9).

6.3.4 Conclusion

In this section, it was shown how a thermodynamic model such as COSMO-RS can be used to perform solvent rankings in order to reduce the tremendous cost of a complete experimental solvent screening. It could be shown that the COSMO-RS ranking shows the right trends concerning the application of available solvents

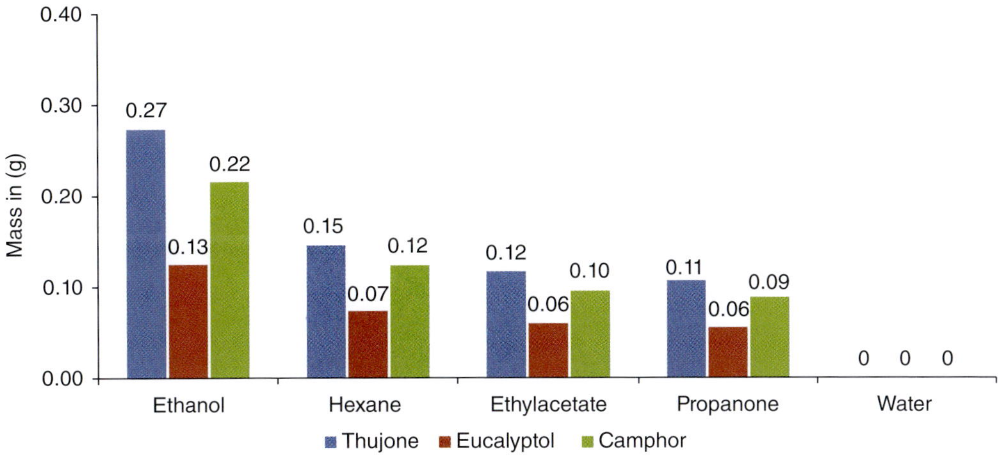

Figure 6.9 Amount of thujone, eucalyptol, and camphor after extraction.

Table 6.4 COSMO-RS solvent ranking: green = good match with experimental data; yellow = possible match with experimental data; and red = no match with experimental data.

	Main components		**side components**		
Solvent ranking	**carnosol**	**carnosic acid**	**thujone**	**camphor**	**eucalyptol**
1	propanone	propanone	propanone	propanone	hexane
2	ethylacetate	ethylacetate	ethylacetate	ethylacetate	propanone
3	ethanol	ethanol	hexane	ethanol	ethylacetate
4	hexane	hexane	ethanol	hexane	ethanol
5	water	water	water	water	water

for extraction of natural products. Thus it is possible to identify the best from a variety of solvents.

However, these reduced solvents must be examined subsequently experimentally, because the calculation of the solubility alone is not sufficient to draw a conclusion on yield in solid–liquid extraction processes (Table 6.4).

The reason is that in solid–liquid extraction processes, the accessibility and the localization of the substances in the plant part, the structure of the raw material and the humidity are critical factors, which are of course not considered by thermodynamic models such as COSMO-RS. Those effects are taken into account by physicochemical modeling, as explained in Chapter 2.

6.4 Alternative Solvents for Green Extraction

6.4.1 Ionic Liquids

In the context of green solvents, ionic liquids (ILs) and, more recently, the so-called DESs are discussed in numerous papers. According to the definition, ionic liquids are pure molten salts with low melting points, usually (and arbitrarily defined) below 100 °C or so. DESs are mixtures of solids that have melting points far below the melting points of each compound, usually also below 100 °C.

ILs are considered to be green mainly because their vapor pressure is very small, often negligible, and hence they are by no means volatile organic compounds. On the other hand, this characteristic excludes their separation from the product by solvent distillation. Many ILs are stable over wide temperature ranges, and as an almost unlimited amount of cation–anion combination seems to be possible, one could expect that for nearly all industrial processes, a proper IL can be found that is better suited than conventional solvents.

Despite these promising properties, there are still only few industrial processes that involve ILs. The most prominent one is still BASF's BASIL process, in which the generated HCl is transformed to an IL (with the help of 1-methylimidazol) instead of a solid-precipitating salt. The advantage is that the IL, as a liquid, can more easily be handled and separated than a solid salt. The catalytic activity of the base is an appreciable side effect.

Concerning separation processes, there are several attempts to integrate ILs, although to our knowledge, none of these concepts have already been upscaled to an industrial level. Among these approaches, we mention their role as potential additives to separate azeotropic mixtures [23] or the use of an IL as solvent in combination with a crown ether to efficiently extract different types of amino acids from acidic aqueous solutions stemming from pharmaceutical samples or fermentation broth [24]. The latter example is a rather exotic one, but it demonstrates that usually ILs are only suitable for the extraction of very pricey products or as extraction solvents in analytical chemistry. In addition, simpler molecules such as butanol [25] or monoethylene glycol [26] have been considered, but the price of the process may still be prohibitive. Among others, Carda-Broch *et al.* determined experimentally the partition coefficient of a significant number of substances between a particular IL and water [27].

A recent review paper, "Application of ionic liquid for extraction and separation of bioactive compounds from plants" [28] gives a nice overview of this topic. The choice of the journal already shows that only analytical applications are targeted. Combinations of ILs with supercritical CO_2, a truly green solvent, have also been proposed, for example, to extract organic solutes from ILs with supercritical CO_2 [29, 30]. In addition, ILs can be induced to separate from organic and aqueous mixtures by applying gaseous CO_2 pressure, an original way to create a different extraction phase [31]. Of course, the ILs should always be commercially

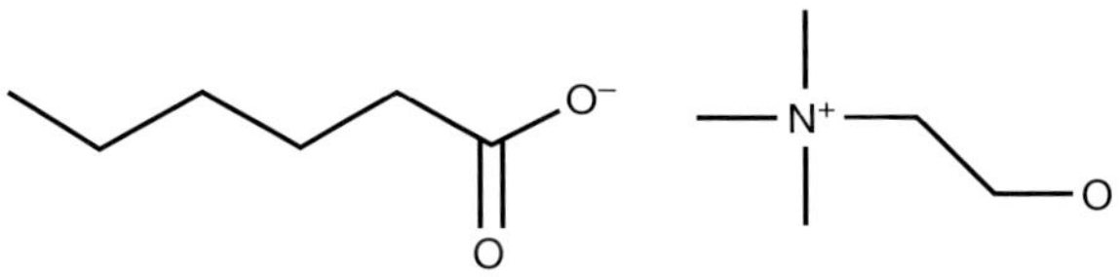

Figure 6.10 Structure of choline hexanoate.

available, relatively cheap, and of low toxicity, such as AMMOENG™ 102 [32], an amphiphilic tetraalkylammonium ethylsulfate.

Except for catalysis [33] and for very special applications, such as in electrochemistry or as functional media in sensor design, ILs do not (yet) seem to be suitable for broad industrial processes. Especially in the case of plant extraction, their use is up to now limited to analytical processes. What are the reasons and is there any hope to see ILs as green solvents for plant extraction in the future?

One major reason is the price of sufficiently pure ILs. While the synthesis of ILs can be quite simple, the purification processes are far from trivial. As a consequence, often ILs still containing a significant amount of by-products are sold, or, if high purity is required, the high cost of the ILs excludes their competitivity with conventional solvents. Another drawback is the fact that they cannot directly replace other solvents, because of the low vapor pressure or the lack thereof. Instead of distillation, alternative separation processes must be designed, which is time-demanding and expensive. Further, ILs nearly always have a high viscosity at room temperature, making their use in extraction processes complicated and also slowing down diffusion processes.

Even if these drawbacks are overcome, there is another major deficiency in most ILs: they are simply not "green" or have only very few green aspects (such as low vapor pressure). Most of the investigated ILs are still imidazolium-based ones, such as [bmim]BF_4 (butyl-methyl-imidazolium tetrafluoroborate), and their synthesis is neither green nor mainly or entirely based on renewable resources [3]. What is more, short-chain imidazolium-based ILs are not readily biodegradable, whereas long-chain ones show a significant toxicity [34, 35].

So, what is the next step? From what has been discussed, it is clear that ILs in plant extraction have a future only if they are nontoxic, relatively pure, and not expensive. A substance class that could fulfill these criteria are choline-based ILs, such as choline butanoate or choline hexanoate (Figure 6.10).

Choline, formerly known as Vitamin B_4, is an essential nutrient for mammals, present in most foods and a popular food additive. It is produced in very high quantity and hence easily available. The same is true for hexanoic acid. It can be easily combined with choline hydroxide to give the desired green IL. It is nontoxic and potentially available in hundreds of tons at a competitive price. The disadvantages of choline-based ILs are their relatively low temperature stability (choline salts should not be heated above 80 °C for a long time) and that they are forbidden in cosmetics, simply for regulation reasons and without real scientific reason.

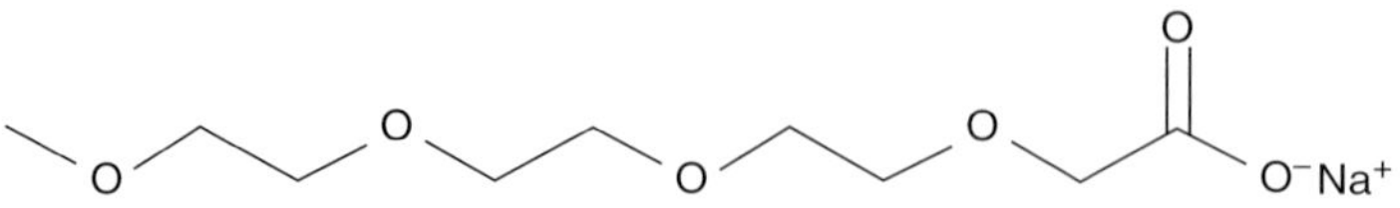

Figure 6.11 Structure of a TOTO salt (sodium 2,5,8,11-tetraoxatridecan-13-oate (TOTO)).

Recently, such choline salts have been successfully used to extract for the first time the important biopolymer suberin from cork without destroying its structure [36].

According to some preliminary and unpublished results, such choline salts are highly specific extraction agents permitting one to extract very selective components from plants. Varying the chain length of the anion allows for a fine-tuning of the selectivity.

Although not entirely green, there is another substance class of ILs that might be promising: the so-called TOTO (sodium 2,5,8,11-tetraoxatridecan-13-oate) salts [37, 38]. In contrast to all other known ILs, their low melting point comes from a high flexibility in the anion chain, whereas the cation can be as simple as a sodium ion (Figure 6.11). The TOTO ions with different alkyl-chain lengths are known as Akypos™ and commercially available from Kao Chemicals at competitive prices.

It can be concluded that only green ILs should be considered for plant extraction, but even so, they have to fulfill further criteria, such as being available at a reasonable price and being part of an optimized extraction process that allows one to recover the IL without its distillation.

6.4.2 Low-Transition-Temperature Mixtures and Deep Eutectic Solvents

Ten years ago, Abbott and coworkers reported about so-called DES [39]. They discovered that liquids with melting points below 100 °C can be obtained by mixing two solids that form an appropriate deep eutectic. The classical example was the mixture of choline chloride (melting point 302 °C) with urea (melting point 133 °C) resulting in a DES with a melting point of only 12 °C at a molar ratio of 1 : 2. In the meantime, his group and many others extended this concept to numerous other combinations including sugars, fatty acids, amino acids, and so on. More generally, these mixtures are called *low-transition-temperature mixtures* (*LTTMs*). The current state of art in this field is summarized in a recent review paper, where they are discussed as promising alternatives to ILs [40]. Indeed, further to their green properties (origin, biocompatibility, biodegradability, nonflammability, etc.), the easy way of producing them by simple mixing without synthesis and further purification steps is a major advantage compared to ILs.

Another recent review paper gives an overview of application domains [41]. Especially for biomass processes and bioconversions including enzymatic reactions, green LTTMs seem to be very promising. However, for separation processes, some drawbacks must not be ignored. One is the extraordinarily high viscosity at room temperature, which often requires the addition of another solvent, usually water. Further, because of the intrinsic properties of LTTMs, it seems that no hydrophobic, water-insoluble mixtures are conceivable and, as a result, liquid–liquid extraction is not possible. Finally, it should be kept in mind that LTTMs are always mixtures of at least two components thus rendering their separation and recovery more demanding.

6.4.3 Ionic Liquids Screening with COSMO-RS

Considering the remarkable increase in ILs research over the last 20 years, it is not astonishing that COSMO-RS has also been frequently applied to predict solubilities of substances in these media. For example, Kahlen *et al.* [42] used COSMO-RS to predict the solubility of cellulose in a combination of 32 anions and 71 cations resulting in 2272(!) potential ILs. Their analysis is based on following equation:

$$x_i^L = \exp\left[\frac{\Delta h_{m,i}}{RT}\cdot\left(\frac{T}{T_i^m}-1\right)\right]\cdot\frac{1}{\gamma_i^L} \tag{6.22}$$

where x_i^L is the mole fraction of the dissolved component in a saturated solution in equilibrium with solid of type i, $\Delta h_{\mathrm{m},i}$ is the enthalpy of melting of component i and T_i^{m} is its melting temperature. The specific interactions of i with the solvent L are given by its activity coefficient at saturation γ_i^L, which contains a combinatorial part and a residual part, which can be calculated by COSMO-RS. As can be seen from the preceding equation, the higher the dissolution is, the more negative is ln γ_i^L, that is, the lower is γ_i^L. To simplify the calculations, the authors determined γ_i^L at infinite dilution, which should reflect the main interactions and thus the solubilization power of the IL. The authors compared several of their predictions to experimentally determined values and they found an overall satisfactory correlation. Especially, the trends are at least qualitatively correctly predicted, which makes COSMO-RS a valuable tool for the estimation of cellulose solubilities in ILs.

As stated in the preceding subsection, the use of ILs for extraction processes has some serious limitations. One way to overcome them is the application of mixtures with other solvents such as alcohols. In this case, LLE and VLE data of such mixtures are important to know. COSMO-RS can also be used to predict them qualitatively, as shown by Freire *et al.* [43]. The same authors also considered IL–water mixtures [44]. However, it should be kept in mind that the description of long-range interactions is not possible in COSMO-RS without additional and adjustable parameters. As a result, the thermodynamic properties of salt solutions (nothing else are IL mixtures with solvents) in polar solvents and especially in water are out of the range of the model and corresponding results should be taken with a grain of salt.

On the other hand, the solubility of valuable compounds in IL is not only possible, but also of major interest, considering what was discussed previously. Guo *et al.* [45] performed an extensive calculation of flavonoid solubilities in more than 1800 ILs and also compared some of the predicted values to experimental results. This comparison is given in Figure 6.12 for the flavonoid esculin at 40

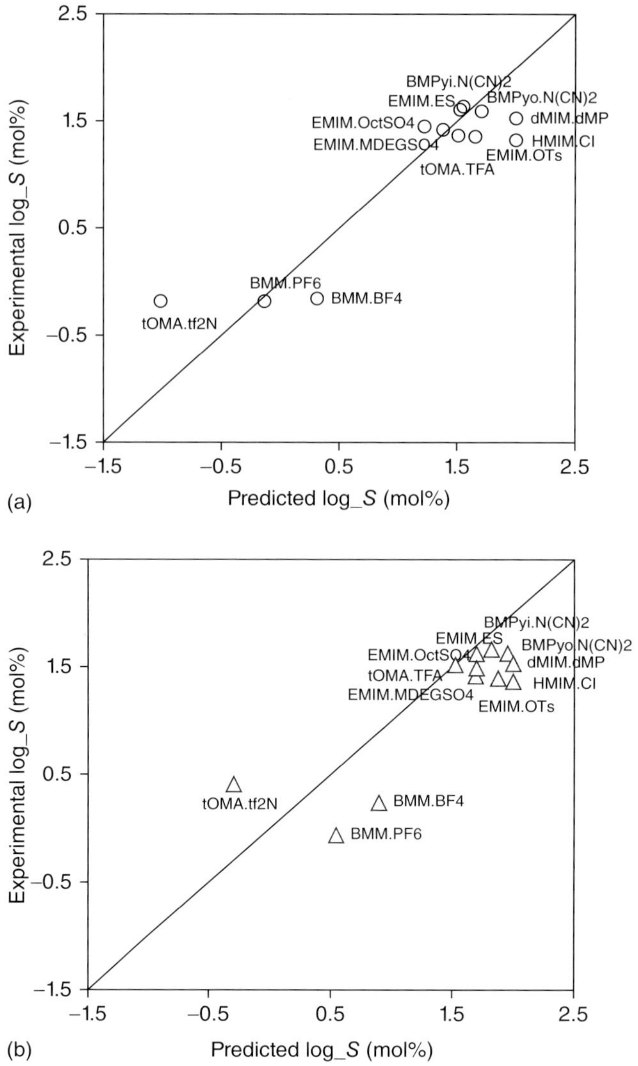

Figure 6.12 COSMO-RS predictions plotted versus experimental values of the solubility of esculin in 12 types of ILs at 313.15 K (a) and 333.15 K (b) according to [45].

and 60 °C. As can be seen, the prediction is satisfactory, so one can be confident that COSMO-RS permits a reasonable prescreening with reliable trends. A few years later, Xu *et al.* published an interesting paper, in which they presented a microwave-assisted extraction (MAE) of flavonoids with various ILs (all imidazolium based) [46] and concluded that "the IL-MAE approach is a green, simple, rapid, highly efficient extraction method for extracting important bioactive compounds from plant materials." However, in the light of the critical discussion in Section 6.4.1 this conclusion might be a little bit too optimistic.

Further to the description and prediction of solubilities in IL, the partitioning of chemical substances between two liquid phases is also of interest for extraction processes. For this purpose, COSMO-RS has also been applied. However, this is intrinsically very complicated. Not only must the LLE be properly described but also the solubility of the considered compound in each of the two phases in equilibrium. As a consequence, it is difficult to find a clear trend for partition coefficients of chemicals between ILs and other organic solvents with any approach and without adjusted parameters [47].

Concerning solubilities in ILs and further phase equilibria involving ILs, further information and data banks can be found on the homepage of COMOlogic (*www.cosmologic.de*).

6.5 Purification Strategies of Natural Products

Selective extractions of active pharmaceutical ingredients from plant-based materials have increased in recent years owing to the high consumer demand. The world trading volume of phytopharmaceuticals was approximately US $100 billion in 2011 [48]. In other areas, such as the cosmetic and food industries, growth rates of up to 6% for the coming years are expected [48]. The benefits of plant extracts compared to synthetic products are, in addition to public acceptance, the opportunity to obtain complex compounds economically. One example is the production of secondary metabolites, such asmono-, di-, and sesquiterpenes, which cannot be produced either synthetically or economically. While the previous sections have dealt with "green extraction processes," the focus in this section is on the purification of plant extracts. The subsequent purification of extracts is not carried out in a target-oriented manner and is mostly empirically driven with only few experimental examinations. To meet economic and environmental requirements for the production of herbal substances in the future, optimized processing is required. Therefore, this section focuses on the purification of plant extracts as representative for efficient handling of complex mixtures.

In general, the design and setup of industrial processes are divided into several development phases such as basic and detail engineering, equipment procurement, installation and start-up, and system operation [49]. The aim is to arrive at optimal process conditions at an early stage, as changes become more expensive the further development has progressed [49]. Therefore, an attempt is made using a conceptual process design (CPD) to arrive at an optimal process design for the separation sequence very early during process development, and in spite of low information content (Figure 6.13).

For this to happen, the process design has to be model based, because this is the only way to find the process optimum quickly and cost-effectively [18, 50]. Depending on the modeling depth, either physicochemical properties or only separation factors are needed. These can either be retrieved from databases (e.g.,

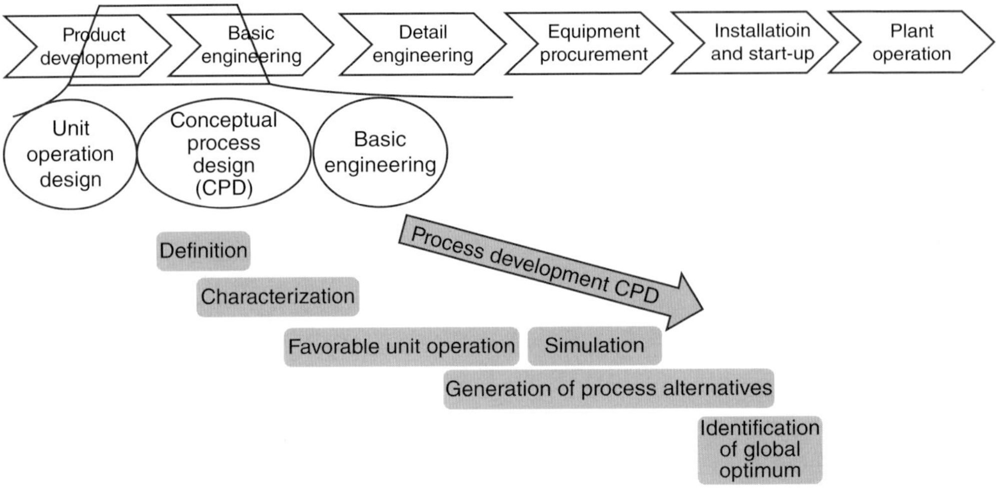

Figure 6.13 Conceptual process design.

Dortmunder Datenbank (DDB), DIPPR, or Reaxys), calculated by means of thermodynamic theories (UNIFAC and COSMO-RS), or determined experimentally.

This approach has already been established for multicomponent mixtures with small molecules and is considered state of the art [51–53]. In this work, it will be assessed how this methodology can be transferred to complex mixtures with the objective to generate an optimal separation sequence relatively early in process design. This task is complicated by the fact that those multicomponent mixtures consist of many compounds from a variety of compound classes, for which at least in part neither substance data nor molecular structures are known. Therefore, a systematic approach is to be developed for characterization and determination of the required physical properties for complex mixtures. This systematic process for determination of physical properties will be performed on technical sample systems from industry. The sample systems including yew needles, fennel fruits, sage leaves, vanillin pods, and black tea were chosen to represent several industries in an optimal way. The target component from yew needles (10-deacetylbaccatin III) provides the starting material for semisynthesis of the breast cancer drug taxol [54]. Fenchone and anethole are target components from the fennel fruit which are mainly used in the cosmetic and phytopharmaceutical industry. The most valuable substances in sage leaves are carnosol and carnosic acid, which are predominantly applied as antioxidants and preservatives in the food industry. Vanillin as a flavoring agent and black tea with different aromas and alkaloids such as caffeine are also used in the food industry.

The challenge in CPD of complex molecules lies in correct determination of the product parameters required for a qualified process simulation. Therefore, in line with a systematic protocol, the approach laid out in Figure 6.14 will be applied.

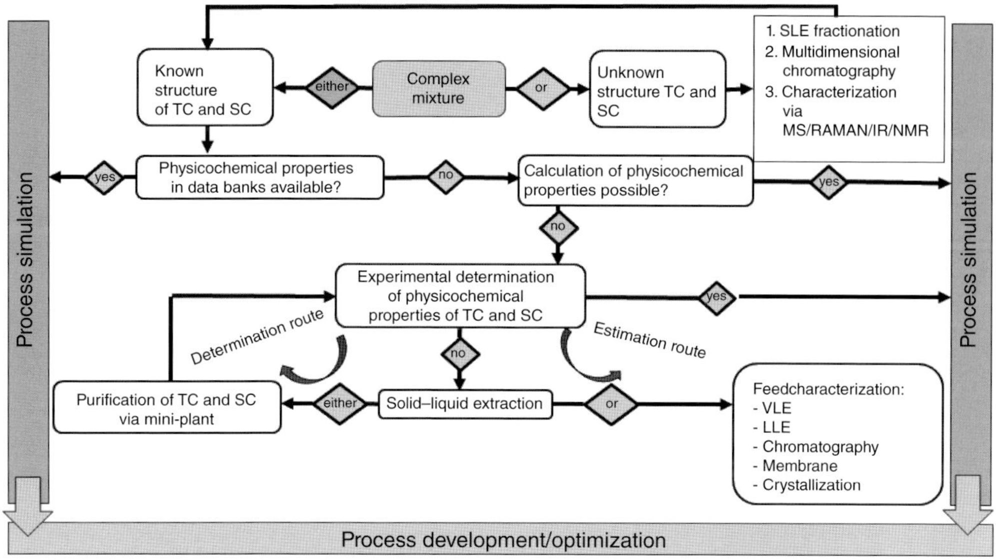

Figure 6.14 Process design of complex mixtures.

Data generation for the process simulation can be conducted in two ways:

1) Feed characterization
2) Product parameter determination of single compounds, where parameter determination is segmented into
   a. Database search
   b. Calculation of product parameters
   c. Experimental parameter measurements.

Feed characterization includes the measurement of characteristic values for the various unit operations, for example, boiling curves, distribution coefficients, and hydrophobicity for the complete complex mixture. It can be used to conduct thermodynamic equilibrium assessments of individual unit operations with relatively small amounts of starting material relatively quickly and effectively. For example, the boiling curve of a mixture is determined to assess the feasibility of a distillation, distribution coefficients are measured for a liquid–liquid extraction, chromatography requires hydrophobicities and isoelectric points, and crystallization or precipitation need solubility properties of mixtures, as can be taken from Table 6.5. It is not necessary to determine the equilibrium parameters for all unit operations, because it can be estimated on the basis of available product parameters as to which unit operations are at all thermodynamically feasible for product separation/purification. For example, the determination of the boiling curve is hardly necessary, if the desired compound is a solid. In the same way, solubility experiments for liquids are useless, as is the determination of partition coefficients for gases. This process is discussed in Section 6.5.2 in detail.

Table 6.5 Feed characterization.

Unit Operation	Equipment	Measuring parameters	
Distillation		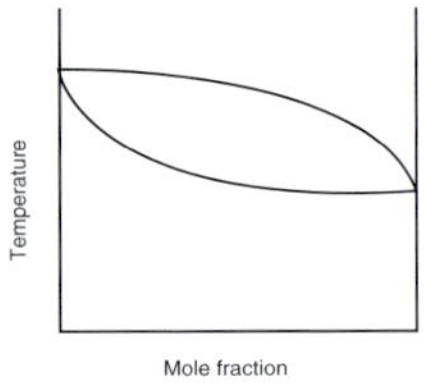	Boiling curve
Liquid-liquid Extraction	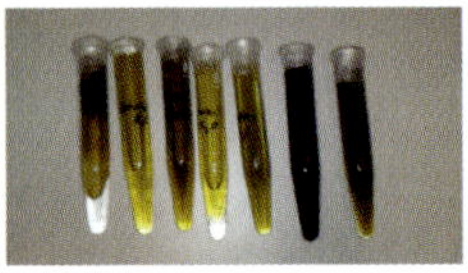	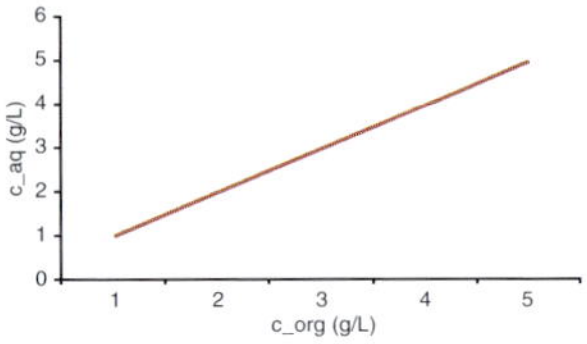	Partition Coefficient
Reversed/normal phase chromatography		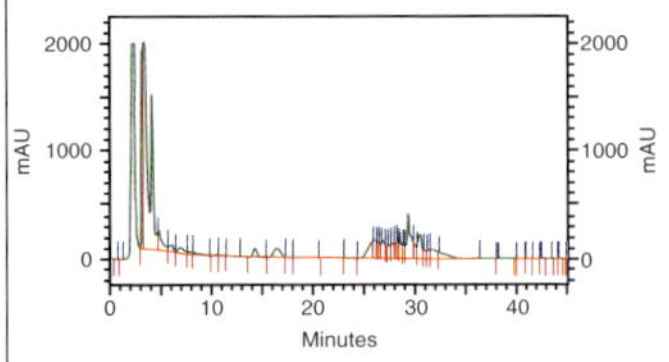	Hydrophobicity
Crystallization	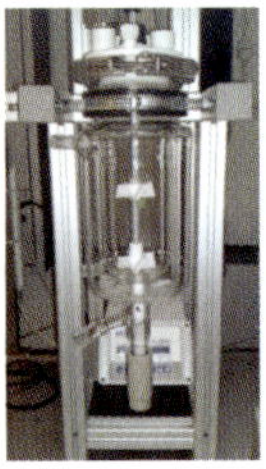	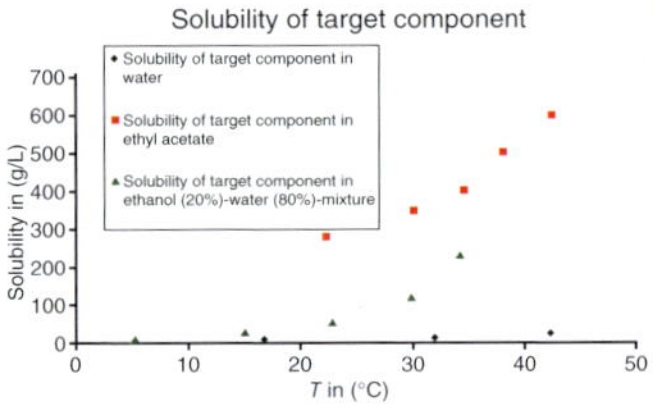	Solubility

For the *experimental parameter determination* of single components, it has to be checked first, whether the structures of the target and side components of the extract are known. If that is not the case, a qualified amount of extract has to be prepared via solid-phase extraction, which then allows to produce fractions of the pure compounds by multidimensional chromatography. In a first step, the extract is pre-purified on a preparative column. Following that, pure compounds can be isolated by further chromatographic fractionation, as is schematically depicted in Figure 6.15 [49, 55].

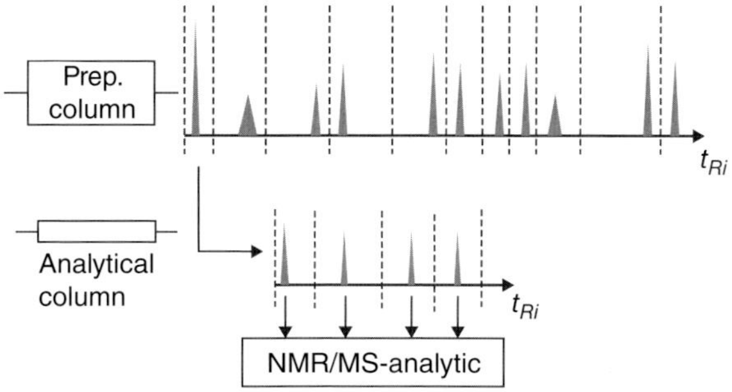

Figure 6.15 Multidimensional chromatography.

These fractions can then be structurally identified via MS, Raman, IR, and NMR spectroscopy [56].

In case the compounds are known, the accessible *databases* are searched for compound data. These can include databases such as Reaxys, DDB, or DIPPR [57–59]. If no compound data can be found, the next step is their calculation on the basis of available thermodynamic models. Typical examples used in this work are group contribution methods and the COSMO-RS-model [14]. A prerequisite for experimental determination of physicochemical properties is access to sufficient amounts of pure target and side compounds. Methods for determination of some important product parameters and required amounts are listed in Table 6.6.

The data in Table 6.6 show that up to 50 g of a pure compound may be necessary for the determination of the physicochemical properties. As many compounds are not commercially available at all or very expensive (>1000 €/kg), an option is presented herewith to generate the reference material from the respective raw material. The starting point in each case is the solid-phase extraction of the raw material with suitable solvents. From the extract, the required amounts

Table 6.6 Methods for determining physicochemical properties.

Physicochemical properties	Methods for determining properties	Required amount of substance
Melting temperature and enthalpy of fusion	Differential scanning calorimetry	100–500 mg
Solubility measurements	Non-isothermal HPLC method	<10 g
Partition coefficient of pure target component	Shake flask experiments	<10 g
Boiling point	Differential scanning calorimetry Distillation process Ebulliomotor	10–15 ml
Enthalpy of evaporation	Differential scanning calorimetry	1 ml
Relative density	Pycnometer	5 ml
Vapor pressure	Isoteniscope	N/A
Viscosity	Capillary viscometer	5 ml

of pure compound can be gained by preparative chromatography. The goal is to get access to the required amounts of high purity material more cost effectively than by purchase from commercial sources, which sometimes lack the necessary purity criteria. It is not the goal at this stage to arrive at an optimal industrial process, which is the purpose of the total development concept, but to provide a good starting base.

Target compound concentrations of 0.2–4% [18] in plant tissue require 20–30 kg raw material to provide the necessary amounts of the pure compounds for the determination of the product parameters. Efficient process design and execution have already clearly shown their worth at this initial stage.

In the following sections, the systematic approaches for product parameter determination will be shown using yew, fennel, and sage as examples. In particular, the objective is to apply specifically the available databases and calculation methods for physicochemical properties on complex molecules, their comparison as well as pointing out their limitations.

The experimental determination of these properties will be described in future publications, to propose a common standard for other working groups, who are interested in expanding physicochemical properties databases.

6.5.1
Databased and Calculated Physicochemical Properties

As described in earlier chapters, physicochemical properties of target and side components are first searched in databases. In this work, the databases Reaxys, DIPPR, and DDB are used, as can be seen in Table 6.7.

Generally, molar mass, dynamic viscosity, critical temperature, critical pressure, critical volume, melting point, boiling point, freezing point, molecular volume of the liquid, polarity (relative as octanol/water partition coefficient), enthalpy of formation, free Gibbs formation enthalpy, entropy, melting enthalpy, enthalpy of evaporation, combustion enthalpy, solubilities (in various solvents), and the virial coefficients were determined.

In order to reduce the number of compound data for the listing in the table, only molecular weight, boiling and melting temperature, evaporation and melting enthalpy, and the octanol/water partition coefficient are considered, as those data are relevant for the design of at least one unit operation, and they have somewhat similar behavior. The Kow value, in this context, is a measure for the polarity of a substance and is defined as the partitioning of a substance in an octanol and a water phase. The shaded areas indicate that melting enthalpies and temperature are determined only for solids, while boiling temperature and enthalpy of evaporation are only determined for liquids and gases,

As can be taken from Table 6.7, sufficient substance parameters exist for the main and side components of fennel. The databases Reaxys and DIPPR are larger and therefore allow finding more substance data than in the DDB. It should be noted that all three databases considered show little information on enthalpy of evaporation and the octanol/water partition coefficient. Another deficiency of the

Table 6.7 Fennel property data from data banks.

Fennel								
Target components	**Substance class**	**Data bank**	**Molecular weight ($kg\,kmol^{-1}$)**	**Boiling temperature (K)**	**Enthalpy of evaporation ($J\,mol^{-1}$)**	**Melting point (K)**	**Enthalpy offusion ($J\,kmol^{-1}$)**	**LogKow**
Anethole	Phenylpropene	DIPPR	148.2	508.5	x			X
		Reaxys	148.2		x			3.33
		DDBST	x	486.6	x			X
Fenchone	Monoterpenoid	DIPPR	152.2	476.0	x			X
		Reaxys	152.2	425.4	51 720			X
		DDBST	x	509.8	X			X
Side components								
α-Pinene	Monoterpenoid	DIPPR	136.2	429.3	X			0.16
		Reaxys	136.2	429.2	X			X
		DDBST	x	x	X			X
Camphene	Monoterpenoid	DIPPR	136.2	433.7	X			0.16
		Reaxys	136.2	431.2	46 900			X
		DDBST	x	x	x			X
Myrcene	Monoterpenoid	DIPPR	x	x	x			X
		Reaxys	136.2	440.2	x			X
		DDBST	x	425.7	x			X
α-Phellandrene	Monoterpenoid	DIPPR	136.2	448.2	x			0.16
		Reaxys	136.2	450.9	x			X
		DDBST	x	450.8	x			X
Estragole	Monoterpenoid	DIPPR	x	x	x			X
		Reaxys	148.2	x	x			X
		DDBST	x	479.2	x			X

Table 6.8 Salvia property data from data banks.

Salvia								
Target components	**Substance class**	**Data bank**	**Molecular weight ($kg\ kmol^{-1}$)**	**Boiling temperature (K)**	**Enthalpy of evaporation ($kJ\ kmol^{-1}$)**	**Melting point (K)**	**Enthalpy offusion ($kJ\ mol^{-1}$)**	**LogKow**
Carnosic acid	Diterpenoid	DIPPR	x			x	X	X
		Reaxys	332.44			458–463	x	X
		DDBST	x			x	39550	X
Carnosol	Diterpenoid	DIPPR	x			x	x	x
		Reaxys	330.42			486–494	x	x
		DDBST	x			x	x	x
Side components								
Thujone	Diterpenoid	DIPPR	x	x	x			x
		Reaxys	152.24	475	x			x
		DDBST	x	x	x			x
Camphor	Diterpenoid	DIPPR	152.23	480.57	x			0.182
		Reaxys	152.24	482.25	0.46			x
		DDBST	x	509.78	x			x
Eucalyptol	Diterpenoid	DIPPR	x	x	x			x
		Reaxys	154.25	449.55	x			x
		DDBST	x	473.18	x			x
Oleanolic acid	Triterpenoid	DIPPR	x			x		
		reaxys	456.71			581.15	x	x
		DDBST	x			x	32 062	x

databases is the lack of information on the experimental conditions applied during the measurements, as no common methods are used. Therefore, differing values exist in the different databases for the same compounds, as can be seen looking at boiling-point data, for example.

For the main and side components of sage, less substance data are available than for fennel (Table 6.8). This can be explained by the fact that the substances from sage belong to the class of diterpenoids, which are considerably more complex than monoterpene substances. On the other hand, sage ingredients are less well researched in literature. The same applies when looking at the substance data for yew compounds.

Table 6.9 shows that, for this system only, the Reaxys database contains molar masses and melting points, all other substance data cannot be found there. Therefore, it can be concluded that for less complex monoterpenoids and phenylpropenes, the corresponding substance data may be found in databases. For more complex molecules, such as diterpenoids, little or no substance data are available. Consequently, the compound data for such molecules either need to be calculated or experimentally determined. The following paragraph shows first results of such calculation of compound data applying different thermodynamic models.

Table 6.10 displays the possibilities for the calculation of substance parameters for fennel. For all target and side components belonging to the compound class of monoterpenoids or phenylpropenes, physicochemical properties can be calculated. With the group contribution method of Lydersen, Joback, and Gani/Constantinou boiling points, enthalpy of evaporation, melting point, and melting enthalpy are calculated. Comparing the calculated values with those from the database DDB shows that the calculated values coincide with the Joback model and the database DDB. On the basis of missing information on the origin of the data in the DDB, it cannot be ruled out that the values have been calculated and have not been found experimentally. To decide which of the three methods for the compound determination of target and side components of sage is better suited, they would also have to be experimentally verified. Owing to the lack of comparable measurement techniques for compound data stored in databases, they cannot be referred to for judging the accuracy of a model.

With the UNIFAC, modified UNIFAC, and COSMO-RS models, octanol/water partition coefficients are calculated. As described in Chapter 4, group contribution methods, such as the UNIFAC model, segment molecules into groups, which then permit the description of thermodynamic properties of a system. This step, requires interaction parameters, which for many molecule groups exist in databases. Simple monoterpenoids, with the isoprene as a core structure can be calculated with UNIFAC. COSMO-RS can be applied here, as owing to the simple molecular structure from the protective charges, activity coefficients and thereby physicochemical properties can be calculated. Thereby, sufficient data for the simulation are available, which can be verified from Table 6.10.

In general, the determination of physicochemical properties for monoterpenoids with UNIFAC, modified UNIFAC, and COSMO-RS is possible. For a

Table 6.9 Yew property data from data banks.

Yew								
Target components	**Substance class**	**Data bank**	**Molecular weight (kg kmol^{-1})**	**Boiling temperature (K)**	**Enthalpy of evaporation (kJ kmol^{-1})**	**Melting point (K)**	**Enthalpy offusion (kJ kmol^{-1})**	**LogKow**
10-Deacetylbaccatin I	Diterpenoid	DIPPR	x			x	X	x
		Reaxys	544.60			517–518	X	x
		DDBST	x			x	X	x
Side components								
Cephalomannine	Diterpenoid	DIPPR	x			x	X	x
		Reaxys	831.91			457–459	X	x
		DDBST	x			x	X	x
Taxin B	Diterpenoid	DIPPR	x			x	X	x
		Reaxys	583.72			109	X	x
		DDBST	x			x	X	x
Baccatin III	Diterpenoid	DIPPR	x			x	X	x
		Reaxys	586.64			510–511	X	x
		DDBST	x			x	X	x

Table 6.10 Calculated property data for fennel.

Target components	Method	Boiling temperature (K)	Enthalpy of evaporation (J mol^{-1})	Melting point (K)	Enthalpy offusion (J mol^{-1})	LogKow
Fennel						
Anethole	Lyderson	510	x			x
	Joback	487	43 160			x
	Gani/Constantinou	507	x			x
	UNIFAC	x	x			3.57
	mod. UNIFAC	x	x			2.59
	COSMO-RS	522	x			3.47
Fenchone	Lyderson	466	x			x
	Joback	510	39 488			x
	Gani/Constantinou	x	x			x
	UNIFAC	x	x			x
	mod. UNIFAC	x	x			x
	COSMO-RS	499	x			2.5
Side components						
a-Pinene	Lyderson	426	x			x
	Joback	446	37 346			x
	Gani/Constantinou	435	x			x
	UNIFAC	x	x			3.94
	mod. UNIFAC	x	x			2.77
	COSMO-RS	436	x			4.18
Camphene	Lyderson	x	x			x
	Joback	x	x			x
	Gani/Constantinou	425	x			x
	UNIFAC	x	x			3.98
	mod. UNIFAC	x	x			2.46
	COSMO-RS	436	x			4.02
Myrcene	Lyderson	532	x			x
	Joback	426	36 632			x
	Gani/Constantinou	452	x			x
	UNIFAC	x	x			4.31
	mod. UNIFAC	x	x			2.59
	COSMO-RS	430	x			4.52
α-Phellandrene	Lyderson	436	x			x
	Joback	451	39 141			x
	Gani/Constantinou	442	x			x
	UNIFAC	x	x			4.14
	mod. UNIFAC	x	x			2.84
	COSMO-RS	440	x			4.23
Estragole	Lyderson	497	x			x
	Joback	479	42532			x
	Gani/Constantinou	492	x			x
	UNIFAC	x	x			3.71
	mod. UNIFAC	x	x			2.51
	COSMO-RS	517	x			3.43

Table 6.11 Calculated property data for salvia.

Salvia						
Target components	**Method**	**Boiling temperature (K)**	**Enthalpy of evaporation (J/mol)**	**Melting point (K)**	**Enthalpy offusion (J/mol)**	**LogKow**
Carnosic acid	Lyderson			x	x	x
	Joback			808	39550	x
	Gani/Constantinou			401	x	x
	UNIFAC			x	x	6.02
	mod. UNIFAC			x	x	4.18
	COSMO-RS			x	x	6.79
Carnosol	Lyderson			x	x	x
	Joback			718	38211	x
	Gani/Constantinou			391	x	x
	UNIFAC			x	x	5.84
	mod. UNIFAC			x	x	x
	COSMO-RS			x	x	4.82
Side components						
Thujone	Lyderson	495	x			x
	Joback	505	40079			x
	Gani/Constantinou	484	x			x
	UNIFAC	x	x			2.95
	mod. UNIFAC	x	x			2.40
	COSMO-RS	510	x			2.72
Camphor	Lyderson	482	x			2.26
	Joback	510	39488			x
	Gani/Constantinou	481	x			x
	UNIFAC	x	x			2.92
	mod. UNIFAC	x	x			2.34
	COSMO-RS	503	x			2.04
Eucalyptol	Lyderson	451	x			x
	Joback	473	39923			x
	Gani/Constantinou	x	x			x
	UNIFAC	x	x			x
	mod. UNIFAC	x	x			2.81
	COSMO-RS	490	x			3.13
Oleanolic acid	Lyderson			x	x	x
	Joback			850	32062	x
	Gani/Constantinou			410	x	x
	UNIFAC			x	x	9.81
	mod. UNIFAC			x	x	8.14
	COSMO-RS			x	x	6.28

statement regarding the precision of models, the calculated and experimental data sets have to be compared. For anethole, K_{OW} values in literature vary between 3.11 and 3.7 [60]. The comparison between calculated and literature data shows the best correlation for the COSMO-RS model. For α-pinene, camphene, myrcene, α-phellandrene, and estragole, K_{OW} values of 4.49, 4.56, 4.50, 4.55, and 3.47 can be found in literature [61]. In all cases, COSMO-RS shows the highest accuracy. However, it should be noted, that different existing COSMO software, such as ARTIST or COSMOtherm, lead to large deviations.

To compare the two g^E models, the original UNIFAC shows better results than modified UNIFAC. The modified UNIFAC model requires other interaction parameters than those for the original model. In case of missing parameters, the use of those from the original model explains the large variations. As is known from literature [62], the quality of the COSMO-RS model depends on the precise capture of the molecule for the calculation of the protective charges. If the geometry of the molecule is not represented correctly, large deviations from reality result. This means that large molecules cannot be described this way.

Table 6.11 shows the target and side components of salvia with the corresponding calculated substance parameters. In addition, in this case, the parameters can be determined for the simple molecules of the monoterpenoid class. Furthermore, the data calculated with the Joback model can be compared with the DDB database. Limitations are found only for the target compounds carnosic acid and carnosol and for the side component oleanolic acid.

For these three compounds, belonging to the diterpenoid class, parameters for sorting the molecules in groups are missing. Only the Joback model can provide physicochemical properties also for these compound classes. The reason for this however, is that the grouping for this method is kept very simple and therefore calculations are also possible for complex molecules. In addition, no interactions between groups are considered, making the method less precise and thereby not applicable to complex molecules. In general, COSMO-RS calculations for complex molecules are possible, but these are associated with high effort and low accuracy. For this reason, the applicability of COSMO-RS for detailed process simulation is limited.

The K_{OW} value of carnosol and carnosic acid, for example, calculated by UNIFAC and modified UNIFAC shows a large deviation. For more complex molecules, such as diterpenoids, the currently available models provide no physicochemical properties at all or only low accuracy data. This can also be shown in the example of yew, as listed in Table 6.12.

For this system, data could only be calculated with the Joback and COSMO-RS models. Because of the low accuracy of the Joback method for complex molecules, the values cannot be used for the simulation. For the target compound 10-deacetylbaccatin III and other diterpenoids found in yew, no properties can be calculated with UNIFAC or modified UNIFAC, because the group contribution methods lack the parameters required for the molecule groups. It is possible to calculate physicochemical properties, such as ocatanol–water partition coefficient with the COSMO-RS model. However, these values must

Table 6.12 Calculated property data for yew.

Yew						
Target components	**Method**	**Boiling temperature (K)**	**Enthalpy of evaporation (J/mol)**	**Melting point (K)**	**Enthalpy offusion (J/mol)**	**LogKow**
10-DAB III	Lyderson			x	x	x
	Joback			x	x	x
	Gani/Constantinou			x	x	x
	UNIFAC			x	x	x
	mod. UNIFAC			x	x	x
	COSMO-RS			x	x	3.1
Side components						
Cephalomannine	Lyderson			x	x	x
	Joback			1300	99879	x
	Gani/Constantinou			x	x	x
	UNIFAC			x	x	x
	mod. UNIFAC			x	x	x
	COSMO-RS			x	x	6.5
Taxin B	Lyderson			x	x	x
	Joback			x	x	x
	Gani/Constantinou			x	x	x
	UNIFAC			x	x	x
	mod. UNIFAC			x	x	x
	COSMO-RS			x	x	5.8
Baccatin III	Lyderson			x	x	x
	Joback			x	x	x
	Gani/Constantinou			x	x	x
	UNIFAC			x	x	x
	mod. UNIFAC			x	x	x
	COSMO-RS			x	x	4.4

be checked experimentally to draw a conclusion on the accuracy. Currently, no thermodynamic models exist to describe such complex molecules with sufficient accuracy, because the required parameters are missing. These might be generated by fitting with experimental data. This means that for such molecules, the physicochemical properties have to be determined experimentally. An advantage of the experimental determination would be that with the data for models such as UNIFAC, additional parameters could be determined so that in future, a calculation might be possible.

In this work, a systematic approach for the determination of physicochemical properties of complex molecules is presented. In particular, it was shown to what extent these properties can be obtained from databases or can be calculated using thermodynamic models. Selected sample systems have been fennel and salvia; and for the different target and side components of these systems boiling and melting temperatures, evaporation and melting enthalpies, and the octanol/water partition coefficients were determined. It could be shown that for all target and side components of fennel physicochemical properties exist in databases. Only the octanol/water partition coefficients could not be found for all substances.

In case of salvia, fewer properties could be found for target and side components. Especially, octanol/water partitioning data were also lacking here. Furthermore, for the yew system, apart from melting temperatures in the Reaxys database, no additional physicochemical properties could be found. From this, it may be concluded that for simple molecules of the monoterpenoid class, physicochemical properties can be found in databases such as Reaxys, DIPPR, and the DDB.

For more complex molecules such as di- and triterpenoids few or even no physicochemical properties can be found. A further disadvantage of the database lies in the lack of information on the measurement methods for the available properties so that consistency in the measurements is not guaranteed. This is shown in the differing values of physicochemical properties from different databases, which limits their comparability to some extent.

The calculation of physicochemical properties showed that these properties can be determined for monoterpenoids and phenylpropenes. The accuracy by comparison with databases could not be verified, because the databases do not provide consistent material data. For the calculation of the K_{OW} value, the COSMO-RS model and the original UNIFAC model performed well. Modified UNIFAC generated large deviations from literature data.

No consistent theory, which allows an accurate calculation of properties, is currently available for diterpenoids. Group contribution methods such as UNIFAC are missing the necessary interaction parameters. Owing to low accuracy the COSMO-RS model is not applicable to complex molecules. Nevertheless, COSMO-RS is the only method, which is applicable very early in process design. COSMO-RS calculations show some trends, which could be used to reduce the experimental effort during process design. But for more proper and accurate simulations, physicochemical properties for such molecules have to be determined experimentally. That way, data for comparison as well as for parameter determination of the models can be generated.

6.5.2 Feed Characterization

The number of products used as agro-chemicals, food additives, flavors, aromas, pharmaceuticals, and nutraceuticals which are made by solid-extraction from plants has increased significantly. Despite this growth, initial predictions for a potential product purification process for these complex mixtures remains entirely experimentally based on empirical knowledge. This section represents an initial study to demonstrate the benefits of a systematic approach for product purification, where no or a few physicochemical properties of target and side components are known.

The current state of research is discussed. On the basis of this evaluation, the most efficient method for conceptual process development has been identified and further developed. The resulting methodology consists of model-based cost accounting accompanied by experimental model parameter determination. The latter is carried out in miniaturized laboratory-scale measurement cells for each unit operation, using the complete original feed. The model parameter determination is accompanied by a comprehensive error analysis. The experimental plan currently includes the determination of thermodynamic equilibrium conditions in the mixture directly from the raw material mixture. Transport kinetics and fluid dynamic parameters are first estimated from known correlations or preexisting knowledge. Later on, these parameters are determined exactly in mini-plant experiments. Furthermore, biological- and botanical-based guidelines are developed to identify thermodynamically favored basic operations. Finally, the developed approaches are successfully validated using two plant extracts. Firstly, it could be proven that the botanical preselection can reduce the experimental plan significantly. Secondly, it was shown that the experimental equilibrium data of the kinetics and fluid dynamics can have a significant impact on the separation costs. Therefore, detailed rigorous modeling approaches have to be chosen instead of shortcut methods in order to make any valid process development conclusions or to further optimize the system.

The fundamental limitation which has to be taken into account is the limited availability of feedstock, usually of the order of a few liters. At this stage of the project, when the probability of project failure is high, there is also limited analytical and experimental resource to be exploited.

Figure 6.16 shows that only improvements which are introduced in the product development or early basic engineering stages can reduce costs significantly. But

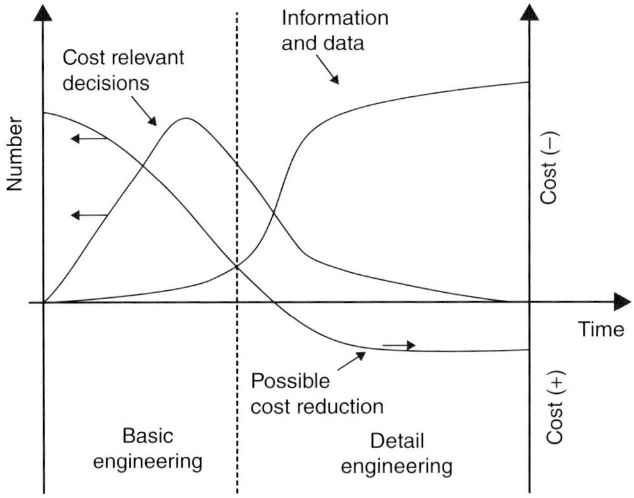

Figure 6.16 Available information and possible cost reduction over the engineering phases.

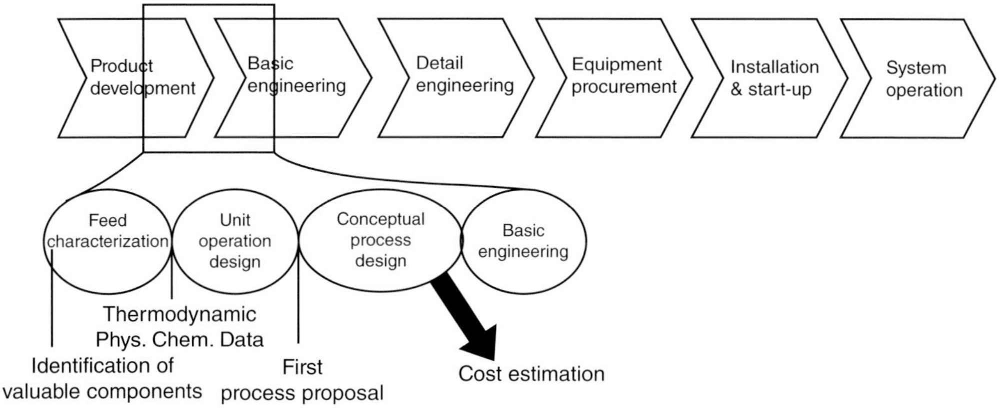

Figure 6.17 Location of feed characterization in process synthesis.

these potential improvements require additional investment in detailed research and development at an early process development stage when there is a higher risk of project failure. Therefore, to justify these increased efforts in process optimization at an early project stage, an extremely efficient approach has to be found. If opportunities for improvements are identified later on, they are often no longer economical. In the context of a CPD, it should therefore be attempted to predict an optimal process at an early stage of process design at the limited basis of available information. These facts result in the demand of a systematic and standardized approach to characterize the regarded feed. The unit operations have to be evaluated in terms of feasibility and economy. The central question to be considered then is this: what approach can be used in order to gain the necessary increased level of information without requiring unacceptable additional experimental or analytical effort?

Figure 6.17 locates the feed characterization step in the overall process development. Individual unit operations have to be modeled before they can be arranged as a first-process proposal in the CPD accompanied by cost estimations. In advance, an identification of key components and their thermodynamic physical–chemical data with regard to the needed modeling depth has to be done. Experiments to determine model parameters in lab scale are inevitable.

The transfer of existing CPD methods to complex mixtures fail. This is because, so far, no efficient way has been found to identify the structure of all relevant substances in these mixtures. This is required to begin the determination of the relevant thermodynamic data:

- Both main tools of the early process development stages, namely, the database queries and quantum mechanical- or group-contribution-based calculations need a structure-determining analysis.
- Experimental determination of thermodynamic properties for both pure components and mixtures requires the isolation of representative sample quantities.

Therefore, to date, it has not been possible to examine the feasibility or accessibility of approaches in CPD for the separation of complex mixtures such as plant extracts. A standardized, systematic approach for the determination of an optimal separation sequence based on modeling of operating and investment costs, but with minimal experimental efforts, currently does not exist. In this section, the state of the art is depicted and discussed.

6.5.2.1 Conceptual Process Design

In academic literature, different approaches have been presented in the last few years. A recent analysis of references using methodical approaches showed the following (Figure 6.18):

- Most often (37%) shortcut models are used.
- More detailed models are used 16% of the time.
- Synthesis tools using heuristic rules are used by 13% of the analyzed literature.
- Least frequently (with 2%) used were purely experimentally driven methods for process synthesis.
- Of the literature-described examples of process synthesis, about 75% of the physical properties needed are known in databases or are calculated from molecular structure, about 16% are unspecified, and only the remaining roughly 9% are experimentally determined.
- The unit operations considered for the process synthesis examples in literature are predominantly distillation, followed by crystallization, chromatography, and membrane separation processes. Absorption, solid–liquid extraction, and liquid–liquid extraction are found to a much lesser extent in these examples.

In the field of process synthesis, model-based methods are well established. Regardless of the different approaches, it is important for model-based process synthesis to have a reliable and comprehensive data basis. For shortcut calculations at the lowest level of detail, the model input parameters used are separation factors. Accuracy of shortcut calculations could be improved by the use of thermodynamic mixture data instead of separation factors, for example, see [63, 64].

In reality, the degree of nonideality in the equipment operated determines the degree to which each unit operation should be optimized. In process design and optimization, only the optima which can be achieved in reality should be compared. For process design and optimization, the detailed model used ought to take into account equilibrium thermodynamic data and both mass transport and fluid dynamics parameters.

In order to define the necessary modeling depth for each unit operation due to the sensitivity of detailed modeling versus shortcut approaches, mass transfer coefficient and axial distribution are varied in typical magnitudes.

Figure 6.19 demonstrates this effect in a typical liquid–liquid extraction column used in the purification of plant extract. It is clear that fluid dynamic and mass transfer nonidealities account for between 20% and 40% of the separation costs. This is a significant contribution and therefore should be taken into account. A

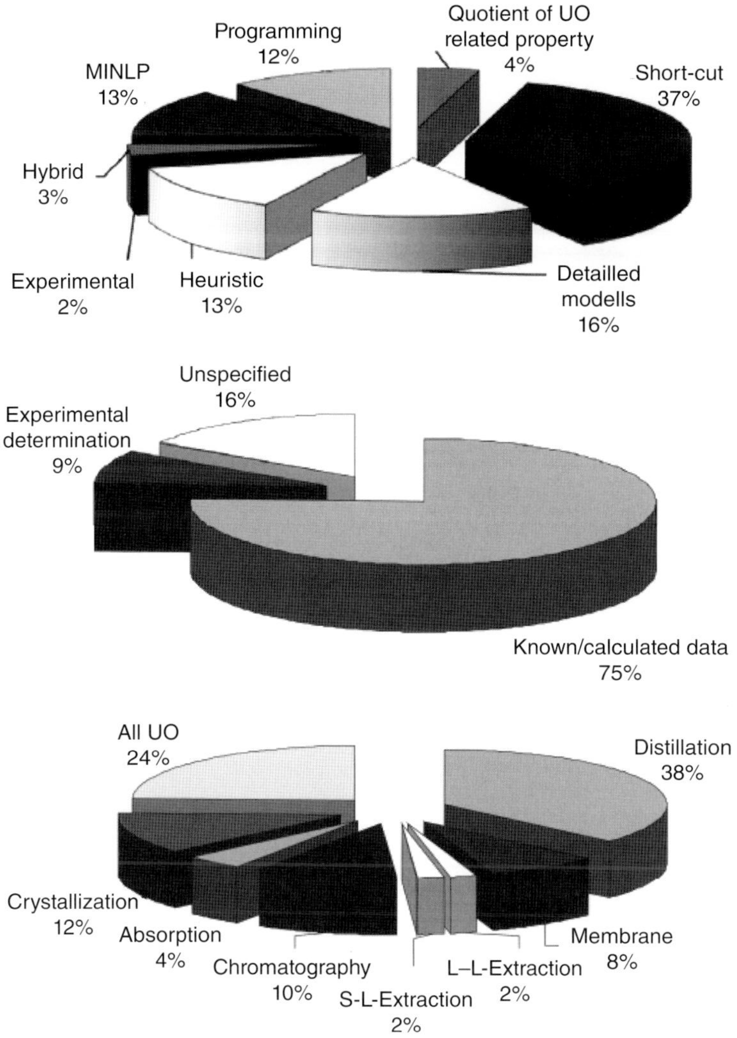

Figure 6.18 Distribution of methods used in process synthesis.

shortcut approach would fail to generate data which is reliable when being realized industrially.

The majority of the data used in academia is obtained from databases or by calculation methods for physical properties based on molecular structure.

- In about 16% of the literature how data is gained is not explained. However, it is likely that it is taken from databases.
- For only about 9% of the references the required data are determined experimentally; this 9% is largely concerned with either process synthesis of chromatographic separations [65–67] or optimization of existing processes [68–70].

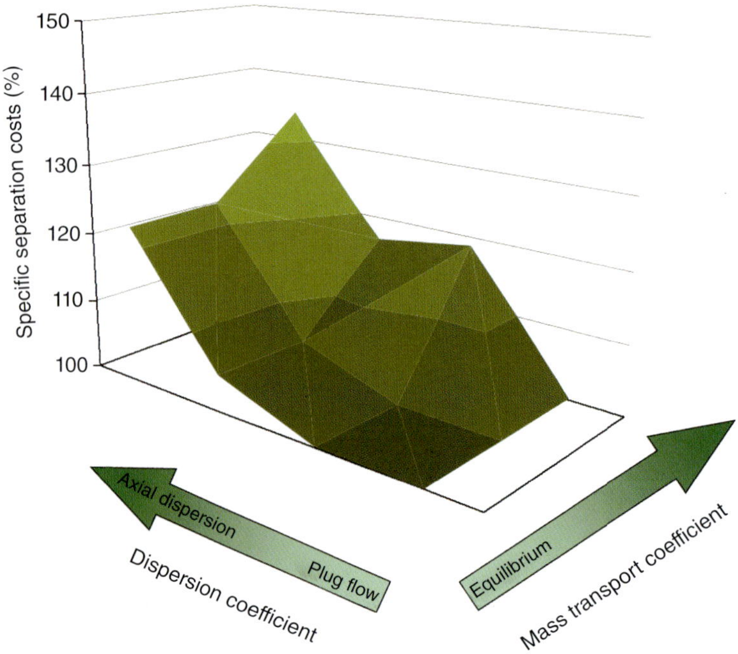

Figure 6.19 Separation costs and as they vary with dispersion and mass transport coefficients.

- The references include all the usual basic unit operations, while the majority deal with distillation operations. Those labeled in figure 6.18 as "all unit operations (all UO)" (24%) include mostly review articles whose statements can be applied to all unit operations.

From this literature research, it can be concluded that currently there are three different ways for obtaining the necessary data for model-based process synthesis:

1) *Databases*: There are extensive databases in which the properties of already known and well-studied substances and systems are described. The development of distillation columns for the separation of well-studied and documented systems is possible without previous experiments with a high degree of accuracy [71]. But there exists none or only a few data for complex mixtures, as described in Section 6.5.1.
2) *Calculation methods*: There are comprehensive methods for the calculation of physicochemical property data based on molecular structure. These include quantum mechanics as well as group contribution methods [12, 66, 72–78]. The applicability for complex mixtures is shown in Section 6.5.1.
3) *Experimental determination*: Equilibrium phase distribution coefficients and separation factors for the individual unit operation are determined in laboratory-scale experiments [67, 70, 79–86].

Table 6.13 Required model parameters for different modeling depth.

Unit operation	Thermodynamical equilibrium for key components parameter	Kinetic parameter	Fluid dynamic parameter	Cost calculation parameter
Size exclusion chromatography	Molecular weight, size	Diffusion coefficient (film, pore, bulk, surface)	Axial dispersion (liquid)	Henry, capacity, productivity, dilution
Ion exchange chromatography	Sequence of elution = f(c(modifier, column)), isoelectric point	Diffusion coefficient (film, pore, bulk, surface), isotherm for target component	Axial dispersion (liquid)	Henry, capacity, productivity, dilution
Reversed/normal phase chromatography	Sequence of elution = f(c(modified, column)), hydrophobicity	Diffusion coefficient (film, pore, bulk, surface), isotherm for target component	Axial dispersion (liquid)	Henry, capacity, productivity, dilution
Distillation	Mixture-VL-equilibrium = f(T, p)	kLa	Axial dispersion (liquid, gas)	cP of components and solvent ΔHV of low boiling component/solvent
Liquid–liquid-extraction	Mixture-LL-equilibrium = f(pH, T, solvent, salt)	Diffusion coefficient (bulk, film)	Axial dispersion (continuous dispersed phase)	Capacity, density, viscosity, surface tension
Membrane	Molecular weight, size	Diffusion coefficient, density of pure solvent	Axial dispersion (permeate, retentate)	Molecular weight cut-off, pressure-flow-curve
Solid–liquid-extraction	Mixture-SL-equilibrium = f(solvent, T, pH, salt)	Diffusion coefficient (film, pore, bulk) solubility	Axial dispersion (liquid)	Capacity, SL-equilibrium line, dilution

6.5.2.2 Modeling Depths and Feed Characterization Approach

To describe a thermodynamically feasible separation as a unit operation, several model parameters are needed.

In Table 6.13 the phase equilibrium, mass transport (kinetic), and fluid dynamic parameters are listed for the main unit operations in separation of plant extracts. Furthermore, parameters important for cost calculations are also listed.

To make decisions between alternative unit operations and their sequencing requires models of these unit operations, which are sufficiently precise. As described before, these decisions are in most cases quite in a narrow range; they cannot be filed on the basis of a rough estimation that it would be if only separation coefficients between target and side components are considered and if so-called short-cut-calculations are applied.

On the other hand, the required model parameters must be accurately determined and with reasonable cost and experimental effort, even at this early stage in the project. Therefore, a compromise between maximization of the model depth and accuracy of the method for model parameter determination has to be found.

Experience shows that inclusion of mass transfer and overall fluid dynamic nonidealities often brings a sizable gain in modeling accuracy and should therefore be included (Figure 6.19). For this reason, in this study, axially dispersed plug models with equilibrium description of mixtures by key component groups and mass transfer resistances are used.

The parameters for the equilibrium behavior have to be determined using one of the above described methods.

Owing to the fact that an exact experimental measurement of the mass transfer and fluid dynamic parameters would result in significant additional experimental work and an increase in the required sample volume, these parameters should be assessed as follows: initially, at this early stage of process development, the parameters are only calculated and/or assumed by using existing correlations, preexisting knowledge, and experience. Subsequently, their significance to the separation should be checked using parameter-sensitivity studies. They should then be improved using experimental results. At a later stage in the project, these efforts are justified by the benefit of continuing with the project.

The feed characterization provides a way to develop a separation sequence, even in the absence of physicochemical property data. The experimentally driven method starts with a characterization of the feed mixture in the individual unit operations at different operating conditions. With the data obtained from these mixture experiments, the simulation and process synthesis will carried out.

For measurement of the equilibrium data, standardized laboratory-scale experiments are built up. An error calculation has to be made to determine the minimal equipment volume to get results with a relative mass balance error of ±5%. The Gaussian error propagation law (Equation 6.23) was used to consider the different error sources.

$$S_F^2 = \sum_{k=1}^{N} \left(\frac{\partial F}{\partial f_k} \cdot \Delta f_k \right)^2 \tag{6.23}$$

The error calculation for the standardized parameter determination experiments for (a) liquid–liquid-extraction and (b) distillation, results in the volume-dependent graph shown in Figure 6.20.

From these calculations, it is possible to determine the minimum required volume for minimized error. Table 6.14 summarizes these minimum required volumes for the experimental model parameter determination of unit operations such as the following:

- Liquid–liquid-extraction
- Distillation

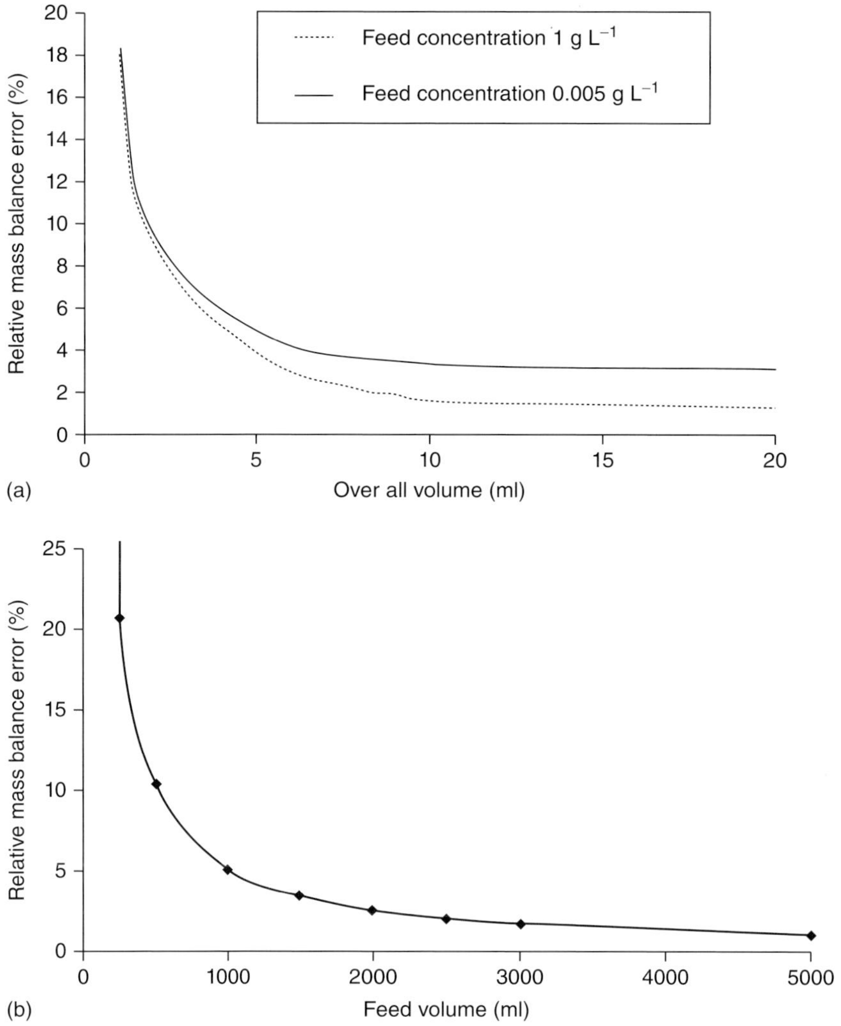

Figure 6.20 Error calculation for the determination of equilibrium data for liquid–liquid extraction (a) and distillation (b).

- membrane separation processes
- Chromatography.

The conclusion of the analysis is that with about 1 l of natural extract broth, all relevant experiments should be possible within the desired accuracy for detailed modeling.

Initially, information from literature and heuristics are used to describe the possible behavior of the target component in the unit operations (Figure 6.21).

A fundamental superstructure is derived from literature data and is shown in the same figure.

Table 6.14 Required volumes for the determination of equilibrium data.

	Parameter variation	Experimental volume (ml)	Repetition	Required minimum volume (ml)
Extraction	5	7	3	105
Distillation	1	1000	3	600
Membrane	1	10	3	30
Chromatography	1	10	3	30
Overall volume				765

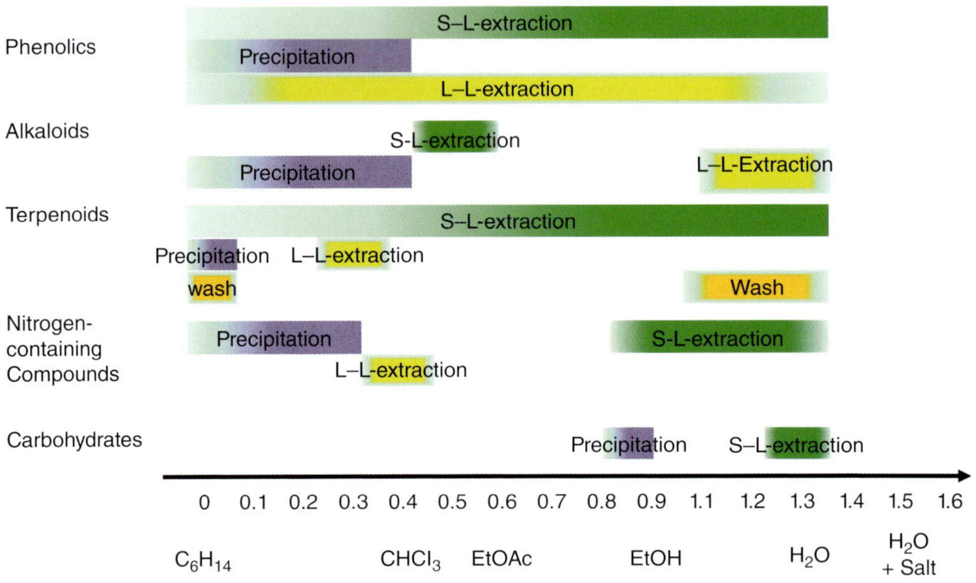

Figure 6.21 Heuristics for secondary metabolites, generated from data available in [87].

Of course, these data are not complete and not sufficiently accurate to achieve precise enough predictions. But, on the basis of this initial, educated guess at potential unit operations, a superstructure of a process flow sheet is the starting point for model-based process development. Therefore, the experimental parameter determination according to the "unit operation–based" approach can be reduced, but not replaced, by this method.

Afterward, the "unit operation–based" laboratory-scale experiments are optimized with the objective that the target component is always preferentially separated. The parameters obtained from these experiments are sufficient for modeling and process synthesis.

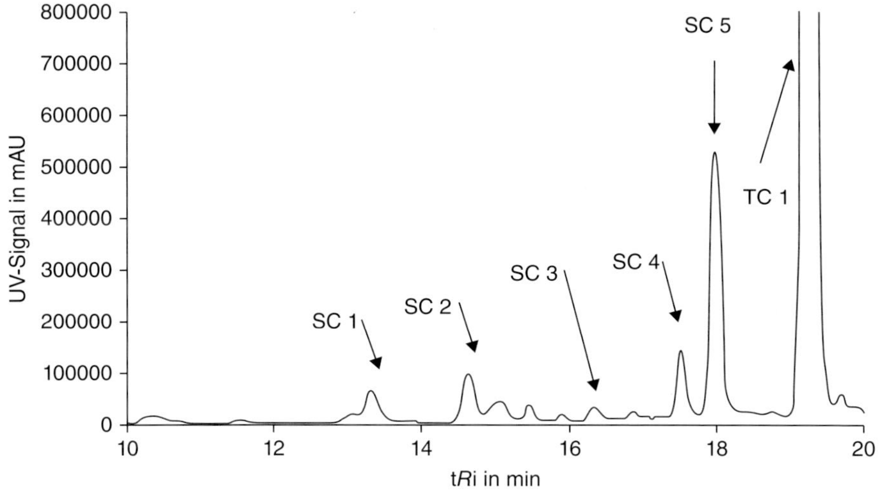

Figure 6.22 Chromatogram of vanillin extract with target and side components.

6.5.2.3 **System 1: Vanillin**

An extract made by solid–liquid extraction with a water/ethanol mixture (1 : 1) of vanilla pods was considered. Analysis is based on an RP-separation with a UV detector. An example of the resulting chromatogram is shown in Figure 6.22. The target component (TC) is vanillin, the five side components (SC) are assumed to be unknown.

The target component is the most apolar in this chromatogram. Any industrial-scale/preparative separation by RP chromatography, or even using the more economical NP chromatography, is therefore an option which must be considered in comparison to the other unit operations.

As analytics, the separation is often based on chromatography. Any transfer to technical chromatography is therefore an option directly on hand. Economy has to be evaluated in process synthesis, of course.

6.5.2.4 **Potential Unit Operations for Product Purification**

Fundamentally, all separation mechanisms, and therefore all unit operations, are possible for the isolation of the target component. Previously known information about the target and side component led to the following restrictions:

- It is known that both target and side components have a molecular weight below 500 g mol^{-1}. Therefore, the use of separation techniques using molecular size, for example, filtration membranes makes no sense. Nanofiltration may be an option, but molecular sizes are too similar.
- Chromatographic separations may not be feasible because of the low value of the target component.

Thus, the appropriate separation operations to be examined are liquid–liquid extraction, distillation, and crystallization. Liquid–liquid extraction and distillation are considered here and a possible process is proposed.

Option: Liquid–Liquid Extraction

1) Solvent choice
   An experimental plan can be derived for solvent combinations which are immiscible in liquid–liquid extraction:
   a. The feed solvent is a 1 : 1 mixture of ethanol and water. Water is immiscible with many organic solvents; however ethanol is soluble in any solvent.
   b. Miscibility tables determine which solvents are immiscible with an ethanol–water mixture. This has been proved in preliminary laboratory-scale tests. Examples of solvents which are not miscible with the system water–ethanol are hexane, heptane, toluene, di-chloromethane, and chloroform.
   c. Especially for flavors and nutrition, most of the named solvents are not applicable by regulations. To exemplify solvent selectivity, they were taken into account in this study.
2) Determination of equilibrium phase distribution coefficient
   The various solvent and feed mixtures were added to 15 ml tubes and their exact volumes measured. The sealed samples were individually weighed. To increase the surface area the reaction vessels were shaken horizontally for 24 h at 20 °C. The emulsion formed is separated into two distinct phases using a centrifuge. At the end of the experiment the volumes of the phases are again determined. The individual phases are withdrawn using syringes and analyzed for their contents by HPLC-UV. The resulting distribution coefficients are given in Table 6.15.
   To evaluate the separation efficiency of the LLE, the distribution coefficients between the two phases and the separation factor relative to the target component were calculated:
   The solubility of all substances in hexane and heptane is small. Even if the separation factor relative to the target component is very good, the absolute solubility of the target component in the extract phase is low.

Table 6.15 Distribution coefficients for LLE.

	Distribution coefficient					
Solvent	**SC 1**	**SC 2**	**SC 3**	**SC 4**	**SC 5**	**TC 1**
Hexane	0.0012	0.0002	0.0060	0.0048	0.0096	0.0185
Heptane	0.0005	0.0001	0.0022	0.0025	0.0052	0.0154
Toluene	0.0621	0.0237	0.0690	0.1055	0.3538	0.7864
Chloroform	0.0181	0.1462	0.5510	1.4141	2.0947	7.0861
Dichloromethane	0.0039	0.3055	0.7061	1.0110	1.9996	6.1227

a. Toluene is not suited for the purification either. Although the relative separability of the secondary components to the target component is good, the distribution coefficient is too small for an economic extraction.
b. Chloroform and dichloromethane are, however, very efficient at separating minor components numbers 1–3. The target component is enriched in the solvent phase more strongly than all the other components. These are ideal conditions for an economic treatment of the target component.

Option: Distillation To test experimentally whether the extract could be separated by distillation, a distillation test is carried out at starting temperatures of 40 and 70 °C. The pressure in the mini-scale apparatus is gradually lowered until the mixture begins to boil. In advance, a DTA analysis for thermal stability of the target components is relevant.

At a fixed temperature, for example, 40 °C, the mixture in the product vessel is kept at first at atmospheric pressure for about 15 min and cycled in the equipment. If a fraction is evaporated and condensed, the fraction is taken out of the recycle until the temperature drops significantly. Than a lower pressure step is adjusted. A clear sign that an equilibrium has been achieved is a constant temperature of each pressure level. Samples are drawn at the top and at the bottom and the pressure is lowered further. The samples are analyzed using HPLC. This results in the chart shown in Figure 6.23.

Figure 6.23a shows the distribution of components between the two phases. In this unit operation, none of the components goes into the vapor phase to a sufficient degree. During distillation, only almost pure solvent evaporates. Side and target components accumulate at the bottom.

Figure 6.23b depicts the overall mass of the individual components in the vapor phase. At 90 mbar, over 50% of the solvent has already been evaporated. Side component 3 is the most volatile substance, but even at 90 mbar, there is only around 0.22% of the total mass in the gas phase. As a consequence, an economic purification of the target component by distillation is not feasible.

6.5.2.5 Data Evaluation

By summing up the knowledge gained from the distillation and liquid–liquid-experiments, it is possible to draw the partition of physicochemical equilibrium parameters shown in Figure 6.24.

By summarizing the measured data as shown in Figure 6.24, the benefit of liquid–liquid extraction can be seen at a glance.

6.5.2.6 Model-Based Process Design and Calculation of Separation Costs

The experimentally determined partition coefficients and estimated model parameters for mass transfer kinetics and fluid dynamics are used for a first cost estimate of the purification of the target component.

Simulations show that by using liquid–liquid extraction, it is possible to separate the first three side components from the target component with an acceptable number of theoretical stages.

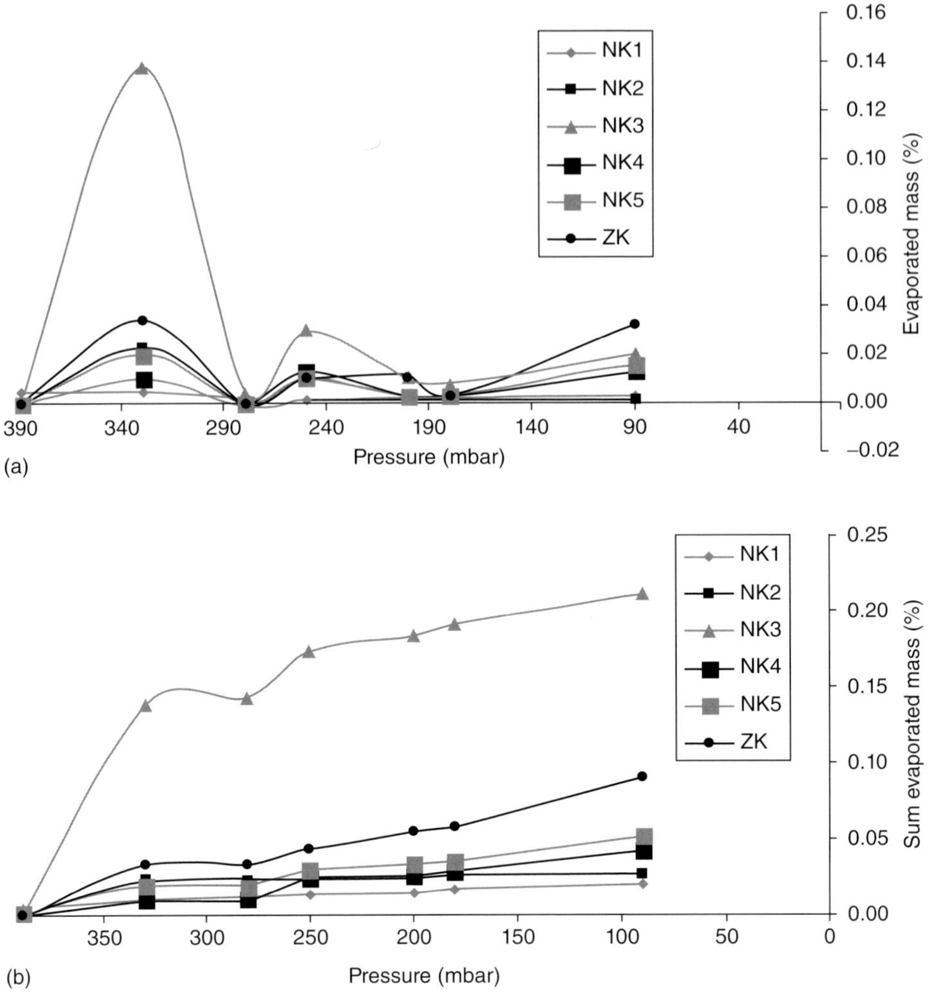

Figure 6.23 Percentage of vaporized mass, (a) sample concentration and (b) cumulative curve.

To understand the influence of mass transfer kinetics and fluid dynamics, the mass transfer coefficient was increased by a factor of 10 to simulate equilibrium conditions and reduced by a factor of 10 to simulate mass transfer limitations. In addition, fluid dynamics was varied by a factor of 10. The results in the change of the separation costs can be seen in Section 6.5.2.7.

Process Proposal and Design For process simulation, the following parameters were used: $10\,m^3\,h^{-1}$ feed with a target and side component concentration of $10\,g\,l^{-1}$, a water phase of $10\,m^3\,h^{-1}$, and organic phase of $10\,m^3\,h^{-1}$. For equilibrium conditions, the experimentally determined values are used, while the kinetics is taken into account by a mass transfer coefficient of $0.05\,m\,h^{-1}$

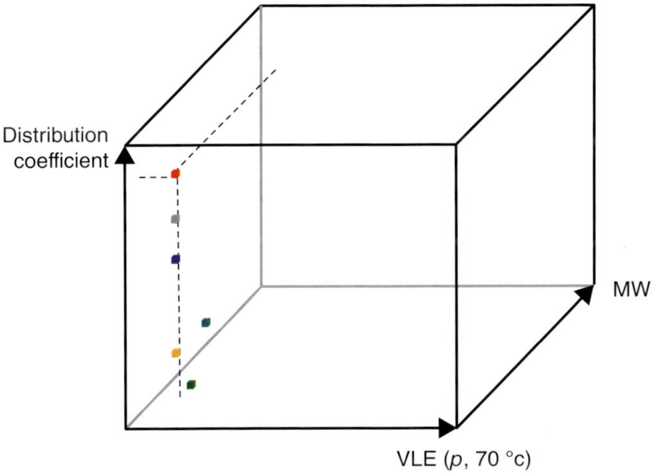

Figure 6.24 Partition of physicochemical equilibrium parameters.

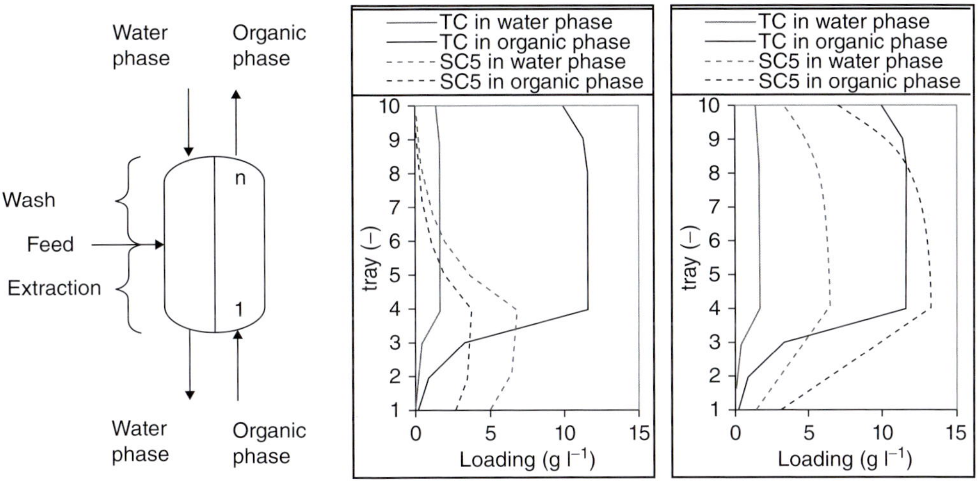

Figure 6.25 Column profile of the target component (TC) as well as for the components 3 (SC3) and 5 (SC5).

calculated according to the Schmidt–Sherwood–Reynolds correlation as a first approximation.

The result of the separation of this mixture is obtained in 10 theoretical stages and the feed inlet in stage 4 as shown in Figure 6.25. From the results, it is clear that a complete separation of the target and side components 1–3 is possible. Although the column is designed to remove the side components 4 and 5 from the aqueous phase using the six stages above the feed inlet, it is only possible to reduce the concentration of these substances. This effect can be well observed in the column profile shown in Figure 6.25.

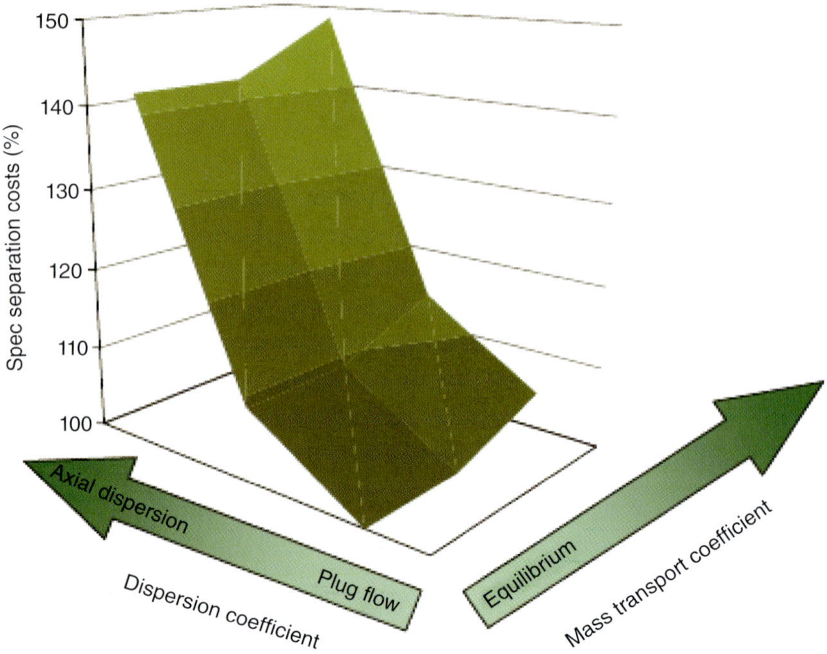

Figure 6.26 Change in separation costs due to variation in model parameters.

6.5.2.7 **Separation Cost Estimation**

On the basis of the target component, the separation costs are estimated to be about 0.17 € kg^{-1}. To examine the sensitivity of the cost to kinetics, the mass transfer coefficient, and axial dispersion are varied.

- The reduction of the mass transfer coefficient by a factor of 100 results in a separation cost increase of 4.5%.
- With a change in fluid dynamic, from plug flow to a flow regime which is dominated by axial dispersion, the separation cost increases by up to 48%, as can be seen from Figure 6.26.

6.5.2.8 **System 2: Tea Aroma**

The second system is a solid–liquid extract made by percolation of a tea with ethyl acetate. The ingredient substances were analyzed by gas chromatography (GC) with a flame ionization detector (FID). For the preparation of this extract, the tea is percolated by ethyl acetate. The extract produced by this percolation consists of a large part of caffeine, matrix components, catechins, chlorophyll and, in smaller parts, of a variety of flavor compounds, which are shown in Table 6.16.

6.5.2.9 **Data for Potential Unit Operation**

For this example, initially a separation by liquid–liquid extraction was investigated. Crystallization was not considered as a first step in this work, because of

Table 6.16 Distribution coefficient and derivations.

	Experiment 1		Experiment 2		
	Distribution coefficient	Derivation mass balance (%)	Distribution coefficient	Derivation mass balance (%)	Derivation distribution coefficient (%)
Linalooloxide Z	37.66	−0.23	32.23	−4.82	14.42
Linalooloxide E	381.41	1.01	119.24	−4.82	68.74
Linaloole	Infinite	1.64	Infinite	−5.18	—
Phenylethanol	122.61	−1.51	80.28	−5.14	34.52
1-Nonanol	Infinite	3.90	Infinite	−4.96	—
Geraniol	Infinite	—	Infinite	—	—
Damascenon	110.38	5.74	80.11	1.03	27.42
Caffeine	0.77	−7.36	0.76	−5.56	0.36
Guaiacol	4.59	−0.30	4.62	1.78	−0.58

its solid-handling complexity. Membranes are not suitable because of the small molecular weights involved.

The preselection system, Figure 6.21, shown above, can be used in this case to limit the number of experiments:

The flavors such as linalool, geraniol, and guaiacol are part of the main groups "terpenoids" and "phenols" and should be soluble – according to Figure 6.21 – in polar to medium polar solvents.

- Caffeine can be extracted as a substance of the main group "nitrogen-containing substances" only in very polar solvents.
- On the basis of this information and the lack of choice of solvents that are immiscible with ethyl acetate, water was chosen as the counterphase. According to the scheme shown in Figure 6.21, caffeine should pass into the aqueous phase, and the flavors tend to remain in the less polar organic ethyl acetate phase.

The experiments for the model parameter determination for the LLE are performed as described above. Water is the only possible solvent which does not dissolve completely in ethyl acetate. The experimental determination of distribution coefficients in the shaking experiments produces the results shown in Table 6.16.

The separation factors indicate a very strong enrichment of the flavors in the organic phase (upper phase, uP), while caffeine passes into the aqueous phase (lower phase, lP).

6.5.2.10 Process Design and Cost Estimation

The cost calculation for a separation of flavors and caffeine in the organic and the aqueous phase results in the following process parameters and structure: with plug flow, a mass transfer coefficient of 0.1 m h^{-1} and for a caffeine purity of 99.99% with a caffeine-yield of 99.98%, the separation costs are about 7 € kg^{-1}. In order

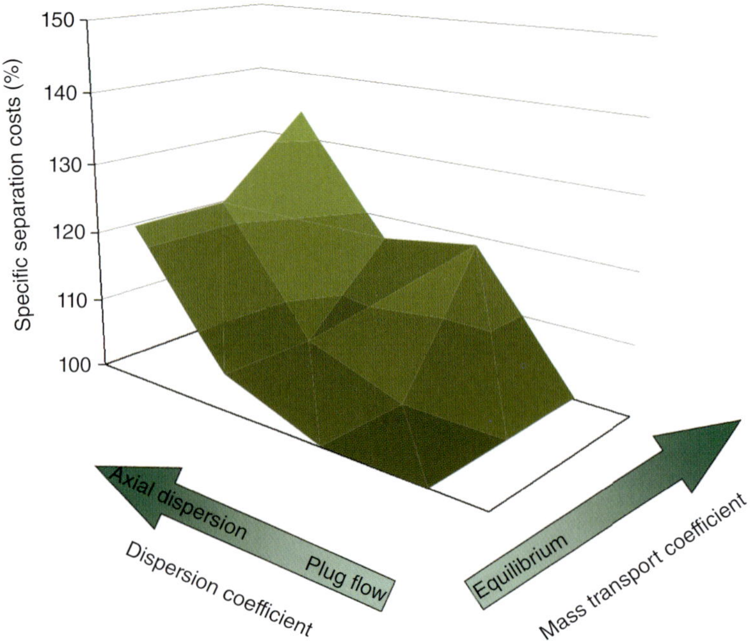

Figure 6.27 Separation cost increase with variation of kinetic and fluid dynamics – tea aroma.

to examine the sensitivity of the result to kinetics and fluid dynamics, the mass transfer coefficient, and fluid dynamic nonidealities have been varied.

- The reduction of the mass transfer coefficient by a factor of 10 from 0.1 to 0.01 $m\,h^{-1}$ results in a separation cost increase of 18%.
- With the increase of fluid dynamic nonidealities, a separation cost increase of up to 34% occurs. This can be seen from Figure 6.27.

Résumé

- The cost calculations for vanilla and for tea extract show very different influences of mass transfer coefficient and fluid dynamics.
- Any change in these parameters does not always have the same influence on the separation costs, but the influence is significant and has to be considered individually for each system.
- Therefore, with the aid of detailed models, both influences must be included in cost accounting. Sensitivity studies have to be carried out individually for each system under consideration.

6.5.2.11 Discussion and Conclusions

In this section, an initial systematic methodology and the necessary tools required for the CPD of complex mixtures have been presented. First, it was shown that a modeling depth is required which does not only take separation coefficients (so called shortcut models) into account but also mixture phase equilibrium as well as mass transfer coefficient and axial dispersion fluid dynamic nonidealities. With the aid of such models of different unit operations, any first process could be proposed, designed, evaluated, and then chosen on the basis of separation cost calculations.

Even in the conceptual process development and design stages of a project, thermodynamic equilibrium data has to be measured experimentally at laboratory scale using minimal volumes and with sufficient accuracy. Mass transfer coefficients, in addition to axial dispersion fluid dynamics parameters, must also be estimated using preexisting knowledge and/or correlations.

The currently known approaches for the determination of the thermodynamic equilibrium data of complex mixtures can be divided into three schemes. An approach for complex mixtures, which is based on molecular structure and databases of physical properties is not feasible. This is because only about 5% of data for plant-based extract components are available. Calculations with methods such as COSMO-RS are possible, if the structures of the components ware known, but the accuracy for detailed modeling is low. Nevertheless, these calculations can be used to reduce the experiments for feed characterization.

On the basis of this evaluation, only an experimental determination of model parameters by means of feed characterization is feasible for complex systems. It reduces the time and cost-intensive laboratory work most effectively. To determine model parameters for all unit operations known in separation engineering, miniaturized laboratory experiments, and standardized experimental procedures were designed and the minimal required sample volumes were determined via error calculations. The experimental plan and the volumes needed could be reduced significantly to a magnitude of about 1 l of broth.

The developed experimental approaches were tested successfully using two different plant extracts. The results show that mass transport and fluid dynamic parameters have a significant impact on the costs. Therefore it is necessary to have good estimates of the parameters and then to proof them in the later process development.

Moreover, the results confirmed that the preselection of unit operations, based on molecular group membership of target and side components, reduces the experimental plan drastically.

Symbols

g (J)	Gibbs enthalpy
h (J)	Enthalpy
p (bar)	Pressure
q_i (−)	Surface parameter
Q_k (−)	Surface of group
r (−)	Volume parameter

R (J mol^{-1}·K)	Gas constant
R_k (-)	Volume of group
S (-)	Dimensionless excess entropy
T (K)	Temperature
$\nu^{(i)}{}_K$ (-)	Frequency of group k in molecule i
V (cm^3 mol^{-1})	Molar volume
x (-)	Mole fraction
X_m (-)	Mole fraction
z (μm)	Spatial coordinate in the solid

Greek Letters

γ (-)	Activity coefficient
Θ (-)	Surface fraction
Θ_m (-)	Surface fraction
μ (J)	Chemical potential
σ (-)	Protective charge
Φ (-)	Volume fraction
Ψ_{mn} (-)	Adjustable parameter of UNIFAC

Indices

B	Boiling
C	Critical
C	Combinatorial
F	Formation
I	Component
M	Melting
R	Residual
S	Saturation

References

1. Capello, C., Fischer, U., and Hungerbühler, K. (2007) What is a green solvent? A comprehensive framework for the environmental assessment of solvents. *Green Chem.*, **9**, 927–934.
2. Estévez, C. (2009) in *Sustainable Solutions for Modern Economies*, Chapter 10 (ed. R. Höfer), RSC Publishing, Cambridge, pp. 407–423.
3. Jessop, P.G. (2011) Searching for green solvents. *Green Chem.*, **13**, 1391–1398.
4. Gmehling, J., Kolbe, B., Kleiber, M., and Rarey, J. (2012) *Chemical Thermodynamics for Process Simulation*, WILEY-VCH Verlag GmbH & Co.KGaA, Weinheim.
5. Lüdecke, D. and Lüdecke, C. (2000) *Physikalische-chemische Grundlagen der thermischen Verfahrenstechnik*, Springer-Verlag, Berlin.
6. Bondi, A. (1968) *Physical Properties of Molecular Crystals, Liquids and*

Glasses, John Wiley & Sons, Inc., New York.
7. Gmehling, J. (2012) Company Consortium for the Revision, Extension and Further Development of the Group Contribution Methods UNIFAC, Mod. UNIFAC (Do) and the Predictive Equation of State PSRK.
8. Koudous, I., Both, S., Gudi, G., Schulz, H., Strube, J., (2014) Process design based on physicochemical properties for the example of obtaining valuable products from plant-based extracts, *C. R. Chim.*. **17**, (3) 218–231.
9. Hildebrand, J.H. and Scott, R.L. (1950) *The Solubility of Nonelectrolytes*, 3rd edn, Reinhold Publishing Corp, 488 pp..
10. Hansen, C.M. (2007) *Hansen Solubility Parameters: A User's Handbook*, 2nd edn, CRC Press, Boca Raton, FL.
11. Klamt, A. (1995) Conductor-like screening model for real solvents: a new approach to the quantitative calculation of solvation phenomena. *J. Phys. Chem.*, **99**, 2224–2235.
12. Klamt, A. (2005) *COSMO-RS – From Quantum Chemistry to Fluid Phase Thermodynamics and Drug Design*, Elsevier.
13. Klamt, A., Eckert, F., and Arlt, W. (2010) COSMO-RS: an alternative to simulation for calculating thermodynamic properties of liquid mixtures. *Annu. Rev. Chem. Biomol. Eng.*, **1**, 101–122.
14. Cosmo Logic *http://www.cosmologic.de*
15. Eckert 1999
16. Abbott, S. (2008–2010) *HSPiP Software*, Steven Abbott TCNF Ltd, Suffolk.
17. OECD SIDS (1994) Report on Indigo Blue, *http://www.inchem.org/documents/sids/sids/482893.pdf* (accessed 12 August 2014).
18. Kassing, M., Jenelten, U., Schenk, J., and Strube, J. (2010) A new approach for process development of plant-based extraction processes. *Chem. Eng. Technol.*, **33** (3), 377–387.
19. Moity, L., Durand, M., Benazzouz, A., Pierlot, C., Molinier, V., and Aubry, J.M. (2012) Panorama of sustainable solvents using the COSMO-RS approach. *Green Chem.*, **14**, 1132–1145.
20. Durand, M., Molinier, V., Kunz, W., and Aubry, J.-M. (2011) Classification of organic solvents revisited by using the COSMO-RS approach. *Chem. Eur. J.*, **17**, 5155–5164.
21. Aichele, A. and Schwegler, H.W. (2008) *Die Blütenpflanzen Mitteleuropas*, **5** Bände, Kosmos (Franckh-Kosmos), Stuttgart.
22. Theophrastus, N.H.V. (Hrsg.) (2004) *Die Natur ist der Arzt – Salbei Salvia officinalis*, Verein zur Förderung der naturgemäßen Heilweise Verlag, München.
23. Seiler, M., Jork, C., Kavarnou, A., Arlt, W., and Hirsch, R. (2004) Separation of azeotropic mixtures using hyperbranched polymers or ionic liquid. *AIChE J.*, **50** (10), 2439–2454.
24. Smirnova, S.V., Torocheshnikova, I.I., Formanovsky, A.A., and Pletnev, I.V. (2004) Solvent extraction of amino acids into a room temperature ionic liquid with dicyclohexano-18-crown-6. *Anal. Bioanal. Chem.*, **378** (5), 1369–1375.
25. Fadeev, A.G. and Meagher, M.M. (2001) Opportunities for ionic liquids in recovery of biofuels. *Chem. Commun.*, **3**, 295–296.
26. Garcia-Chavez, L.Y., Schuur, B., and de Haan, A.B. (2012) COSMO-RS assisted solvent screening for liquid-liquid extraction of mono ethylene glycol from aqueous streams. *Sep. Purif. Technol.*, **97**, 2–10.
27. Carda-Broch, S., Berthod, A., and Armstrong, D.W. (2003) Solvent properties of the 1-butyl-3-methylimidazolium hexafluorophosphate ionic liquid. *Anal. Bioanal. Chem.*, **375**, 191–199.
28. Tang, B., Bi, W., Tia, M., and Row, K.H. (2012) Application of ionic liquid for extraction and separation of bioactive compounds from plants. *J. Chromatogr. B*, **904**, 1–21.
29. Blanchard, L. A., Hancu, D., Beckman, E. J., Brennecke, J. F. (1999) Green processing using ionic liquids and CO2. *Nature*, **399**, 28-29.
30. Blanchard, L.A. and Brennecke, J.F. (2001) Recovery of organic products from ionic liquids using supercritical

carbon dioxide. *Ind. Eng. Chem. Res.*, **40**, 287–292.
31. Scurto, A.M., Aki, S., and Brennecke, J.F. (2002) CO_2 as a separation switch for ionic liquid/organic mixtures. *J. Am. Chem. Soc.*, **124**, 10276–10277.
32. Pereiro, A.B. and Rodriguez, A. (2009) Application of the ionic liquid Ammoeng 102 for aromatic/aliphatic hydrocarbon separation. *J. Chem. Thermodyn.*, **41**, 951–956.
33. Werner, S., Szesni, N., Kaiser, M., Haumann, M., Wasserscheid, P., (2012) A Scalable Preparation Method for SILP and SCILL Ionic Liquid Thin-Film Materials. *Chem. Eng. Tech.*, **35**, (11), 1962–1967.
34. Abbott, A.P., Capper, G., Davies, D.L., Rasheed, R.K., and Tambyrajah, V. (2003) Novel solvent properties of choline chloride/urea mixtures. *Chem. Commun.*, **1**, 70–71.
35. Stolte, S., Abdulkarim, S., Arning, J., Blomeyer-Nienstedt, A.K., Bottin-Weber, U., Matzke, M., Ranke, J., Jastorff, B., and Thöming, J. (2007) Primary biodegradation of ionic liquid cations, identification of degradation products of 1-methyl-3-octyl-imidazolium chloride and electrochemical waste water treatment of poorly biodegradable compounds. *Green Chem.*, **10** (1), 214–224.
36. Garcia, H., Ferreira, R., Petkovic, M., Ferguson, J.L., Leitão, M.C., Gunaratne, H.Q.N., Seddon, K.R., Paulo, L., Rebelo, N., and Pereira, C.S. (2010) Dissolution of cork biopolymers in biocompatible ionic liquids. *Green Chem.*, **12**, 367–369.
37. Zech, O., Kellermeier, M., Thomaier, S., Maurer, E., Klein, R., Schreiner, C., and Kunz, W. (2009) Alkali oligoether carboxylates – a new class of ionic liquids. *Chem. Eur. J.*, **15**, 1341–1345.
38. Zech, O., Hunger, J., Sangoro, J.R., Iacob, C., Kremer, F., Kunz, W., and Buchner, R. (2010) Correlation between polarity parameters and dielectric properties of [Na][TOTO] – a sodium ionic liquid. *Phys. Chem. Chem. Phys.*, **12**, 14341–14350.
39. Abbott, A.P., Capper, G., Davies, D.L., Rasheed, R.K., and Tambyrajah, V. (2003) Novel solvent properties of choline chloride/urea mixtures. *Chem. Commun.*, 70–71.
40. Francisco, M., van den Bruinhorst, A., and Kroon, M.C. (2013) Low-Transition-Temperature Mixtures (LTTMs): a new generation of designer solvents. *Angew. Chem. Int. Ed.*, **52**, 3074–3085.
41. Zhang, Q., Vigier, K.D.O., Royer, S., and Jérôme, F. (2012) Deep eutectic solvents: syntheses, properties and applications. *Chem. Soc. Rev.*, **41**, 7108–7146.
42. Kahlen, J., Masuch, K., and Leonhard, K. (2010) Modelling cellulose solubilities in ionic liquids using COSMO-RS. *Green Chem.*, **12**, 2172–2181.
43. Freire, M.G., Santos, L.M.N.B.F., Marrucho, I.M., and Coutinho, J.A.P. (2007) Evaluation of COSMO-RS for the prediction of LLE and VLE of alcohols + ionic liquids. *Fluid Phase Equilib.*, **255** (2), 167–178.
44. Freire, M.G., Ventura, S.P.M., Santos, L.M.N.B.F., Marrucho, I.M., and Coutinho, J.A.P. (2008) Evaluation of COSMO-RS for the prediction of LLE and VLE of water and ionic liquids. *Fluid Phase Equilib.*, **268**, 74–84.
45. Guo, Z., Lue, B., Thomasen, K., Meyer, A.S., and Xu, X. (2007) Predictions of flavonoid solubility in ionic liquids by COSMO-RS: experimental verification, structural elucidation, and solvation characterization. *Green Chem.*, **9**, 1362–1373.
46. Xu, W., Chu, K., Li, H., Zhang, Y., Zheng, H., Chen, R., and Chen, L. (2012) Ionic liquid-based microwave-assisted extraction of flavonoids from bauhinia championii (Benth.) benth. *Molecules*, **17**, 14323–14335.
47. Kuhlmann, E. (2007) Entwicklung von Extraktions- und Absorptionssystemen auf Basis ionischer Flüssigkeiten für die Entschwefelung von Kohlenwasserstoffen. Dissertation. Universität Erlangen.
48. Bart, H.J., Hagels, H.J., Kassing, M., Jenelten, U., Johannisbauer, W., Jordan, V., Pfeiffer, D., Pfennig, A., Tegtmeier, M., Schäffler, M., and Strube, J. (2012)

Positionspapier der Fachgruppe Phytoextrakte – Produkte und Prozesse, DECHEMA.
49. Josch, J.P. and Strube, J. (2012) Characterization of feed properties for conceptual process design involving complex mixtures. *Chem. Ing. Tech.*, **84** (6), 918–931.
50. Kassing, M., Jenelten, U., Schenk, J., Hänsch, R., and Strube, J. (2012) Combination of rigorous and statistical modeling for process development of plant-based extractions based on mass balances and biological aspects. *Chem. Eng. Technol.*, **35**, 109–132.
51. Manski, R. and Bart, H.J. (2004) Gleichgewichtsmodellierung bei der Trennung von Cobalt und Nickel durch Reaktivextraktion. *Chem. Ing. Tech.*, **76** (7), 924–929.
52. Franke, M., Górak, A., and Strube, J. (2004) Auslegung und Optimierung von Hybriden Trennverfahren. *Chem. Ing. Tech.*, **76** (3), 199–210.
53. Franke, M.B., Nowotny, N., Ndocko, E.N., Gorak, A., and Strube, J. (2008) Design and optimization of a hybrid distillation/melt crystallization process. *AIChE J.*, **54** (11), 2925–2942.
54. Dewick, P.M. (2009) *Medical Natural Products: A Biosynthetic Approach*, John Wiley & Sons, Ltd, Chichester.
55. Josch, J.P., Both, S., and Strube, J. (2012) Characterization of feed properties for conceptual process design involving complex mixtures, such as natural extracts. *Food Nutr. Sci.*, **3**, 836–850.
56. Schulz, H. (2008) in *Modern Techniques for Food Authentication* (ed. D.W. Sun), Elsevier, Amsterdam.
57. DDBST *ddbonline.ddbst.com/* (accessed 12 August 2014).
58. DIPPR *http://dippr.byu.edu/students/* (accessed 12 August 2014).
59. Reaxys (2014) Reaxys *https://www.reaxys.com/reaxys/secured/search.do*.
60. Shojaei, Z.A., Linforth, R.S.T., and Taylor, A.J. (2007) Estimation of the oil water partition coefficient, experimental and theoretical approaches related to volatile behaviour in milk. *Food Chem.*, **103** (3), 689–694.
61. Schmid, C., Steinbrecher, R., and Ziegler, H. (1992) Partition coefficients of plant cuticles for monoterpenes. *Trees*, **6**, 32–36.
62. Kontogeorgis, G.M. and Folas, G.K. (2010) *Thermodynamic Models for Industrial Applications: From Classical and Advanced Mixing Rules to Association Theories*, John Wiley & Sons, Ltd, Chichester.
63. von Watzdorf, R., Bausa, J., and Marquardt, W. (1999) Shortcut methods for nonideal multicomponent distillation: 2. Complex columns. *AIChE J.*, **45** (8), 1615–1628.
64. Urdaneta, R., Bausa, J., Brüggemann, S., and Marquardt, W. (2002) Analysis and conceptual design of ternary heterogeneous distillation processes. *Ind. Eng. Chem. Res.*, **41**, 3849–3866.
65. Lienqueo, M.E., Leser, E.W., and Asenjo, J.A. (1996) An expert system for the selection and synthesis of multistep protein separation processes. *Comput. Chem. Eng.*, **20** (1), 189–194.
66. Lienqueo, M.E., Mahn, A., Navarro, G., Salgado, J.C., Perez-Acle, T., Rapaport, I., and Asenjo, J.A. (2006) New approaches for predicting protein retention time in hydrophobic interaction chromatography. *J. Mol. Recognit.*, **19** (4), 260–269.
67. Scherpian, P. and Schembecker, G. (2009) Scaling-up recycling chromatography. *Chem. Eng. Sci.*, **64** (18), 4068–4080.
68. Faber, R., Li, P., and Wozny, G. (2003) Sequential parameter estimation for large-scale systems with multiple data sets. 1. Computational framework. *Ind. Eng. Chem. Res.*, **42** (23), 5850–5860.
69. Faber, R., Li, P., and Wozny, G. (2004) Sequential parameter estimation for large-scale systems with multiple data sets. 2. Application to an industrial coke-oven-gas purification process. *Ind. Eng. Chem. Res.*, **43** (15), 4350–4362.
70. Faber, R., Arellano-Garcia, H., Li, P., and Wozny, G. (2007) An optimization framework for parameter estimation of large-scale systems. *Chem. Eng. Process.*, **46** (11), 1085–1095.

71. Gmehling, J. (2009) Present status and potential of group contribution methods for process development. *J. Chem. Thermodyn.*, **41** (6), 731–747.
72. Fredenslund, A., Jones, R.L., and Prausnitz, J.M. (1975) Group-contribution estimation of activity coefficients in non- ideal liquid mixtures. *AIChE J.*, **21** (6), 1086–1099.
73. Eckert, F. and Klamt, A. (2002) Fast solvent screening via quantum chemistry. COSMO-RS approach. *AIChE J.*, **48** (2), 369–385.
74. Mahn, A., Zapata-Torres, G., and Asenjo, J.A. (2005) A theory of protein-resin interaction in hydrophobic interaction chromatography. *J. Chromatogr. A*, **1066** (1-2), 81–88.
75. Salgado, J.C., Rapaport, I., and Asenjo, J.A. (2006) Predicting the behaviour of proteins in hydrophobic interaction chromatography 1: using the hydrophobic Imbalance, No. HI) to describe their surface amino acid distribution. *J. Chromatogr. A*, **1107** (1-2), 110–119.
76. Salgado, J.C., Rapaport, I., and Asenjo, J.A. (2006) Predicting the behaviour of proteins in hydrophobic interaction chromatography 2. Using a statistical description of their surface amino acid distribution. *J. Chromatogr. A*, **1107** (1-2), 120–129.
77. Salgado, J.C., Andrews, B.A., Ortuzar, M.F., and Asenjo, J.A. (2008) Prediction of the partitioning behaviour of Proteins in aqueous two-phase systems using only their amino acid composition. *J. Chromatogr. A*, **1178** (1-2), 134–144.
78. Siebert, K.J. (2003) Modeling protein functional properties from amino acid composition. *J. Agric. Food Chem.*, **51** (26), 7792–7797.
79. Hachem, F., Andrews, B.A., and Asenjo, J.A. (1996) Hydrophobic partitioning of proteins in aqueous two-phase systems. *Enzyme Microb. Technol.*, **19** (7), 507–517.
80. Andrews, B.A., Schmidt, A.S., and Asenjo, J.A. (2005) Correlation for the partition behavior of proteins in aqueous two-phase systems: effect of surface hydrophobicity and charge. *Biotechnol. Bioeng.*, **90** (3), 380–390.
81. Haskard, C. A. and Li-Chan, E. C. (1998) Hydrophobicity of Bovine Serum Albumin and Ovalbumin Determined Using Uncharged, and Anionic, ANS, Fluorescent Probes. *J. AGR. FOOD CHEM.*, **46** (7), 2671–2677.
82. Mazza, C.B., Sukumar, N., Breneman, C.M., and Cramer, S.M. (2001) Prediction of protein retention in ion-exchange systems using molecular descriptors obtained from crystal structure. *Anal. Chem.*, **73** (22), 5457–5461.
83. Ladiwala, A., Rege, K., Breneman, C.M., and Cramer, S.M. (2005) A priori prediction of adsorption isotherm parameters and chromatographic behavior in Ion-exchange systems. *Proc. Natl. Acad. Sci. U.S.A.*, **102** (33), 11710–11715.
84. Malmquist, G., Nilsson, U. H., Norrman, M., Skarp, U., Stromgren, M. and Carredano, E. (2006) Electrostatic Calculations and Quantitative Protein Retention Models for Ion Exchange Chromatography. *J. CHROMATOGR. A.*, **115** (1-2) 164-186.
85. Xu, L. and Glatz, C.E. (2009) Predicting protein retention time in Ion-exchange chromatography based on three-dimensional protein characterization. *J. Chromatogr. A*, **1216** (2), 274–280.
86. Gu, Z. and Glatz, C.E. (2007) A method for three-dimensional protein characterization and its application to a complex plant, extract. *Biotechnol. Bioeng.*, **97** (5), 1158–1169.
87. Stahl, E. and Schild, W. (1986) *Isolierung und Charakterisierung von Naturstoffen*, Gustav Fischer Verlag, Stuttgart.

7
Water as Green Solvent for Extraction of Natural Products

Loïc Petigny, Mustafa Zafer Özel, Sandrine Périno, Joël Wajsman, and Farid Chemat

7.1
Introduction

Water is the most abundant molecule on the earth, covering 70% of its surface. Life processes on this planet are crucially dependent on the presence of water. It is often recognized that water has tremendous benefits as a green extraction solvent because it is not only inexpensive and environmentally benign but it is also nonflammable and nontoxic, providing opportunities for clean processing and pollution prevention. The water molecule contains three nuclei, one modestly heavy (oxygen), and two light ones (hydrogen). Because of water's electronic structure, the oxygen atom has a slight negative charge on it and the hydrogen atoms have a slightly positive charge [1]. When water molecules are closed together, their positive and negative regions are attracted, these attractive forces are known as *hydrogen bonds*. The energy associated with a hydrogen bond is 8–40 kJ/mol. The O–H bond lengths are 0.9572 Å and the average HOH bond angle is 104.52°, slightly smaller than the tetrahedral angle (109.5°). The molecule is very small with a hard sphere diameter of 2.75 Å. The smallness of the water molecule has important consequences for the hydration of solutes. The contours of the total electron density in the HOH plane of the water molecule indicate a shape not far from spherical. The water molecule has a dipole moment of 1.85 D with two partial positive charges on the hydrogen atoms and a single zone of negative charge on the oxygen atom.

It is well known that water molecules are strongly associated in the liquid phase by H-bonding. Hydrogen bonds result dominantly from electrostatic interactions with a light contribution of covalent bond formation. The H-bond interactions in liquid water constitute a 3-D H-bond network with localized and structured clustering. Therefore, the liquid structure is not determined by hard-core repulsions between the individual water molecules but rather by intermolecular and directional H-bonding. The broad variation of H-bond angles further demonstrates the noncrystalline arrangement of the water molecules and appears to indicate that some bond angle distortion in the H-bonds is rather easily allowed in the liquid. It is important to recognize that intermolecular interaction

Green Extraction of Natural Products: Theory and Practice, First Edition.
Edited by Farid Chemat and Jochen Strube.

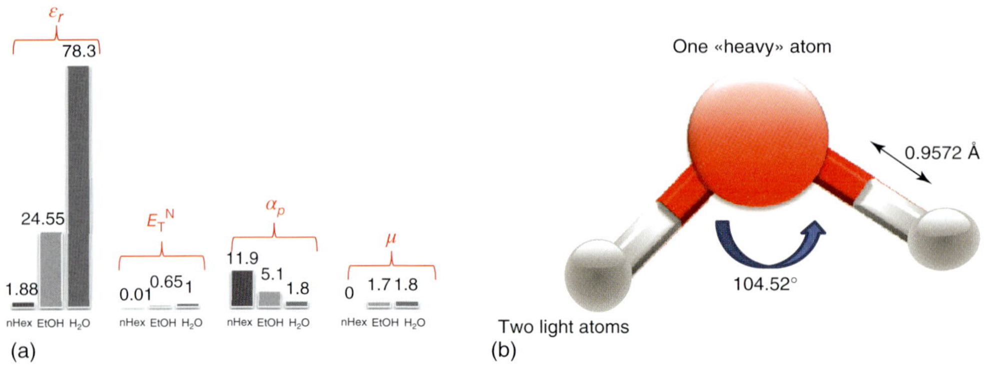

Figure 7.1 (a) Selected solvent properties of water and others extraction solvents and (b) single water molecule and its average geometry.

between water molecules is aided by the substantial dipole polarizability of water. While the dipole moment of a single water molecule is 1.85 D, the dipole moment in the liquid is significantly larger, ranging from about 2 to 4 D, depending on the nature of the water–water interaction [2]. Moreover, water is amphoteric. At 298 K and 1 atm, the ionization constant of water is 14. The proton and hydroxide concentrations are so small that the water activity is almost unity. In Figure 7.1, solvent parameters are described for the most used solvents in extraction. The macroscopic dielectric constant of a solvent (εr) is the most popular parameter to characterize the polarity of the medium and controls the ionic dissociation of salts. The high dielectric constant of water, related to the dipole orientations in the H-bond network, makes the liquid a highly polar solvent. The polarity of water falls off at high temperatures and pressures as anticipated, because the H-bond network breaks down. This means that when the temperature increases, the solubility of organic compounds increases much more than expected for the natural effect of water temperature. For example, at 250 °C, water exhibits a density and polarity similar to those of acetonitrile at room temperature. The dielectric constant drops rapidly with temperature, and it falls from 78.5 at 25 °C to 27.5 at 250 °C [3].

In addition to the environmental advantages of using water instead of organic solvents, purification of products is normally facilitated because, once cooled, the organic products are not soluble in ambient temperature water. Moreover, Dimroth–Reichardt $E_T{}^N$ values are regarded as the parametric characteristics of the polarity of the solvent molecules; actually, they reflect the acceptor strength of the solvent. α_p represents the polarizability of the molecule, that is, the ability of the electron cloud to deform easily. Finally, dipole moment μ has already been discussed in the preceding. Comparison of the μ values for water and ethanol shows how little the ethyl group contributes to charge separation.

The properties of water are reflected in the thermodynamics of hydration of ionic, polar, and apolar molecules. Recently, it was concluded that in absence of solute–water H-bonding, the water molecules keep their 3-D H-bonding network as intact as possible, which, for not too big solutes, results in a preference for tangential orientation of the water OH-bonds relative to the apolar surface. Of course, this direction preference of the water molecules decreases with increasing temperature because of the breakdown of the H-bonds. In fact, large heat capacity changes are characteristic for water and find their origin in large fluctuations of the H-bond network, associated with compensating enthalpy and entropy effects [4].

At a critical concentration, it is unavoidable that the large hydrophobic hydration shells start to overlap, leading to mutually destructive breakdown of these water arrangements. These H-bonding interactions result in a solvent-induced sticking together of hydrophobic surface areas of the two solutes, primarily driven by the gain in entropy that comes from the release of water molecules into bulk water from the hydrophobic hydration shells. Hydrophobic hydration shells are quite voluminous. Depending on the nature of the hydrophobic molecules, one can distinguish pairwise interactions, the formation of small aggregates ("moving units," e.g., in the case of hydrotropes), and the formation of larger aggregates (bulk hydrophobic interactions) as in the case of surfactant aggregates such as micelles and vesicles. The ultimate state of aggregation will, of course, involve phase separation.

Using water as an alternative solvent during synthesis, purification, or extraction is a research area that has an impact in several fields of modern chemistry. The advantages of using water as alternative solvent for extraction includes reduced environmental impact, selective extraction, use of simple equipment, no hazards, faster start-up, and simplification of process steps.

This chapter presents a complete picture of current knowledge on using water as alternative solvent for natural product extraction. It provides the necessary theoretical background and some details about extraction using water, the technique, the mechanism, some applications, and environmental impacts.

7.2 Maceration

7.2.1 Principle and Process

Maceration, the soaking of a vegetal matrix in a solvent, is a very old and simple method of extraction. Used every day, from a simple cup of tea to the industrial production of several thousand liters, this method presents a great variety of uses and conditions that we discuss in this section.

Maceration consists in intimate interaction and mixing of a solvent, here water, and a solid matrix. It aims to transfer a selection of compounds from the matrix inside plant roots, bark, wood, or leaves to solvent bulk. It is then followed by separation of water from the solid matrix by filtration or centrifugation. Compounds such as antioxidants, anthocyanins, tannins, aromas, pigments, and so on, can be extracted by maceration. The overall mechanism of solid–liquid extraction can be described in four steps: interaction between solvent and matrix surface, solvent diffusion inside the matrix, solute diffusion in solvent, solute transfer outside the matrix surface toward the bulk solvent. The plant is a living organism. It is composed of cells surrounded by a solid cellular wall composed of cellulose, hemicellulose, pectin, and often, lignin. The cell wall is one of the borders that solvent and solutes must cross. The more porous and broken the cell walls are, the easier the transfer can be done. The solvent, water, is polar, possesses a density, a viscosity, and an activity which can impair solubilization, the transfer, or the contact with the matrix. However, by changing parameters and physical conditions of extraction, they can be overcome. Changing pH can help solubilization and solvent diffusion inside the matrix. Reducing the particle size gives an important factor increase in solvent–matrix interaction. A good stirring process not only ensures matrix and solvent homogenization but also final solute transfer. Higher temperature of extraction can increase solvent–matrix interaction but also transfer, solubilization, and diffusion. It can, however, accelerate or trigger the degradation of thermosensitive molecules. It becomes then a competition between extraction and degradation.

Maceration can be defined in different ways, as temperature and process change.

- Maceration is done at room temperature.
- Decoction is done while the water is boiling.
- Infusion, such as teas, is a controlled and high-temperature extraction.
- Batch process: the matrix is set in a vat with water; a stirring device is often added to the apparatus. The method can only be used up to a certain amount of matrix/solvent ratio, as its higher ratios tend to impair the homogenization of the solvent. In addition, solubility and thermodynamic phase equilibrium of solute between solvent and matrix can set an upper limit to the extraction.
- Percolation: similarly to coffee brewing, percolation is a semicontinuous process which fixes the vegetal matrix in a bed. Fresh solvent is flowed through it at different temperatures. This method can circumvent the solubility or thermodynamic equilibrium limits as new solvent is fed, thus enabling the complete solute stripping of the matrix.
- Continuous extraction: this brings two flows, one of solvent and one of plant matrix, in contact This method is best used for large-scale process of single plant-based products. They can be compact in design; however, it is the rate of flow process that makes the process high capacity. They are often built so that the flows of matrix and solvent are in counter current. This can be achieved in different ways from an endless screw to the use of pumps and conveyor belts for true continuous flows of both solvent and solid material, or for a semi continuous mode.

7.2.2
Applications

Maceration feasibility in extractpreparation has been investigated on different matrices, as shown in Table 7.1. *Litchii chinensis* flowers cannot produce fruits

Table 7.1 Applications and experimental conditions – examples of plant material maceration.

Matrix	Analytes	Experimental conditions	References
Litchi flowers *Litchi chinensis* Sonn.	Polyphenols, anthocyans, flavonoids, tanins	Conventional maceration 100 °C, 30 min, 2.5–5% (w/v)	[5]
Meadowsweet Filipendula ulmaria L.	Quercetin and salicylic acid	Conventional maceration 90 °C, 15 min, 2.5% (w/w)	[6]
Cistus incanus leaves	Polyphenols, flavonoids	Conventional maceration 90 °C, 5 min, water Milli-Q, 1.5% (w/w)	[7]
Green tea leaves *Camellia sinensis* L.	Polyphenols	Microwave extraction, 80 °C, 30 min, 5% (w/v), 600 W	[8]
Flame tree flowers *Delonix regia.*	Anthocyans, flavonol, and phenolic acid	Ultrasound extraction, 1 h, 1%(w/v) whole flower, water, 25 kHz, 150 W	[9]

when they are not pollinated and fall. They can represent a significant mass to be discarded. However, these flowers contain several molecules of interest such as polyphenols, flavonoids, condensed tannins, and proanthocyanidins. Yang *et al.* [5] studied the extraction of such molecules by maceration. At 100 °C, litchi water extracts showed important amounts of polyphenols and flavonoids. Anthocyanins and proanthocyanidins were found as well as condensed tannins and ascorbic acid. Such water extract showed a good protective effect on cardiovascular health. Meadowsweet, a European plant has reported anti-inflammatory effects. Water extracts from this plant can be used for incorporation in beverages. A study and optimization of this aqueous extraction was done by Harbourne *et al.* [6]. They analyzed the antioxidants, tannins, quercetin, and salicylic acid contents of such extracts. In addition to extract composition, they measured a color factor, as color and its intensity impacts the final product's "desirability." Extraction parameter optimization was conducted on temperature, time of extraction, and pH for an increase in antioxidants, quercetin, and salicylic acid, and avoiding tannin extraction. pH change did not significantly improve the yields of extraction. However, a large increase in total phenolic content was observed when temperature was increased from 20–40 °C to 90–100 °C. This increase was attributed to cell breakdown due to high temperature, thus releasing the cell contents in water. As tannins were slow to extract, an extraction time of 15 min was enough to extract the suitable compounds, keeping a low tannin content.

Riehle *et al.* [7] encountered another kind of competition in infusion of *Cistus incanus*. In traditional folk medicine, this Mediterranean plant is used for its anti-inflammatory and antimicrobial properties. The plant's properties are mainly associated with high antioxidant activity from phenolic acids and flavonol compounds. However, these flavonoids were suspected to be thermally sensitive. A temperature influence was therefore expected and antioxidant stability in the extraction process was studied. Further optimization was done on plant extraction by selection of water quality and extraction time. Even though it was found that higher temperature increased overall antioxidant yield, the amount of some polyphenols, such as myricitin, decreased and these were lost by thermal degradation. Water quality played a role as lower water hardness accelerated the yield of extraction. Parameter optimization was the key to a good extract quality and to avoid undesirable components or degradation of desirable compounds. However, a change of technology is needed to further improve extraction efficiency. Such technological change was studied by Nkhili *et al.* [8]. Polyphenol extraction from tea leaves was done through conventional heating of water and compared to extraction by microwave heating. At the same temperature and with a shorter time of extraction, microwave heating showed increased extraction yields of total polyphenols and individual compounds. In addition, microwave heating accelerated the extraction, preserved thermosensitive components and saved on energy. Cold maceration intensification and activation is possible as studied by Adjé *et al.* [9]. Polyphenol extraction from the *Delonix regia* flower was studied

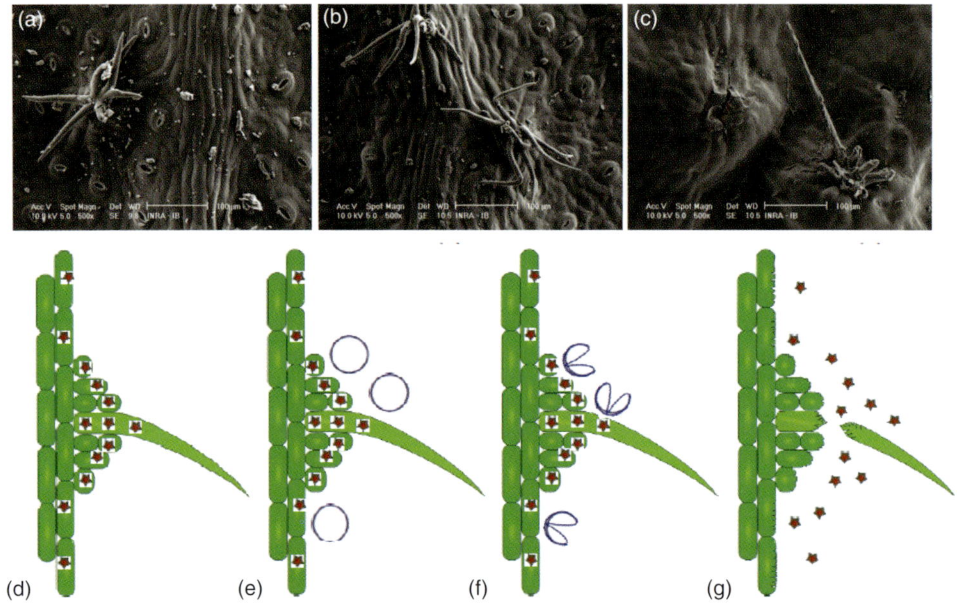

Figure 7.2 (a) Boldo leaf, magnification 500 on the original picture scale; (b) Boldo leaf after 2 h of maceration, magnification 500 of the original picture scale; (c) Boldo leaf after UAE at optimized conditions, magnification 500, on the original picture scale; (d) plant profile with a trichome at the surface of the leaf; (e) generation of a cavitation bubble; (f) collapse of the cavitation bubble which generates a micro-jet directed toward the surface; and (g) abrasion of the surface, breaking of the trichome, and release of soluble material in the surrounding medium.

under various conditions of maceration. Ultrasound-assisted extraction allowed a decrease in extraction time to one-third of conventional extraction time. Ultrasound effects are illustrated in Figure 7.2. As the process used water, it could not increase the total level of polyphenols extracted.

7.3 Subcritical Water Extraction

7.3.1 Principle and Process

Subcritical water extraction (SWE) is also known as superheated water extraction, hot water extraction, pressurized hot water extraction, hot liquid water extraction, pressurized low polarity water extraction, and high-temperature water extraction. SWE is a green extraction technique. Similar to maceration by its batch or percolation dynamics, SWE uses water in conditions of pressure and temperature that are quite different, and therefore needs specific equipment (Figure 7.3). The effects and uses of such conditions on water properties are described in the following section. Table 7.2 shows organic solvent properties, Hansen parameters, and subcritical water temperature equivalents. Water possesses many thermodynamic properties. These can be influenced by a number of physical parameters, among them temperature and pressure. Together, they can modify its thermal behavior, density, and viscosity, thus defining the physical state of water: solid, liquid, or gas. With higher temperatures and pressures, these properties bring water over a point known as the *critical point*. This point is defined at 221 bar and 374 °C. Over this pressure and temperature, water is neither a liquid nor gas, but in an in-between

Figure 7.3 Tournaire, solid–liquid extractor EF2 150 l, *http://www.tournaire-equipement.com.*

Table 7.2 Organic solvents properties, Hansen parameters, and subcritical water temperature equivalent.

Solvent	Dielectric constant ε	Hansen parameter δ_d (MPa$^{1/2}$)	Hansen parameter δ_p (MPa$^{1/2}$)	Hansen parameter δ_h (MPa$^{1/2}$)	Subcritical water temperature (°C)
DMSO	48	18.4	16.4	10.2	150
Acetonitrile	38	15.3	18.0	6.1	175
Methanol	33	14.7	12.3	22.3	200
Ethanol	24	15.8	8.8	19.4	275
Acetone	21	15.5	10.4	7.0	300
Chloroform	5	17.8	3.1	5.7	<374 T_C

state called a *supercritical* state. With pressure and temperature under the critical point, transition between liquid and gas is discontinuous; over the critical point, however, the transition is continuous. What we focus on is the domain of liquid water under this critical point and over what we can achieve in usual conditions: between 100 and 374 °C; 1 and 221 bar. This domain is called *subcritical* water. Keeping water in liquid state at high temperature is possible thanks to an appropriate pressure, thus avoiding its vaporization. Water solvent properties change considerably [10].

Among the varying properties of water, the dielectric constant ε can heavily impact on water solvent power. At 25 °C and atmospheric pressure, water presents a high value of $\varepsilon = 80$. This high value reflects the presence of extensive hydrogen bonds formed by water molecules. As temperature rises, this value decreases steadily along with viscosity and density. Also, water diffusivity increases with temperature. At 250 °C and 50 bar, water has the ε of ethanol = 27, close to that of methanol ($\varepsilon = 33$) at 25 °C and 1 bar. Under these conditions, water is able to solubilize more apolar molecules. Therefore, by tuning water temperature, we can adapt its polarity and target molecules of interest for extraction [10, 11]. Miller and Hawthorne [12] studied the change in the solubility of molecules with increase in water temperature. It was shown that D-limonene solubility in water increased by a factor of 57 from 20 °C –and 1 bar to 200 °C and 65 bar. Carvone solubility under similar change in conditions was increased by a factor of 25. Miller and Hawthorne [13] looked at the solubilities of hydrophobic organic compounds such as chlorothalonil in subcritical water at temperatures from 25 to 250 °C and pressures (30–70 bar). When the temperature of chlorothalonil was raised from 25 to 250 °C, the solubility increased by a factor of 130 000. Pressure increase is the key to obtaining subcritical water; however, evaluation of its influence on water polarity properties shows limited relevance [14, 15]. Therefore, the minimum pressure to apply on the system can be established from the temperature/pressure diagram of water phase changes [15].

SWE was first achieved in laboratories, by assembling a system of HPLC pumps and a GC oven with an empty HPLC column. Typically, an SWE system consists of a syringe or HPLC pump, oven, valves, water reservoir, extraction vessel, and collection vial. The basic SWE system setup is similar to that in accelerated solvent extraction (ASE) and supercritical fluid extraction systems [16]. In 1995, Dionex commercialized a laboratory apparatus able of providing subcritical water conditions, termed *accelerated solvent extraction*. This apparatus can offer two ways of extraction: static or dynamic. Firstly, in static extraction, plant material is loaded into the column, along with a neutral filler such as diatomaceous earth. Then, the column is sealed in the oven and filled with water at an appropriate pressure and temperature, using volumetric pumps. The water and matrix stay in static contact for a set time in these conditions before being emptied by the opening of a valve and a purging stream of gas, often nitrogen. The extract stream is cooled before pressure release and is collected [10, 17]. Another way of static SWE is to use a closed vessel to undergo the process. The vessel is a high-pressure-resistant reactor, pressurized and heated to the required conditions. The second way of SWE is dynamic extraction. This method is similar to percolation and uses the apparatus described previously. Water flows continuously into a plant-filled column and depletes the matrix under subcritical conditions. A new factor is added with pressure and temperature: subcritical water flow. In addition to extraction composition, concentration is also affected by the flow rate. Low flow rates and long extraction times can give higher yields and solutes provided with higher concentrations are not thermal sensitive. The choice between low or high flow rates depends on the balance between the thermal sensitivity of solutes, economic costs in terms of time and solvent used, as well as concentration of the final extract [10, 18, 19]. Larger-scale SWE exists, especially in industry, as seen in Figure 7.4.

7.3.2 Applications

SWE is still a small-scale process. SWE is used in research and development with potential industrial applications. Subcritical water began to be used in the extraction of nonpolar to mid-polar groups of compounds such as polycyclic aromatic hydrocarbons (PAHs) [20] and essential oils [21]. Its use gained popularity and now there are more uses such as the extraction of flavor and fragrance compounds, essential oils, and phenolic compounds. The literature on toxic solvents has been increasing from year to year. The use of toxic solvents in chemical laboratories and the chemical industry is considered a significant problem affecting the health and safety of workers and causing environmental pollution. Most solvents are organic molecules with hazardous and toxic properties and can be costly in themselves and/or in their disposal. Their use, storage, and disposal also cause environmental problems.

Barbero *et al.* [22] developed a reliable way to extract capsaicin from hot peppers using SWE for analytical purposes. Ground pepper was loaded into a steel cell of an ASE apparatus from Dionex. They studied the extraction process with temperatures up to 200 °C with a pressure of 10 MPa in static mode. The stability of the target molecule was verified at high temperature and no degradation was

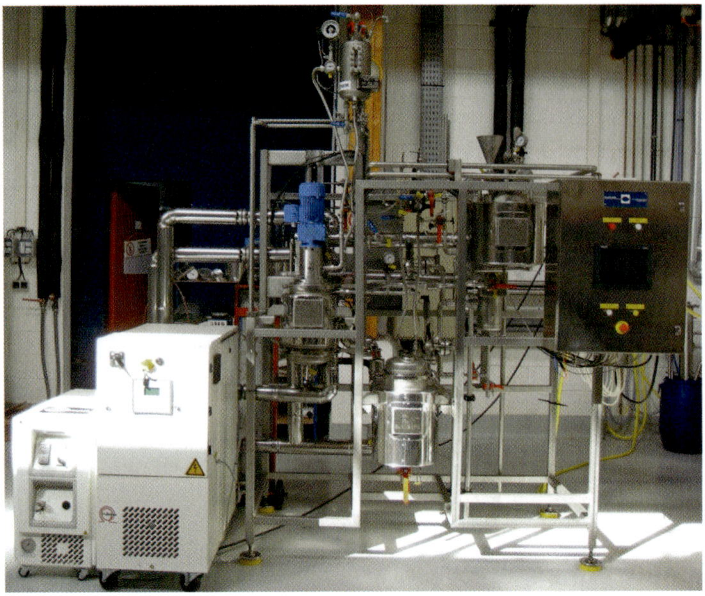

Figure 7.4 Tournaire, subcritical water extractor, *http://www.tournaire-equipement.com*.

detected after extraction. This is likely due to the absence of oxygen in the cell and water, which can greatly accelerate oxidation degradation at high temperature. They found that the best temperature to activate the extraction was at 200 °C. Then, they compared it to various mixes and solvents such as water, ethanol, methanol, and water mixtures of alcohols at room temperature. Of these, only methanol was suitable for capsaicin extraction from peppers at room temperature.

Kumar *et al.* [23] compared not only different solvents to SWE but also different extraction techniques. Maceration and Soxhlet extraction were compared to SWE for polyphenol recovery from sea buckthorn leaves. Leaf maceration in water was carried out for 24 h, the Soxhlet in 70% ethanol–water used as references for comparison. The SWE was done using a similar apparatus described previously, a Dionex ASE in static mode. The studied temperatures ranged from 25 to 200 °C for an extraction tie of 15 min and a pressure of 100 bar. A temperature of 150 °C showed to give the best yields of total polyphenols and of individual molecules such as quercetin -3-galactoside, kaempferol, or isohamnnetin compared to the conventional extractions. However, at a temperature of 20 °C, polyphenol yields decreased considerably, under the yield levels of maceration and Soxhlet extraction, showing a possible thermal degradation of target molecules.

Mannitol extraction from olive leaves can be difficult because of its possible conversion into sorbitol. Ghoreishi and Shahrestani [24] designed its extraction with SWE. As previously, an ASE system is used. This time a static and dynamic approach was followed: a period of static SWE was done followed by a flow of subcritical water. Temperature, pressure, initial static extraction time, and flow rate were studied. With a 5 min initial static time, the following flow rate had little influence on the extraction yield. Temperature increase from 60 to 100 °C also increased the yield. However, temperatures over 100 °C decreased the yield mainly due to thermal degradation. Pressure change was revealed to have an influence on yield. An optimum of extraction was reached at 50 bar. Such variations can be explained by changes in the partition coefficient induced by pressure, a phenomenon also described by David *et al.* [25]. Compared to Soxhlet extraction, SWE is much faster, safer, solvent saving, and cost effective.

A comparison between SWE and supercritical CO_2 was done by Fernández-Ponce *et al.* [26] for antioxidant extraction from mango leaves. On the same system, two dynamic methods were used on crushed leaves. A steady flow, either supercritical CO_2 or SWE, passed through the matrix. With a temperature of 100 °C and 4 MPa, use of SWE substantially increased the extraction yield compared to supercritical CO_2. A higher antioxidant activity was also noted, as well as higher concentrations of individual molecules such as mangiferin and quercetin. SWE was found more suitable for the extraction of these molecules than supercritical CO_2.

Lang and Wai [27] compared the SWE method to different solvent extraction methods for recovery of terpene trilactones. Ginkgo biloba leaves contains such compounds, ginkgolide and bilobalides. On an ASE system, with a temperature range between 20 and 140 °C and pressure up to 100 bar, they studied a static and dynamic extraction mode. It was revealed that the solutes can be extracted safely for temperatures up to 60 °C. Higher temperatures and, therefore, subcritical water, were unsuitable for the recovery of terpene trilactones due to degradation. However, pressure increase showed an increase in extraction yield. They explained this increase not because of solvent block or slow diffusion rates, but because of the natural protection against water possessed by the ginkgo leaves. Higher pressure seemed to overcome the leaf surfaces' strong hydrophobic structure, which protect the tree from drought. Effectively, SWE method is best suited for this matrix and solutes, not for its solvent power tuning properties, but as physical assistance for the extraction.

A comparative study of volatiles extracted from *Rosa damascena* using various techniques, namely SWE, water distillation, and direct thermal desorption was carried out with the analysis being performed using comprehensive two-dimensional gas chromatography mass spectrometry (GC × GC-TOF/MS). It was found that SWE gave a slightly higher oil yield than water distillation. The percentages of highly fragrant and desirable phenylethylalcohol content was found to be significantly higher using the SWE (38.14%) and direct thermal desorption (36.52%) techniques compared to the water distillation (1.92%) technique [28].

Essential oils from the leaves of *Thymbra spicata* L. were extracted using subcritical water. Different temperatures (100, 125, 150, and 175 °C) pressures (20, 60, and 90 bar) and flow rates (1, 2, and 3 ml min^{-1}) were used. Pressure did not show any difference on the extraction yield. A flow rate of 2 ml min^{-1} and 150 °C gave a better extraction yield. The kinetic study showed that SWE of essential oils from *Thymbra spiacata* was mostly completed within 15 min [16].

Zou *et al.* [29] extracted hydrophobic bio-oil from *Danaliella tertiolecta* using SWE. It is a new application for SWE. The temperatures that can be safely used depend on the type of material being from which the oil is extracted. A high amount of bio-oil (37%) was obtained in 30 min at a temperature of 360 °C from marine microalgae using SWE [29].This study suggested that SWE is a very good extraction technique for bio-oil production.

7.4 Enzymatic Assistance

7.4.1 Principles and Process

With softer conditions of temperature and pressure, use of water additives to help the extraction of natural product is reviewed in this section. The first assisted additives reviewed are enzymes. They are macromolecules, proteins synthesized by bio-organisms that act as biocatalysts for most chemical reactions happening in processes sustaining life. Most of the industrially used enzymes are from either fungus or microorganisms. Only a very small number of enzymes originate from plants or animals [30]. Each enzyme is highly specific for one particular kind of reaction, with specific reactive compounds. One enzyme molecule is capable of catalyzing a reaction million times, having a very high turnover rate. These large proteins present an active site acting similarly to a key and lock system. It is in this site that the reaction occurs through a catalytic group in the active site or by close encounter between two reacting molecules. Enzymes are classified into six categories: oxidoreductases catalyze oxidoreduction reactions; transferases transfer a chemical group from one molecule to another; hydrolases concern the hydrolytic cut of molecular bounds; lyases cut molecular bounds through elimination; isomerases modify the structure or geometry of molecules; and ligases bind two molecules together.

As enzymes are highly specific, each application, reaction, or matrix, requires a specific enzyme. They need particular pH and temperature conditions to be operational, and extreme condition leads to a denaturation of the enzyme, rendering it either inactive or destroyed [31, 32]. Enzymes such as amylases, cellulases, and proteases help open cell walls of the vegetal matrix and they also can reduce particle size. Cell wall disruption and size reduction are important in the release of oils and analytes in the solvent [33, 34]. Enzymatic treatment is also a matrix preparatory step, to denature its structure in order to activate the final extraction. Essential oils containing matrixes can be treated by enzyme to facilitate their extraction by hydrodistillation [35]. As enzymes possess a high turnover rate, some enzymes can be expensive or difficult to obtain in large quantities. As with catalysts or solvents, it would be beneficial to recycle them. In the case of aqueous enzyme extraction of oils, enzymes stay in water phase and can be reused with minimal treatment; this is the basis of homogeneous enzymatic catalysis. Enzymes can also be adsorbed on a substrate in order to recover it after an extraction step; or the substrate can be recovered from the solid residual treated by enzymes. This method is called *heterogeneous enzymatic catalysis* [36]. Both homogeneous and heterogeneous enzymatic catalysis can be used with equipment already available for conventional maceration. This include continuous and batch processes.

7.4.2 Applications

Enzymes are used for various applications, a selection of these are shown in Table 7.3. Bayberry kernels were used as source of oil for an enzymatic extraction by Zhang *et al.* [37]. A system with temperature and pH control is needed to operate at the right conditions for enzymatic treatment. They optimized a process of extraction with an experimental design, with enzyme amount, solid–liquid ratio, temperature, and time of extraction. On kernels with a 61% maximum yield, the aqueous enzymatic extraction reached 31%. Rosenthal *et al.* [38] studied oil and protein extraction from soy bean flour with combined effects of four enzymes: cellulose, hemicellulose, pectinase, and protease. They used a process similar to bacteria culture by employing a fermenter unit reactor. As protease is a basic pH optimum enzyme and the three other enzymes are acid pH enzymes, the process

Table 7.3 Applications and experimental conditions – examples of plant material extraction by subcritical water.

Matrix	Analytes	Experimental conditions	References
Sea buckthorn leaves *Hippophae rhamnoides*	Antioxidants	Static extraction Dionex ASE-350, 2 g, 33 ml cell, 20–200 °C, 10 MPa, 15 min, pure water	[20]
Peppers *Capsicum*	Capsaicinoids	Static extraction Dionex ASE-200, 0.7 g, 11 ml cell, 50–200 °C, 10 MPa, 30 min, pure water	[19]
Olive leaves *Olea europaea*	Manitol	Static and dynamic extraction 2.5 g crushed leaves with crushed glass (60 : 40 w/w), 60–100 °C, 3–11 MPa, 300s static then 2–33 μl s^{-1}, pure water	[18]
Ginkgo biloba leaves	Terpene trilactones	Static and dynamic extraction, 1 g, 20–140 °C, 10 MPa, 15 min static, then 1 ml min^{-1} to collect 20 ml, water with 0.2% acetic acid	[23]
Mango leaves *Mangifera indica*	Polyphenols, mangiferin, quercetin	Dynamic extraction, SF100 Thar Tech, 15 g, 100 °C, 4 MPa, pure water, 3 h, 10 ml min^{-1} flow	[22]

must be done in two steps with two distinct pH ranges, one for each enzyme to be able to work properly. The first hour of extraction was done at pH 5 for cellulose, hemicellulose, and pectinase to react, at a temperature of 37 °C. Then, the pH was adjusted to pH 8 for 15 min, with protease addition. It was found that, mainly due to protease, and on high-granulometric flour, proteins, and oils yields increased. Two proteases were used sequentially by Nyam *et al.* [39] for aqueous extraction of Kalahari melon seed oil. Neutrase® and Flavourzyme® were used separately. They were studied along with temperature, pH, time of extraction, and enzyme concentration as the parameters of an experimental design. While both of them benefit from longer extraction time and concentration, the two enzymes behaved differently with temperature and pH: Neutrase benefited from higher pH and temperature, whereas Flavourzyme was not significantly affected by temperature and pH modifications. Comparing optimum conditions of extraction for both enzymes, it was found that Flavourzyme is more suitable for oil recovery. It is noted that aqueous phase mixing is important as higher speeds lead to emulsion formation and yield decrease. A more suitable and slower mixing allows oils droplets to coalesce and be separated more easily from water.

Incomplete solvent extraction of lycopene from extruded tomato products was increased by Dehghan-Shoar *et al.* [40] by using pancreatin to digest the product before extraction, thus freeing lycopene from tomato peels. This study shows the use of enzymes not as a mean of extraction, but as an intensification process. The incorporation of this additional step enhanced extracted lycopene proportion from solvents by more than 2.5-fold between products extracted only by solvents.

Jiao *et al.* [35] investigated the extraction process enhancement of *Forsythia suspense* fruit essential oil by an enzymatic pretreatment leading up to microwave-assisted hydrodistillation. The whole process was done with microwaves as the medium of heat source. Three different enzymes were studied: cellulase, hemicellulose, β-glucosidase; their concentration, incubation temperature, microwave power, and incubation time were investigated. Cellulase was considered the best choice for pretreatment. The increase in the concentration of enzymes led to a notable increase in yield until 0.5% concentration. A study of the incubation temperature showed 40 °C optimum temperature for enzyme efficiency. Microwave power investigations showed optimum heating power of 500 W. Yield decrease over this optimum power was seen, affected by temperature homogeneity and denaturation from a part of the enzymes. An obvious increase in extraction yield was observed up to 25 min of digestion and additional time kept yield at a steady level. Comparisons among conventional hydrodistillation, microwave hydrodistillation, and enzymatic-assisted microwave extraction showed clearly that the microwave process was much faster than hydrodistillation. Furthermore, enzymatic-assisted method with the same overall time of extraction as microwave hydrodistillation, showed the hydrodistillation step to be much faster with a higher yield of essential oil.

7.5
Micellar Extraction

7.5.1
Principle and Process

Enzymatic extraction is not the only method using water additive to enhance natural product extractions in water. The method of micellar-assisted extraction makes uses of surfactant molecules. This technique aims to use surfactants properties to form micelles, creating apolar phases dispersed in water. This allows solubility enhancing of more hydrophobic molecules in water [41]. Surfactants are organic molecules presenting two parts: one side of the molecule is apolar, lipophilic, and the other is polar, hydrophilic. They exist in different shapes and electrical properties: nonionic, cationic, anionic, and zwitterionic. This influences greatly the way these molecules self-assemble. At low concentration, surfactants are freely solubilized in water and populate the different interfaces of the aqueous medium. Over a concentration called the *critical micellar concentration* (*CMC*), surfactant molecules self-assemble, and form clusters called *micelles*. The nature of surfactant molecules and their concentration in the medium can influence the structure and of these micelles, which take different geometries. In water, the simplest micelle forms a sphere, bordered by surfactant molecules. They are positioned with their polar end outside and apolar end turned inside. Other structures are based on layers, cylinders, 3D networks, liposome, and also inverse micelles with polar ends turned inside the micelle [42, 43]. All these structures are thermodynamically stable under defined conditions involving pH, temperatures, other solutes, and nature of surfactants. They provide a pseudo-organic phase inside water, making hydrophobic solutes transfer inside the micelle, similar to a partition. Their effects can be seen in the extraction of hydrosoluble rosmarinic acid, and hydrophobic carnosic acid from rosemary in Figure 7.5. The macroscopic effects of surfactants on the rosemary surface are seen in Figure 7.6.

Changes in micelles stability are more noticeable with nonionic surfactants. Above a set temperature termed the *cloud point* (*CP*), micelles formed by surfactant molecules are no longer stable and they break down. Two phases are formed: an aqueous solution of surfactant close to CMC and a phase rich in surfactants [44]. This is however not possible with ionic surfactants; instead they exploit an effect of salting-out with addition of sodium chloride in solution [45]. The cloud point is not a fixed value and can change with surfactants nature, pH, and salt presence in solution [46]. This property of separation is used as a preconcentration method, first to isolate the heavy metal complex and then for purification of biological molecules [42, 46]. Cloud-point extraction (CPE) uses this property: a first step of micellar extraction below the cloud point of hydrophobic molecules is done. Then, the surfactant solution is separated from the matrix and brought over the cloud point. Hydrophobic molecules stay in the surfactant-rich phase and a centrifugation step can be added in order to enhance phase separation.

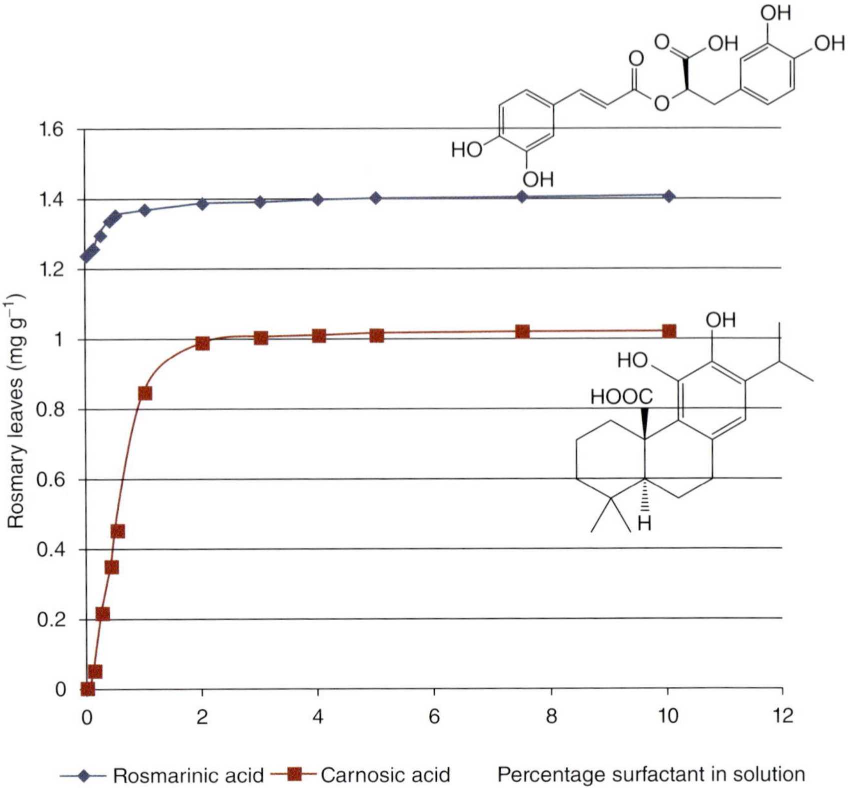

Figure 7.5 Rosmarinic acid and carnosic acid extraction with various surfactant concentrations.

7.5.2
Applications

Various surfactants that were used and documented with different methods of extraction can be seen in Table 7.4. Bi *et al.* [47] compared three surfactants to extract cryptotanshinone (CT) and tanshinone (TA) from *Salvia miltiorrhiza* Bunge. They used a surfactant-assisted maceration followed by a CPE. This study aimed to compare the extracted amounts of CT and TA in sodium dodecyl sulfate (SDS) Tween 20, and Triton X-100 solutions. This last surfactant was proven to be the most effective among the surfactants used. Extraction yield increased with concentration of surfactant up to a maximum, where further increase of Triton X-100 concentration did not increase extracted amounts of CT and TA. A solid liquid ratio of 1 : 70 was chosen mainly for economic reasons, as the optimum yield was for a higher amount of solvent needed for the extraction. Kinetics of extractions up to 160 min was done, showing a maximum yield of extraction after 100 min. Then a cloud-point step was studied. Parameters such as pH, sodium chloride concentration, temperature, and time equilibrium were optimized.

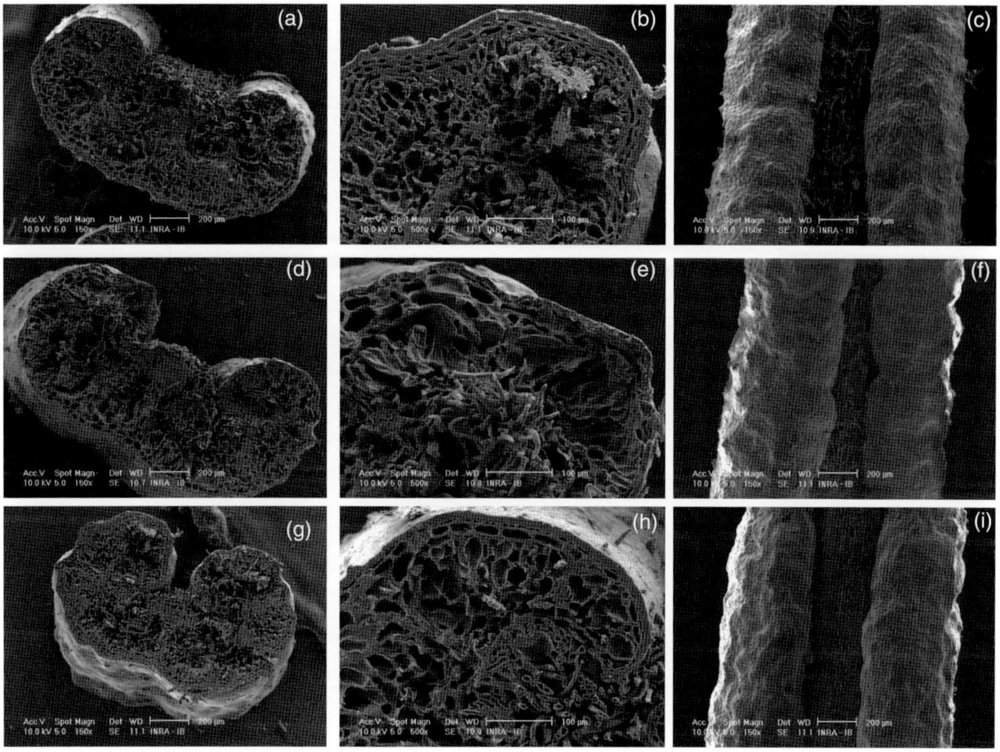

Figure 7.6 (a–c) Magnification 150, 500, and 150 respectively, on the original picture scale, rosemary leaves unprocessed; (d–f) magnification 150, 500, and 150 respectively, on the original picture scale, rosemary leaves after 1 h maceration in water at 40 °C; and (g–i) magnification 150, 500, and 150 respectively, on the original picture scale, rosemary leaves after 1 h maceration in a 1% solution of surfactant at 40 °C.

A variation of pH from the natural pH of extraction reduced the cloud-point step efficiency. Reaching a 15% sodium chloride concentration was enough to reach a maximum separation of TA and CT in the surfactant phase by using a salting in/out phenomenon and reducing compound solubility in water. Higher temperature than 65 °C decreased the yield of extraction, as the compounds are sensitive to temperature. Cloud-point equilibrium was reached in 10 min. Further analysis of aqueous phase originating from CPE shows that no detectable TA and CT remains, effectively concentrating the molecules of interest in the surfactant phase.

The recovery ability of a surfactant, Genapol X-080, for individual and mixtures of polyphenols and tocopherols from olive mill waste water was studied using CPE by Gortzi *et al.* [48]. It was shown that each polyphenol was extracted with different efficiency. Their recovery appeared to be proportional to surfactant concentrations with the observation of a plateau reached for several molecules. A link between surfactant concentration and polyphenol recovery was also observed in

Table 7.4 Applications and experimental conditions examples of enzymatic plant material extraction.

Matrix	Analytes	Experimental conditions	References
Soy bean flour *Glycine max*	Oil, proteins	0.1% Hemicellulase, 0.45% pectinase, 2% protease enzymes concentration, 5–20% (w/w) flour/water, 50 °C, pH 5, 1 h then pH 8, 15 min	[32]
Bayberry kernels *Myrica rubra*	Oil, fatty acids	3.17% (w/w) mixed enzymes (cellulose/neutral protease = 1/2, w/w), 4.91 ml g^{-1} water/kernel, 51.6 °C, 4 h	[31]
Kalahari melon seed *Citrullus Lanatus*	Oil	Neutrase 0.8 l at 25 g kg^{-1} in water, 58 °C, pH 7, 31 h	[33]
Tomato peels *Solanum lycopersicum* L	Lycopene	Pancreatine 10 mg.ml^{-1}, 1 : 5 w/l peels/water, 37 °C, pH 6.5, 20 min	[34]
Forsythia suspense fruits	Essential oil	Cellulase, hemicellulase β-glucosidase 0.5% concentration, 1 : 10 w l^{-1} fruit/water, 50 °C, pH 5, 25 min, 500 W microwave assistance	[29]

the polyphenol mixture. The tocopherols recovery was studied in the same way and the results showed a quantitative recovery from 5% of surfactant with no significant increase for higher concentration. With a second step of CPE, total recovery of tocopherols was possible. Multiple-step CPE was applied on olive mill wastewater, studying surfactant concentration impact and the number of CPE steps. A high recovery of molecules could be reached with three CPE steps and 2% of surfactants, with individual phenol recoveries higher than 40% in one step with a low concentration of surfactant. As Genapol X-080 allows a cloud point at 55 °C, phenol degradation could be avoided, making a single or multistep CPE an interesting method compared to conventional liquid–liquid extraction.

Katsoyannos *et al.* [49] also used multistep CPE, for both olive mill wastewater and red-flesh orange juice. Several surfactants were studied for phenols and carotenoid recovery. Span 20 and poly ethylene glycol 400 (PEG 400) showed similar low yields: below 20% in phenols and close to 10% in carotenoids. Much higher recoveries were demonstrated by the Tween surfactant. 92% and 96% of phenols could be recovered by Tween-20 and Tween-80, respectively. The major difference between the two Tween surfactants was in carotenoid recovery: 64% and 40% for Tween-80 and Tween-20, respectively. After a surfactant concentration optimization of a two-step CPE using Tween 80, 79.8% of carotenoids were recovered by the use of 7 + 7% of surfactants. However, CPE parameters were not optimized in this study, and it is implied that such optimization would allow a higher recovery of carotenoids. Kiathevest *et al.* [50] investigated micellar extraction parameters followed by CPE for concentration of anthraquinones from the roots of *Morinda citrifolia,* and optimized them at room temperature and pressure. They used Genapol X-080 and Triton X-100 and studied concentration of surfactant effects in anthraquinone recovery. They expected a better extraction by Genapol X-080 surfactant use; however a slightly higher yield obtained compared to Triton X-100 extraction is explained by its lower viscosity, thus suggesting a better mass transfer. They further improved anthraquinone recovery by ultrasound assistance. Ultrasound cavitation effects in surfactant solution with solid samples improved their interactions and anthraquinones mass transfer. It is interesting to note that, from 1% to 16% for Triton X-100 and 4% to 16% of Genapol X-080 with ultrasound-assisted micellar extraction, anthraquinone recovery was at the maximum, with no significant increase. At 20%, the recovery was lower. This is explained by an increase in micelles formations induced by ultrasound cavitation, increasing solution viscosity, and decreasing mass transfer from solid to liquid.

Sun *et al.* [51] combined microwaves and CPE surfactant for glycyrrhizic acid and liquiritin extraction from licorice root. They aimed for solvent-free extraction and preconcentration. Various concentrations of surfactants were studied under different temperature of extraction, heated by microwave irradiation. Both liquiritin and glycyrrhizic acid yields were maximum with 5% of surfactant. No significant increase in yield was observed with triton X-100 concentration over 5%. Higher temperature gave higher extraction yields: 100 °C gave a maximum yield after 5 min. The extracts were concentrated afterward with a cloud-point operation. They added sodium chloride in solution, activating a cloud point at 50 °C. It was observed that higher concentration of NaCl decreased the volume of surfactant phase, reaching a maximum preconcentration factor of 13.5. A higher salt concentration also allowed better recoveries (>98%) of glycyrrhizic acid and liquiritin. The concentrated surfactant phase was then analyzed by high performance liquid chromatography (HPLC). Surfactant presence did not disturb the chromatographic analysis, making this approach an effective extraction process for analysis of liquorice root.

7.6 Hydrotropes

7.6.1 Principles and Process

While micellar extraction is similar to a two-liquid-phase extraction method, hydrotropic extraction only involves one continuous liquid phase. This method uses organic molecules called *hydrotropes*. Hydrotropes are short organic molecules with a structure comparable to surfactants: they have a lipophilic and a hydrophilic part. They were first described by Neuberg in 1916 [52] as organic salts able to increase the solubility in water of an organic compound with low solubility. However, unlike surfactants, they cannot form micelles. Hydrotrope molecules can be anionic, cationic, zwitterionic, and nonionic. Hydrotropy describes the increase in solubility of hydrophobic molecules in water by addition of water-soluble organic molecules. Hydrotropes may not form micelles but they can, however, aggregate following their affinity areas [53]. In addition, they have a concentration threshold at which the solubility of hydrophobic compounds increases significantly. This threshold is called the *minimum hydrotropic concentration* (*MHC*). It is also at this point that surface tension changes suddenly. MHC, unlike typical CMC value of of 1 – 10 mM, is more usually of the order of 1 M, which makes the mass of hydrotropes in water tens or hundreds grams per liter. MHC is specific to hydrotropic compounds and nonsensitive to organic compounds it helps to solubilize.

Hydrotrope-assisted solubilization is done through aggregation of hydrotropes around hydrophobic molecules. The hydrotropes orient themselves with their hydrophobic area toward organic molecules and their hydrophilic part toward bulk solvent. Hydrotropic extraction relies mainly on the addition of enough hydrotropes in the solution. The different hydrotropes can be identified by their formula – sodium salicylate, cumene sulfonate, vitamin C, sodium benzoate, and sodium xylene sulfonate being some examples of hydrotropes [54–59] – and they are characterized by their MHC. Different methods of hydrotropic extraction exist. Usually, plant extracts are made and they take part in a formulation, adding in the last steps a formulating agent which can be considered as a hydrotrope. However, many hydrotropes are used in final formulation in cosmetic industries. Therefore, it is possible to change the order of addition of the hydrotrope and use it in the first step for plant extraction. In theory, the amount of hydrotrope used in the extraction process is the amount needed in the final formulation and no additional hydrotrope may be needed. A second method involves preparation of a hydrotropic solution at a selected concentration, followed by plant material extraction. Once filtered, the pH and temperature are changed to specifically crystallize the desired compound and thereby obtain it in pure form. A third method uses ternary models between water, hydrotropes, and the target organic molecule. The organic molecule is extracted with a solution of hydrotropes. To recover the compound, water is added until dephasing or crystallization [60] of the solute occurs. The organic compound is recovered by decantation or centrifugation. Once again, under suitable conditions, we can selectively extract and separate the compounds.

7.6.2 Applications

A selection of several uses of hydrotropes and their extraction methods are shown in Table 7.5.

Dandekar *et al.* [61] used two different hydrotropes to extract limonin from sour orange seeds. They studied sodium cumene sulfonate (Na-CuS) and sodium salicylate (Na-Sal) as hydrotropes. The design of experiment included temperature, hydrotrope concentration, and raw material loading as the varying parameters. A

Table 7.5 Applications and experimental conditions examples of plant material micellar extraction.

Matrix	Analytes	Experimental conditions	References
Sage *Salvia miltiorrhiza* Bunge	Cryptotanshinone, tanshinone	Cloud point extraction, 0.8 mol l^{-1} Triton X-100 in water, 1/70 S l^{-1} (g ml^{-1}) water, 65 °C, 10 min	[41]
Olive mill wastewater	Antioxidants	Cloud point extraction, Genapol-X080 5% in water, 5% NaCl, 20 min, 50 °C, pH 2.5–3.5	[42]
Red flesh orange juice	Phenols, carotenoids	Cloud point extraction, Tween-80 5%, 20% NaCl, pH = 2.5–3.5, 55 °C 30 min	[43]
Morinda citrifolia roots	Anthraquinones	Pressurized hot water-assisted micellar extraction, Triton X-100 1%, 0.5 g/50 ml, 30 °C, 2 h, 35.8 kHz, 270 W	[44]
Licorice root *Glycyrrhiza glabra* L.	Glycyrrhizic acid, liquiritin	Microwave-assisted micellar extraction, 2 g roots, 250 ml 5% (volume) Triton X-100 in water, 3–5 min, 100 °C by focused microwaves	[45]

temperature of 45 °C, 10% raw material loading, and a concentration of 2 M for both hydrotropes was needed to achieve the maximum extraction yield of extraction. A slightly better extraction of limonin was done with Na-CuS solution. They were able to recover limonin by three different methods: extract dilution under MHC, dilution under MHC in acidic water, partition using dichloromethane. The dilution below MHC was simple but slow. In acidic water, it was fast, but did not work with Na-Sal as it was converted into salicylic acid. Dichloromethane partition was efficient for the salicylate method but inefficient with Na-CuS. Desai and Parikh [62] also used Na-Sal and Na-CuS as hydrotropes, to extract citral from lemongrass. In this study, they found that Na-Sal allowed a better solubilization of citral than Na-CuS, indicating that a good affinity is needed between the hydrotropes and the target compounds. A Taguchi method was used to optimize citral yield through hydrotrope concentration, amount of lemongrass, temperature, and particle size. After extraction, citral was recovered by dilution under MHC. Desai and Parikh also considered hydrotrope recycling after citral recovery; by dilution under MHC, the solution is reconcentrated by evaporation until the desired hydrotrope concentration is reached. They managed, with their method of extraction, recovery, filtration, and reconcentration, to retain 95% of the initial hydrotropes. They also observed an unchanged yield in citral extraction with reused hydrotrope solution (Table 7.6).

Mishra and Gaikar [63] studied forskolin extraction from *Coleus forskohlii* roots by using Na-Sal, Na-CuS, and p-toluen sulfonate (Na-pTS). They first assessed

Table 7.6 Applications and experimental conditions – examples of plant material hydrotropic extraction.

Matrix	Analytes	Experimental conditions	References
Sour orange *Citrus aurantium* L. seeds	Limonin	10% seed (w/w) in water, 2 M sodium salicylate or sodium cumene sulfonate, 60 °C, 6 h	[55]
Cymbopogon flexuosus (Steud.) Wats.	Citral	5% w/w in water, 1.75 M sodium salicylate or sodium cumene sulfonate, 30 °C, 8 h	[56]
Coleus Forskohlii Roots	Forskolin	5% w/w in water, 2 M sodium cumene sulfonate, or *p*-toluene sulfonate, 20–90 °C, 3 h	[57]
Black pepper *Piper nigrum*	Piperine	10% $w\,l^{-1}$ ground pepper in water, sodium *n*-butyl benzene sulfonate 0.05–3.4 $mol\,l^{-1}$, 30 °C, 2 h	[58]
Boswellia serrata Resin	Boswellic acid	10% $w\,l^{-1}$ ground resin, sodium *n*-butyl benzene sulfonate 0.05–2 $mol\,l^{-1}$, 25 °C, 2 h	[54]

the solubility of pure forskolin in hydrotropic solutions at various concentrations and temperature. Increase of temperature or hydrotrope concentration also increased the solubility of forskolin, which is nearly insoluble in pure water, regardless of temperature. From 50 $mg\,l^{-1}$ in pure water at 90 °C, forskolin solubility reached close to 3 $g\,l^{-1}$ in 90 °C 2 M Na-CuS solution, the hydrotrope displaying the highest solubility enhancement. A compromise between pulverized roots amount with solid particle size was studied, as stirring operation could become inefficient. Larger particles gave lower yield but higher purity of forskolin recovery, and smaller particles gave higher yield, with lower purity. Extraction yield and recovery purity were also studied with different hydrotropes and also with methanolic extraction. Solutions of Na-pTS and Na-Sal gave lower yields, but managed to obtain a purity close to 60%. Methanol gave the highest forskolin extraction yield of 75% with the lowest purity of 8%. However, the Na-CuS solution helped to reach 70% yield with 85% purity.

Raman and Gaikar [64] studied piperin extraction from black pepper. Parameters such as hydrotrope choice, its concentration, temperature of extraction, and particle size were optimized. They compared hydrotropic extraction and a petroleum ether Soxhlet method. It was observed that they obtained a high purity of piperin extract with hydrotropic process. They studied cell degradation process as well as the way hydrotropic solution penetrated membranes. Microscopic studies showed that piperin was mainly contained in pericarp cells. The different hydrotropes used had different effects on each cell structure, changing extraction efficiency of piperin, but also on other molecules extracted and affecting piperin extraction selectivity. Raman and Gaikar [60] used hydrotropic solutions on *Boswellia serrata* resin in order to extract boswellic acids. Crushed gum resin was mixed with different solutions of hydrotropes. Solubility studies showed a selectivity of hydrotrope, where Na-pTS is less efficient than Na-CS and sodium *n*-butyl benzene sulfonate. This was confirmed by modeling molecular interactions. The aggregation of hydrotrope molecules around boswellic acid molecules depended on the geometry and affinity of the molecules to the hydrotropes. The extraction kinetics study showed a second-order process, with a slow first step of extraction due to hydrotrope penetration resistance inside the gum structure of the resin; and a second fast extraction step which slowed down to a plateau showing the end of extraction process.

7.7 Conclusion

Advantages and drawbacks of using water as a solvent with variable geometry have been compared: conventional maceration using water, subcritical water, micellar-aqueous extraction, hydrotropic-aqueous extraction, and enzymatic-aqueous

Table 7.7 Summary of extraction processes.

Name	Maceration	Subcritical water	Enzymatic assistance	Micellar extraction	Hydrotropes
Description of system	Contact between solvent and matrix	Water at high temperature and pressure to keep it liquid	Addition of enzymes in specific conditions to denature the matrix	Addition of surfactants to create micelles	Addition of organic salts to increase solubility
Investment	Low	High	Medium	Low to medium	Low to medium
Ease of operation	High	Medium	Low	Medium	High
Solubility	Only hydrophilic	Hydrophilic to mild lipophilic	Hydrophilic and lipophilic release	Hydrophilic to mild lipophilic	Hydrophilic to light lipophilic
Sample size (l)	>1000	50	>1000	>1000	>1000
Extraction time	High	Medium	High	Medium	Low
Main disadvantages	Limited by solubility	Not suitable for thermosensitive molecules, high pressure	Price of enzyme, precise conditions necessary	Elimination of surfactants	High concentration of hydrotropes needed
Main advantages	Well mastered, large scale	Tunable solvent power	Matrix pretreatment	Simultaneous extraction of polar and apolar molecules	Hydrotropes can be part of the next step of formulation

(Table 7.7). Water as potential alternative solvent for extraction but also for purification of natural products is increasingly efficient at directly transferring knowledge into technology for commercial development. Using water as alternative solvent with variable geometry makes use of physical and chemical phenomena that are fundamentally different from those applied in conventional extraction techniques. The systems developed to date clearly indicate that water offers net advantages in terms of environmental impact, selectivity, operating time, energy input, and preservation of thermolabile compounds.

References

1. Finney, J.L. (2004) Water? What's so special about it? *Philos. Trans. R. Soc. Lond. B Biol. Sci.*, **359** (1448), 1145–1165. doi: 10.1098/rstb.2004.1495
2. Silvestrelli, P.L. and Parrinello, M. (1999) Water molecule dipole in the gas and in the liquid phase. *Phys. Rev. Lett.*, **82**, 3308–3311. doi: 10.1103/PhysRevLett.82.3308
3. Marcus Lindström, U. (2007) *Organic Reactions in Water: Principles, Strategies and Applications*, Blackwell Publishing, Oxford, Ames, IA.
4. Southall, N.T., Dill, K.A., and Haymet, A.D.J. (2002) A view of the hydrophobic effect. *J. Phys. Chem. B*, **106** (3), 521–533. doi: 10.1021/jp015514e
5. Yang, D.-J., Chang, Y.-Y., Hsu, C.-L., Liu, C.-W., Wang, Y., and Chen, Y.-C. (2010) Protective effect of a litchi (Litchi chinensis Sonn.)-flower-water-extract on cardiovascular health in a high-fat/cholesterol-dietary hamsters. *Food Chem.*, **119** (4), 1457–1464. doi: 10.1016/j.foodchem.2009.09.027
6. Harbourne, N., Jacquier, J.C., and O'Riordan, D. (2009) Optimisation of the aqueous extraction conditions of phenols from meadowsweet (Filipendula ulmaria L.) for incorporation into beverages. *Food Chem.*, **116** (3), 722–727. doi: 10.1016/j.foodchem.2009.03.017
7. Riehle, P., Vollmer, M., and Rohn, S. (2013) Phenolic compounds in Cistus incanus herbal infusions—Antioxidant capacity and thermal stability during the brewing process. *Food Res. Int.*, **53** (2), 891–899. doi: 10.1016/j.foodres.2012.09.020
8. Nkhili, E., Tomao, V., El Hajji, H., El Boustani, E.-S., Chemat, F., and Dangles, O. (2009) Microwave-assisted water extraction of green tea polyphenols. *Phytochem. Anal.*, **20** (5), 408–415. doi: 10.1002/pca.1141
9. Adjé, F., Lozano, Y.F., Lozano, P., Adima, A., Chemat, F., and Gaydou, E.M. (2010) Optimization of anthocyanin, flavonol and phenolic acid extractions from Delonix regia tree flowers using ultrasound-assisted water extraction. *Ind. Crops Prod.*, **32** (3), 439–444. doi: 10.1016/j.indcrop.2010.06.011
10. Mustafa, A. and Turner, C. (2011) Pressurized liquid extraction as a green approach in food and herbal plants extraction: a review. *Anal. Chim. Acta*, **703** (1), 8–18. doi: 10.1016/j.aca.2011.07.018
11. Teo, C.C., Tan, S.N., Yong, J.W.H., Hew, C.S., and Ong, E.S. (2010) Pressurized hot water extraction (PHWE). *J. Chromatogr. A*, **1217** (16), 2484–2494. doi: 10.1016/j.chroma.2009.12.050
12. Miller, D.J. and Hawthorne, S.B. (2000) Solubility of liquid organic flavor and fragrance compounds in subcritical (hot/liquid) water from 298 K to 473 K. *J. Chem. Eng. Data*, **45** (2), 315–318. doi: 10.1021/je990278a
13. Miller, D.J. and Hawthorne, S.B. (1998) Method for determining the solubilities of hydrophobic organics in subcritical water. *Anal. Chem.*, **70** (8), 1618–1621. doi: 10.1021/Ac971161x
14. Kronholm, J., Revilla-Ruiz, P., Porras, S.P., Hartonen, K., Carabias-Martínez, R., and Riekkola, M.-L. (2004) Comparison of gas chromatography–mass

spectrometry and capillary electrophoresis in analysis of phenolic compounds extracted from solid matrices with pressurized hot water. *J. Chromatogr. A*, **1022** (1-2), 9–16. doi: 10.1016/j.chroma.2003.09.052

15. Deng, C., Li, N., and Zhang, X. (2004) Rapid determination of essential oil in Acorus tatarinowii Schott. by pressurized hot water extraction followed by solid-phase microextraction and gas chromatography–mass spectrometry. *J. Chromatogr. A*, **1059** (1-2), 149–155. doi: 10.1016/j.chroma.2004.10.005.
16. Ozel, M.Z., Gogus, F., and Lewis, A.C. (2003) Subcritical water extraction of essential oils from Thymbra spicata. *Food Chem.*, **82** (3), 381–386. doi: 10.1016/S0308-8146(02)00558-7
17. Roger, M.S. (2002) Extractions with superheated water. *J. Chromatogr. A*, **975** (1), 31–46. doi: 10.1016/S0021-9673(02)01225-6
18. Dawidowicz, A.L., Rado, E., and Wianowska, D. (2009) Static and dynamic superheated water extraction of essential oil components from Thymus vulgaris L. *J. Sep. Sci.*, **32** (17), 3034–3042. doi: 10.1002/jssc.200900214
19. Eikani, M.H., Golmohammad, F., and Rowshanzamir, S. (2007) Subcritical water extraction of essential oils from coriander seeds (Coriandrum sativum L.). *J. Food Eng.*, **80** (2), 735–740. doi: 10.1016/j.jfoodeng.2006.05.015
20. Miller, D.J., Hawthorne, S.B., Gizir, A.M., and Clifford, A.A. (1998) Solubility of polycyclic aromatic hydrocarbons in subcritical water from 298 K to 498 K. *J. Chem. Eng. Data*, **43** (6), 1043–1047. doi: 10.1021/Je980094g
21. Kutlular, O. and Ozel, M.Z. (2009) Analysis of essential oils of origanum onites by superheated water extraction using GCxGC-TOF/MS. *J. Essent. Oil Bear. Plants*, **12** (4), 462–470.
22. Barbero, G.F., Palma, M., and Barroso, C.G. (2006) Pressurized liquid extraction of capsaicinoids from peppers. *J. Agric. Food. Chem.*, **54** (9), 3231–3236. doi: 10.1021/jf060021y
23. Kumar, M.S.Y., Dutta, R., Prasad, D., and Misra, K. (2011) Subcritical water extraction of antioxidant compounds from Seabuckthorn (Hippophae rhamnoides) leaves for the comparative evaluation of antioxidant activity. *Food Chem.*, **127** (3), 1309–1316. doi: 10.1016/j.foodchem.2011.01.088
24. Ghoreishi, S.M. and Shahrestani, R.G. (2009) Subcritical water extraction of mannitol from olive leaves. *J. Food Eng.*, **93** (4), 474–481. doi: 10.1016/j.jfoodeng.2009.02.015
25. David, M.D., Campbell, S., and Li, Q.X. (2000) Pressurized fluid extraction of nonpolar pesticides and polar herbicides using in situ derivatization. *Anal. Chem.*, **72** (15), 3665–3670. doi: 10.1021/ac000164y
26. Fernández-Ponce, M.T., Casas, L., Mantell, C., Rodríguez, M., and Martínez de la Ossa, E. (2012) Extraction of antioxidant compounds from different varieties of Mangifera indica leaves using green technologies. *J. Supercrit. Fluids*, **72**, 168–175. doi: 10.1016/j.supflu.2012.07.016
27. Lang, Q. and Wai, C.M. (2003) Pressurized water extraction (PWE) of terpene trilactones from Ginkgo biloba leaves. *Green Chem.*, **5**, 415. doi: 10.1039/b300496c
28. Ozel, M.Z., Gogus, F., and Lewis, A.C. (2006) Comparison of direct thermal desorption with water distillation and superheated water extraction for the analysis of volatile components of Rosa damascena Mill. using GCxGC-TOF/MS. *Anal. Chim. Acta*, **566** (2), 172–177. doi: 10.1016/j.aca.2006.03.014
29. Zou, S.P., Wu, Y.L., Yang, M.D., Li, C., and Tong, J.M. (2010) Bio-oil production from sub- and supercritical water liquefaction of microalgae Dunaliella tertiolecta and related properties. *Energy Environ. Sci.*, **3** (8), 1073–1078. doi: 10.1039/C002550j
30. Chemat, F., Abert Vian, M. (2014) Alternative Solvents for Natural Products Extraction, Springer, NewYok.
31. Iyer, P.V. and Ananthanarayan, L. (2008) Enzyme stability and stabilization–Aqueous and non-aqueous environment. *Process Biochem.*, **43** (10), 1019–1032. doi: 10.1016/j.procbio.2008.06.004

32. Schmid, R.D. (1979) *Advances in Biomedical Engineering*, Springer, Berlin, Heidelberg, pp. 41–118, *http://link.springer.com/chapter/10.1007/3540092625_7* (accessed 19 August 2014).
33. Wijesinghe, W.A.J.P. and Jeon, Y.-J. (2012) Enzyme-assistant extraction (EAE) of bioactive components: a useful approach for recovery of industrially important metabolites from seaweeds: a review. *Fitoterapia*, **83** (1), 6–12. doi: 10.1016/j.fitote.2011.10.016
34. Puri, M., Sharma, D., and Barrow, C.J. (2012) Enzyme-assisted extraction of bioactives from plants. *Trends Biotechnol.*, **30** (1), 37–44. doi: 10.1016/j.tibtech.2011.06.014
35. Jiao, J., Fu, Y.-J., Zu, Y.-G., Luo, M., Wang, W., Zhang, L., and Li, J. (2012) Enzyme-assisted microwave hydro-distillation essential oil from Fructus forsythia, chemical constituents, and its antimicrobial and antioxidant activities. *Food Chem.*, **134** (1), 235–243. doi: 10.1016/j.foodchem.2012.02.114
36. Klibanov, A.M. (1983) in *Advances in Applied Microbiology*, Vol. 29 (ed. A.I. Laskin), Academic Press, *http://www.sciencedirect.com/science/article/pii/S0065216408703526* (accessed 19 August 2014), pp. 1–28.
37. Zhang, Y., Li, S., Yin, C., Jiang, D., Yan, F., and Xu, T. (2012) Response surface optimisation of aqueous enzymatic oil extraction from bayberry (Myrica rubra) kernels. *Food Chem.*, **135** (1), 304–308. doi: 10.1016/j.foodchem.2012.04.111
38. Rosenthal, A., Pyle, D., Niranjan, K., Gilmour, S., and Trinca, L. (2001) Combined effect of operational variables and enzyme activity on aqueous enzymatic extraction of oil and protein from soybean. *Enzyme Microb. Technol.*, **28** (6), 499–509. doi: 10.1016/S0141-0229(00)00351-3
39. Nyam, K.L., Tan, C.P., Lai, O.M., Long, K., and Man, Y.B.C. (2009) Enzyme-assisted aqueous extraction of kalahari melon seed oil: optimization using response surface methodology. *J. Am. Oil Chem. Soc.*, **86** (12), 1235–1240. doi: 10.1007/s11746-009-1462-8
40. Dehghan-Shoar, Z., Hardacre, A.K., Meerdink, G., and Brennan, C.S. (2011) Lycopene extraction from extruded products containing tomato skin. *Int. J. Food Sci. Technol.*, **46** (2), 365–371. doi: 10.1111/j.1365-2621.2010.02491.x
41. Yazdi, A.S. (2011) Surfactant-based extraction methods. *TrAC, Trends Anal. Chem.*, **30** (6), 918–929. doi: 10.1016/j.trac.2011.02.010
42. Paradkar, R.P. and Williams, R.R. (1994) Micellar colorimetric determination of dithizone metal chelates. *Anal. Chem.*, **66** (17), 2752–2756. doi: 10.1021/ac00089a024
43. Nandini, K.E. and Rastogi, N.K. (2009) Reverse micellar extraction for downstream processing of lipase: effect of various parameters on extraction. *Process Biochem.*, **44** (10), 1172–1178. doi: 10.1016/j.procbio.2009.06.020
44. Paleologos, E.K., Giokas, D.L., and Karayannis, M.I. (2005) Micelle-mediated separation and cloud-point extraction. *TrAC, Trends Anal. Chem.*, **24** (5), 426–436. doi: 10.1016/j.trac.2005.01.013
45. Van de Pas, J.C., Buytenhek, C.J., and Brouwn, L.F. (1994) The effects of salts and surfactants on the physical stability of lamellar liquid-crystalline systems. *Recl. Trav. Chim. Pays-Bas*, **113** (4), 231–236. doi: 10.1002/recl.19941130411
46. Xie, S., Paau, M.C., Li, C.F., Xiao, D., and Choi, M.M.F. (2010) Separation and preconcentration of persistent organic pollutants by cloud point extraction. *J. Chromatogr. A*, **1217** (16), 2306–2317. doi: 10.1016/j.chroma.2009.11.075
47. Bi, W., Tian, M., and Row, K.H. (2011) Extraction and concentration of tanshinones in Salvia miltiorrhiza Bunge by task-specific non-ionic surfactant assistance. *Food Chem.*, **126** (4), 1985–1990. doi: 10.1016/j.foodchem.2010.12.059
48. Gortzi, O., Lalas, S., Chatzilazarou, A., Katsoyannos, E., Papaconstandinou, S., and Dourtoglou, E. (2007) Recovery of natural antioxidants from olive mill wastewater using Genapol-X080. *J. Am. Oil Chem. Soc.*, **85** (2), 133–140. doi: 10.1007/s11746-007-1180-z
49. Katsoyannos, E., Gortzi, O., Chatzilazarou, A., Athanasiadis, V.,

Tsaknis, J., and Lalas, S. (2012) Evaluation of the suitability of low hazard surfactants for the separation of phenols and carotenoids from red-flesh orange juice and olive mill wastewater using cloud point extraction. *J. Sep. Sci.*, **35** (19), 2665–2670. doi: 10.1002/jssc.201200356

50. Kiathevest, K., Goto, M., Sasaki, M., Pavasant, P., and Shotipruk, A. (2009) Extraction and concentration of anthraquinones from roots of Morinda citrifolia by non-ionic surfactant solution. *Sep. Purif. Technol.*, **66** (1), 111–117. doi: 10.1016/j.seppur.2008.11.017
51. Sun, C., Xie, Y., and Liu, H. (2007) Microwave-assisted micellar extraction and determination of glycyrrhizic acid and liquiritin in licorice root by HPLC. *Chin. J. Chem. Eng.*, **15** (4), 474–477.
52. Neuberg, C. (1916) Hydrotropy. *Biochemistry*, **76**, 107.
53. Holmberg, K., Shah, D.O., and Schwuger, M.J. (2002) *Handbook of Applied Surface and Colloid Chemistry*, Vol. 1, Wiley-VCH Verlag GmbH, Chichester, New York.
54. Yu, W., Zou, A., and Guo, R. (2000) Hydrotrope and hydrotrope-solubilization action of Vitamin C. *Colloids Surf. A Physicochem. Eng. Asp.*, **167** (3), 293–303. doi: 10.1016/S0927-7757(99)00506-3
55. Theneshkumar, S., Gnanaprakash, D., and Nagendra Gandhi, N. (2009) Enhancement of solubility and mass transfer coefficient of salicylic acid through hydrotropy. *J. Zhe. Univ.- Sci. A*, **10** (5), 739–745. doi: 10.1631/jzus.A0820516
56. Hodgdon, T.K. and Kaler, E.W. (2007) Hydrotropic solutions. *Curr. Opin. Colloid Interface Sci.*, **12** (3), 121–128. doi: 10.1016/j.cocis.2007.06.004
57. Kumar, S., Prakash, D., and Nagendra Gandhi, N. (2009) Effect of hydrotropes on solubility and mass transfer coefficient of lauric acid. *Korean J. Chem. Eng.*, **26** (5), 1328–1333. doi: 10.1007/s11814-009-0219-2
58. Dharmendira Kumar, M. and Nagendra Gandhi, N. (2000) Effect of hydrotropes on solubility and mass transfer coefficient of methyl salicylate. *J. Chem. Eng. Data*, **45** (3), 419–423. doi: 10.1021/je9901740
59. Prakash, D.G., Kumar, S.T., and Gandhi, N.N. (2009) Effect of hydrotropes on solubility and mass transfer coefficient of p-nitrobenzoic acid. *J. Appl. Sci.*, **9** (16), 2975–2980. doi: 10.3923/jas.2009.2975.2980
60. Raman, G. and Gaikar, V.G. (2003) Hydrotropic solubilization of boswellic acids from Boswellia serrata Resin. *Langmuir*, **19** (19), 8026–8032. doi: 10.1021/la034611r
61. Dandekar, D.V., Jayaprakasha, G.K., and Patil, B.S. (2008) Hydrotropic extraction of bioactive limonin from sour orange (Citrus aurantium L.) seeds. *Food Chem.*, **109** (3), 515–520. doi: 10.1016/j.foodchem.2007.12.071
62. Desai, M.A. and Parikh, J. (2012) Hydrotropic extraction of citral from cymbopogon flexuosus(Steud.) Wats. *Ind. Eng. Chem. Res.*, **51** (9), 3750–3757. doi: 10.1021/ie202025b
63. Mishra, S.P. and Gaikar, V.G. (2009) Hydrotropic extraction process for recovery of forskolin from coleus forskohlii roots. *Ind. Eng. Chem. Res.*, **48** (17), 8083–8090. doi: 10.1021/ie801728d
64. Raman, G. and Gaikar, V.G. (2002) Extraction of piperine from piper nigrum (Black Pepper) by hydrotropic solubilization. *Ind. Eng. Chem. Res.*, **41** (12), 2966–2976. doi: 10.1021/ie0107845

8
Coverage Exploitation of By-Products from the Agrofood Industry

Carlos A. Ledesma-Escobar and María D. Luque de Castro

8.1
Introduction

Agriculture in Mediterranean countries is dominated by extensive crops of olives, grapes, and citrus. This region is the first in the world in production of olive oil and wine, and one of the leading producers of citrus juices [1]. In addition to the overall economic importance of these products for the region, they are the main contributors to the consideration by UNESCO of the Mediterranean diet a heritage of humanity [2, 3]. However, residues from the crops and by-products from the derivate industries represent serious environment damage, if not properly treated, both because of the huge amounts thus produced and their seasonal character. There are different treatments for either exploitation or safe disposal of the diverse type of residues generated by the Mediterranean agrofood industry. For many years, the only goal of treatments was to reduce pollution from the residues. Nowadays, research on the characteristics of the residues from these crops and derivate industries has shown they are key sources of raw materials for pharmaceutical, cosmetics, and food industries, among the most important, with a final use, for both the surpluses after supplying the industries and the residues from the industries after extracting the compounds of interest, as an overall source of energy [4]. Proper exploitation of these residues allows a comprehensive utilization of crops and reduces polluting effects on the environment. The different treatments for safe disposal of the residues from the production of olive oil, wine, and citrus juices, and from the related agriculture, as well as their present and potential total or partial exploitation are discussed in this chapter.

8.2
Treatments for Safe Disposal/Exploitation of Agrofood Wastes or Residues

A number of treatments for safe disposal of wastes from the agrofood industry have so far been developed. Simple physical processes such as evaporation,

Green Extraction of Natural Products: Theory and Practice, First Edition.
Edited by Farid Chemat and Jochen Strube.

sedimentation, filtration, centrifugation, or dilution have been used for many years to treat wastes – either liquid wastes or those containing a high percentage of water; nevertheless, none of the treatments alone is able to reduce the chemical oxygen demand (COD) and toxicity to acceptable limits. Usually, a combination of other physical processes coupled with coagulation, flocculation, or adsorption steps is more efficient for removing organic matter, the main reason of high COD and toxicity of wastes. Besides, chemical oxidation is a common treatment to reduce the toxicity of wastewaters. On the other hand, thermal processes such as incineration, pyrolysis, or gasification are effective to reduce the volume and toxicity of solid wastes and also to recover energy that can be used as fuel. Finally, biological treatments allow both pollution caused by waste to be reduced and reuse of the products from microbial degradation. This type of treatments is commonly known as *bioremediation*.

As most physical and chemical treatments are shared by the wastes from the agrofood areas, the object of this chapter, they are discussed in this section, then emphasizing the differences characteristic of their use in each area, if any, in the corresponding section.

8.2.1 Physical Processes

The main physical processes for safe disposal of agrofood wastes are sedimentation, filtration, and centrifugation.

Sedimentation consists of separation by gravity of the solid particles with higher density from the liquid phase in a suspension. The sedimentation velocity depends on the concentration, the weight, and size of the solid particles as well as the viscosity of the liquid phase. To increase sedimentation velocity or allow sedimentation of small or low-density particles, strategies such as addition of salts or application of an electric field are used. The agglomerates thus formed are referred to as *floccules* and their characteristics largely depend on the type of waste; therefore, they are described in each case [5–7].

Filtration is an operation by which solid particles are separated from a liquid or a gas by forcing the mixture through a porous medium that retains the particles. Filtration can be carried out through a porous surface (filter medium) or through a packed bed (mainly glass wool, rock wool, or sand). The purpose of the filter may be to remove undesirable particulate matter from a liquid or to recover a solid product from a liquid–solid system [8].

Centrifugation is an accelerated way to separate solid particles from a liquid or two immiscible liquids of different densities. This separation is made possible by the effect of centrifugal forces as a result of applying a rotational movement to the heterogeneous system thus increasing the gravitational force and allowing precipitation of small particles [9].

8.2.2 Physicochemical Processes

The most common physicochemical processes for safe disposal of wastes are coagulation–flocculation and adsorption.

Coagulation–flocculation processes are mainly used for removal of colloidal material that causes color and turbidity in wastewater; therefore, they have the aim of eliminating suspended solids and as much organic material as possible [10]. In dealing with colloids, each particle is stabilized by a number of charges of the same sign on their surface, causing them to repel each other, thus preventing the formation of larger masses (flocs) and consequently the particles do not sediment. Coagulation destabilizes colloids and allows formation of sedimentable flocs [11], generally achieved by addition of chemicals or application of energy [12]. The terms coagulation and flocculation are often used interchangeably to refer to the formation of aggregates, probably because both phenomena often occur simultaneously; however, conceptual differences exist between these two operations: coagulation is the destabilization of the colloidal suspension and flocculation is limited to transport of the coagulated particles to cause collisions between them, thus allowing agglomeration [13].

Adsorption is the phenomenon by which the molecules of a fluid (gas or liquid) are retained on a solid surface. Depending on the type of intermolecular attraction, adsorption can be physical or chemical. Physical adsorption has low selectivity and is irreversible, whereas chemical adsorption is reversible and can be highly selective. Adsorption processes are widely used in the food industry to ensure the quality and/or appearance of the products; they are also the key in combating environmental pollution, as purification of drinking water, wastewater treatment, or removal of gaseous pollutants from smokestacks [14, 15]. Most adsorbents utilized in technological applications of adsorption are porous solids with large specific surfaces resulting in very high adsorption capacities. Common adsorbents are activated materials such as silica, clay, or charcoal. Some residues from the agrofood industry have recently been used as adsorbents for dyes from aqueous solutions [16, 17].

8.2.3 Advanced Oxidation Processes

Advanced oxidation processes (AOPs) are defined as those that generate hydroxyl radicals in quantities sufficient for oxidizing the majority of the complex chemicals present in wastewater effluents. Hydroxyl radicals are powerful oxidants which lead to very efficient oxidations, the kinetics of which – between hydroxyl radicals and most organic and many inorganic compounds – are very fast [18]. Several studies have shown that AOPs can be used to convert compounds such as insecticides, dyes, surfactants, or organochlorines into relatively harmless products such as carbon dioxide and water [19]. Conventional sources for generation of hydroxyl radicals are oxidants such as hydrogen peroxide (H_2O_2), ozone (O_3), UV (ultraviolet) irradiation, and ultrasound. Among single oxidants, O_3 generally appears to be the most effective for pollutants oxidation, while H_2O_2 alone is inefficient for destruction of high concentrations of organic pollutants. Similarly, UV photolysis alone has also been reported to ensure decomposition of most organic molecules at very slow rates [20]. Combined systems have been developed to enhance oxidation and mineralization rates by generation of hydroxyl radicals. In most cases, combination of UV irradiation, oxidants (H_2O_2 or O_3), and catalysts (Fe^{2+} or TiO_2) can further increase the production of hydroxyl radicals [21]. Basic reactions to generate these radicals by combined systems are as follows:

1) Ozone and hydrogen peroxide: $O_3 + H_2O_2 \rightarrow OH^\bullet + O_2 + HO_2^\bullet$
2) Iron salts and hydrogen peroxide: $Fe^{2+} + H_2O_2 \rightarrow Fe^{3+} + OH^\bullet + OH^-$
3) UV-light and hydrogen peroxide: $H_2O_2\ [+UV] \rightarrow 2OH^\bullet$.

The use of AOPs, mainly as a pretreatment stage for industrial wastewater remediation, is considered a highly competitive water treatment technology for the removal of organic pollutants not treatable by conventional techniques owing to their high chemical stability and/or low biodegradability. In recent years, many authors have developed combined AOPs and biological systems for treatment of a diversity of industrial wastewaters [22].

8.2.4 Thermal Processes

Thermal treatment of wastes aims at reducing the volume and mass of the given waste, inertizing its hazardous components by heat and reducing pollutant emissions into air and/or water. It is considered an environmental and economically friendly way to diminish residues disposal. The process may involve energy recovery by exploiting the produced fuel gas, which may be used to generate electricity or heat. Thermal treatments of wastes can be divided into two groups, within each of which (i) waste is combusted in the presence of oxygen (incineration) and (ii) waste is heated in the presence of little or no oxygen so that there is no direct combustion but pyrolysis or gasification [23]. In modern European waste management, waste incineration plays the absolute dominant role. The processes result in residual products from the waste as well as products from flue gas–cleaning additives, which have to be deposited later at controlled sites such as landfills or mines. After thermal treatment, ferrous and nonferrous metals can be recovered and recycled. In addition, the grate ash or slag can be recovered for building purposes. Nutrients and organic matter are destroyed and cannot be recovered after thermal treatment [24]. A special type of thermal treatment is caused by weather conditions, particularly by the sun.

Incineration involves combustion of waste in a controlled manner to convert it into less harmful components that are less bulky and more controllable. The energy content of waste can be recovered and used as fuel for some industrial applications. Its main limitations are that not all waste can be incinerated and some combustion products of the waste may affect the quality of the industrial products [24]. Compared to other treatments, incineration is very expensive because of the high cost of combustors and the associated air pollution control systems [25]. Stoichiometric or theoretical combustion is the ideal combustion process where fuel is completely burned, as expressed by the following reaction:

$$C_aH_b + \left(a + \frac{1}{4}b\right) O_2 \rightarrow aCO_2 + \frac{1}{2}bH_2O$$

Pyrolysis is the thermal decomposition of organic material through the application of heat without extra air or oxygen [24]. It is an alternative to reduce waste volume and a process to obtain energy from agricultural wastes [26] that has the significant advantage of zero net emissions. In pyrolysis carried out at temperatures around 450 °C and under no additional air, the hydrocarbons in the waste react and generate fuel gas. This gas mainly consists of carbon monoxide and hydrogen, which can be used either for electricity generation or for providing heat in boiler applications. The gas has a calorific value between 22 and 30 $MJ\,m^{-3}$, depending on the waste material being processed. The solid residue from pyrolysis, known as *pyrolysis coke*, is constituted by the residual carbon that is not converted into gas during the process. To improve pyrolysis, dolomite is used as catalyst, the presence of which causes a high increase in the gas fuel yield giving place to a vast rise in hydrogen production [26]. The pyrolysis reaction is described by the following equation [27]:

$$C_aH_bO_c + \text{heat} \rightarrow H_2O + CO_2 + H_2 + CO + CH_4 + C_2H_6 + CH_2O + \text{tar} + \text{char}$$

Gasification of biomass is the thermochemical conversion of a solid or liquid carbon-based waste into a combustible gaseous product (syngas) by the supply of a gasification agent (other gaseous compound) [28]. The gasification agent (e.g., oxygen, carbon dioxide, hydrogen, or water steam) allows conversion of waste into gas by different heterogeneous reactions under high pressure, generally at temperatures above 800 °C. The main chemical reactions of biomass gasification include the following [29]:

1) $C + ½O_2 \rightarrow CO$
2) $C + O_2 \rightarrow CO_2$
3) $C + 2H_2O \rightarrow CO_2 + 2H_2$
4) $C + H_2O \rightarrow CO + H_2$
5) $C + CO_2 \rightarrow 2CO$
6) $C + 2H_2 \rightarrow CH_4$.

The combustible gas, containing H_2, CO, CO_2, CH_4, and light hydrocarbons, can be used in fuel cell units for electricity production at high efficiency levels (40% electrical efficiency and 85% overall thermal efficiency) [30].

Generally, gasification is the process subsequent to pyrolysis where the residual carbon is oxidized from the glowing embers of the pyrolysis coke [31]. Depending on the gasification agent used, the syngas has different heating values: gasification with air 4–7 MJ Nm^{-3}; with oxygen 10–12 MJ Nm^{-3}; with steam 15–20 MJ Nm^{-3} [28].

The goal of evaporation is to reduce the volume of liquid wastes by removing water by evaporation. This is the most common process for liquid elimination from wastes in Mediterranean countries: evaporation in storage ponds in the open air because of the low investment required and the favorable climatic conditions. However, this method needs large areas and involves several problems such as bad odor, infiltration, and insect proliferation [32]. Another option is the use of multistage evaporation systems that reduce the time of the process; however, they require larger investment and high energy demand. The concentrate from evaporation has to be treated before disposal, generally by biological treatments such as aerobic digestion or an activated sludge process – aerobic and anaerobic treatments, respectively – [25].

8.2.5
Biological Treatments

Generally speaking, biological treatments involve degradation of organic matter by microorganisms. Depending on the type of microorganism used and the conditions under which degradation is carried out, biological treatments are divided into aerobic and anaerobic degradation. In addition to helping to reduce pollution from agroindustrial wastes, biological treatments offer a solution for soil amendment and reduce consumption of fossil fuels.

Anaerobic digestion is the mechanism by which the organic molecules of the biomass are decomposed by the action of microorganisms. This degradation is characterized by production of a gas phase and a liquid or solid phase. The former mainly consists of methane and carbon dioxide, commonly known as *biogas*, which can be recovered and used as fuel [33, 34]. The liquid or solid phase, containing the components difficult to degrade plus nitrogen, phosphorus, and other elements originally present in the biomass, can be used as fertilizer [35]. The main factors that affect anaerobic degradation are temperature, pH, nitrogen-to-carbon ratio (C/N), and the content of toxic compounds or inhibitors present in the biomass [36]. Therefore, the type of microorganisms, as well as the operating conditions must be previously optimized according to the residue to be treated.

Aerobic digestion consists of degradation of organic matter under aerobic conditions. This process has for many years been used for soil amendment by producing biofertilizers as a result of composting. The main factors affecting aerobic degradation is the content of air and water and the pH, temperature, and C/N ratio [37]. Because this treatment requires a high intake of air and water – at present not recommended for wastes with high organic load – this type of treatment has in many cases been replaced by anaerobic digestion, which, in addition to reducing pollution from organic waste, produces biogas [36].

8.3 Exploitation of By-products from Olive Trees and Olive Oil Production

8.3.1 Generalities

Olives are one of the major crops in the Mediterranean basin, with a harvested area of 9 million of hectares and approximated production of 15 million of tons per year (Table 8.1) [1]. Olive fruit has been especially recognized because of its high content of polyunsaturated fatty acids and phenols, which is why olive oil is considered the edible oil with better nutritional properties. This oil has for many years been a part of the Mediterranean diet which is famous for low incidence of atherosclerosis, certain cancers, and cardiovascular and neurodegenerative diseases. Some of the healthy properties of the Mediterranean diet have been strongly associated with olive oil consumption [2]. This is the reason why the consumption of olive oil has increased over the world by about 40% in the last decade, a trend that is growing at present [38]. Both the excellent nutritional properties of olive oil

Table 8.1 Total production of olive drupes and olive oil in Mediterranean countries [1].

Mediterranean country	Olive drupes		Olive oil production (tons)
	Production (tons)	Harvested area (ha)	
Spain	3 626 600	2 300 000	992 000
Italy	2 992 330	1 056 005	570 000
Greece	2 100 000	900 000	351 800
Turkey	1 820 000	805 500	206 300
Syrian Arab Republic	1 095 043	700 000	200 000
Tunisia	963 000	1 800 000	192 600
Morocco	1 315 794	968 123	130 000
Algeria	39 3840	328 884	32 000
Libya	139 091	216 013	15 000
Israel	63 000	33 700	12 300
Lebanon	90 307	56 529	11 300
Egypt	465 000	55 000	8 800
France	27 969	17 055	3 600
Cyprus	14 865	10 852	2 400
Albania	12 5000	48 000	800
Croatia	50 900	19 000	600
Slovenia	2 000	1 000	600
Montenegro	2 888	2 350	187
Malta	8	8	3
Bosnia and Herzegovina	160	110	a)
Total Mediterranean production	15 287 795	9 318 129	2 730 290

a) Datum no provided by FAO.

and its healthy properties are accepted worldwide and are now being recognized by official organizations [39].

Olive oil is classified into eight types according to the physicochemical characteristics and the method used for oil production [40]: (i) extra virgin olive oil (EVOO); (ii) virgin olive oil (VOO); (iii) lampante olive oil; (iv) refined olive oil; (v) olive oil composed of refined and VOO; (vi) crude olive pomace oil; (vii) refined olive pomace oil; and (viii) olive pomace oil (Table 8.2 summarizes the characteristics of each oil).

EVOO and VOO are the most valued olive oils, and they result from the fruit of the olive tree (*Olea europaea* L.) using only mechanical processes; thus excluding oils obtained using solvents, reesterification, or any mixture with other oils [41]. They consist of 98% of saponifiable compounds (triacylglycerols, diacylglycerols, monoacylglycerols, and free fatty acids), and 2% of unsaponifiable or minor fraction (aliphatic and triterpenic alcohols, sterols, hydrocarbons, volatile compounds, and antioxidants such as carotenes, tocopherols, and phenols) [42]. As can be seen in Table 8.2, the main difference between EVOO and VOO is lower acidity of the former.

Olive oil quality can be influenced by different factors. Cultivar, environment, and agronomic practices affect the fruit physiology, whereas processing and storage conditions alter oil composition (V. Sánchez de Medina, personal communication, December 10, 2013). The ripening index (RI) has been studied as a key factor affecting the composition of olive oil, showing that the oleic/linoleic acid ratio is inversely proportional to the ripening processes. The RI is determined by the color of the fruits, classified into the following categories: 0 = with deep or dark green color; 1 = with yellow or yellowish–green color; 2 = with yellowish color and reddish spots; 3 = with reddish or light violet color; and 4 = with black color [43]. Table 8.3 shows the differences in fatty acids profile between VOO (RI = 0) and olive oil from ripe olive drupes (RI = 4).

Over 95% of the world's olive oil is produced in the Mediterranean basin, being Spain, Italy, and Greece the producers of around 70% of the olive oil in this region [46].

The overall mechanical process to extract VOO from olive drupes includes three steps: (i) crushing of the fruit to obtain a homogenous paste; (ii) malaxation to increase the percentage of free oil and help small oil droplets to coalesce and agglomerate thus facilitating separation of the oil and water phases; and (iii) separation of the oily phase by pressing (discontinuous process) or centrifugation (continuous process), as shown in Figure 8.1.

Pressing extraction is the oldest method to obtain olive oil. The olive paste after crushing is placed in a uniform layer on fiber disks on top of each other and then placed in the press. Pressure is applied on the disks to compact the solid phase of the olive paste and to percolate the liquid phases (oil and water). To facilitate oil separation from the other phases, a small quantity of water is added and VOO is separated by decantation. This process produces two residues: (i) a solid waste, known as *olive cake* or "*orujo*," containing the olive pulp, skin, stones, and small amounts of water and oil and (ii) a liquid waste, known as *olive mill*

Table 8.2 Olive oil characteristics (as regulated by the Official Journal of the European Union) [40].

Category	Fatty acid methyl esters (FAMEs) and fatty acid ethyl esters (FAEEs)	Acidity (%) (*)	Peroxide index (mEq 0_2/kg) (*)	Waxes (mg kg^{-1})	2-Glyceril-monopalmitate (%)	Stigma-stadiene (mg kg^{-1})[a]	Difference: ECN42 (HPLC) and ECN42 (theoretical calculation)	K232 (*)	K270 (*) "K 270 or K 268[b]"	Delta-K (*)[b]	Organoleptic evaluation median defect (Md) (*)	Organoleptic evaluation fruity median (Mf) (*)
Extra virgin olive oil	Σ FAME + FAEE ≤ 75 mg kg^{-1} or 75 mg kg^{-1} < Σ FAME + FAEE ≤ 150 mg kg^{-1} and (FAEE/FAME) ≤ 1.5	≤0.8	≤20	≤250	≤0.9 if total palmitic acid % ≤14% ≤1.0 if total palmitic acid % >14%	≤0.10	≤0.2	≤2.50	≤0.22	≤0.01	Md = 0	Mf > 0
Virgin olive oil	—	≤2.0	≤20	≤250	≤0.9 if total palmitic acid % ≤14% ≤1.0 if total palmitic acid % >14%	≤0.10	≤0.2	≤2.60	≤0.25	≤0.01	Md ≤ 3.5	Mf > 0
Lampante olive oil	—	>2.0	—	≤300[c]	≤0.9 if total palmitic acid % ≤14% ≤1.1 if total palmitic acid % >14%	≤0.50	≤0.3	—	—	—	Md > 3.5[d]	—
Refined olive oil	—	≤0.3	≤5	≤350	≤0.9 if total palmitic acid % ≤14% ≤1.1 if total palmitic acid % >14%	—	≤0.3	—	≤1.10	≤0.16	—	—
Olive oil composed of refined and virgin olive oils	—	≤1.0	≤15	≤350	≤0.9 if total palmitic acid % ≤14% ≤1.0 if total palmitic acid % >14%	—	≤0.3	—	≤0.90	≤0.15	—	—

(continued overleaf)

Table 8.2 (*Continued.*)

Category	Fatty acid methyl esters (FAMEs) and fatty acid ethyl esters (FAEEs)	Acidity (%) (*)	Peroxide index (mEq O_2/kg) (*)	Waxes (mg kg^{-1})	2-Glyceril-monopalmitate (%)	Stigma-stadiene (mg kg^{-1})[a]	Difference: ECN42 (HPLC) and ECN42 (theoretical calculation)	K232 (*)	K270 (*) "K 270 or K 268[b] "	Delta-K (*)[b]	Organoleptic evaluation median defect (Md) (*)	Organoleptic evaluation fruity median (Mf) (*)
Crude olive-pomace oil	—	—	—	>350[e]	≤1.4	—	≤0.6	—	—	—	—	—
Refined olive-pomace oil	—	≤0.3	≤5	>350	≤1.4	—	≤0.5	—	≤2.00	≤0.20	—	—
Olive-pomace oil	—	≤1.0	≤15	>350	≤1.2	—	≤0.5	—	≤1.70	≤0.18	—	—

a) Total isomers which could (or could not) be separated by a capillary column.
b) K 270 if solvent is cyclohexane, K 268 if solvent is iso-octane.
c) Oils with a wax content between 300 and 350 mg kg^{-1} are considered to be lampante olive oil if the total aliphatic alcohol content is less than, or equal to, 350 mg kg^{-1} or if the erythrodiol and uvaol content is less than, or equal to, 3.5%.
d) Or where the median defect is less than, or equal to, 3.5 and the fruity median is equal to 0.
e) Oils with a wax content between 300 and 350 mg kg^{-1} are considered to be crude olive-pomace oil if both the total aliphatic alcohol content is above 350 mg kg^{-1} and the erythrodiol and uvaol content is greater than 3.5%.

Table 8.3 Effect of the ripening index on the fatty acid profile of olive oil.

Fatty acid	Abbreviated formula	VOO[a)] (RI = 0)	Olive oil[b)] (RI = 4)
Myristic	C14:0	0.01	Nd
Palmitic	C16:0	10.5	1.24
Palmitoleic	C16:1	0.76	0.04
Margaric	C17:0	0.1	0.03
Margaroleic	C17:1	0.14	3.13
Stearic	C18:0	3.53	65.12
trans-Oleic	C18:1n9t	0.00	Nd
Oleic	C18:1n9c	77.49	16.55
trans-Linoleic	C18:2n9,12tt	0.00	Nd
Linoleic	C18:2	6.05	0.45
Linolenic	C18:3	0.61	0.49
Arachidic	C20:0	0.39	0.01
Gadoleic	C20:1n9	0.25	0.11
Behenic	C22:0	0.11	0.04
Lignoceric	C24:0	0.06	3.98
Oleic/linoleic	18:1/18:2	12.81	4.27

a) Provided by Carbonell (SOS Cuétara S.A. Madrid) [44].
b) Olive oil obtained from Tunisian cultivar [45].
nd = no determinate

wastewater (OMWW) or "alpechín," which is the residual water after oil decantation [47]. The process is cheap, technically simple, and requires a small quantity of water (40–60 l water/100 kg olives) [48]. However, it is a batch process with high manpower cost. The traditional pressing process gave way to continuous processes by centrifugation; first, to a three-phase system and, most recently, to a two-phase system. Both processes consist of a decanter to separate phases by centrifugation, based on the different densities of olive oil, water, and insoluble solids.

Three-phase centrifugal decanters generate three fractions at the end of the process: a solid residue (olive cake or "orujo") and two liquid phases (olive oil and OMWW). This process requires warm water to be added to dilute the olive paste. Natural antioxidants in the oil are thus reduced and large amounts of OMWW (80–120 l/100 kg olives) are generated, resulting in worldwide production of more than 30 million m^3 per year of OMWW [49]. This system has some advantages over traditional pressing extraction: it allows complete automation and reduces the area needed; however, it has as main disadvantages the high consumption of water and energy, higher OMWW production, and more expensive installation [32]. Worldwide annual OMWW production is estimated to be over 3 million m^3 [50], which has very high organic loads with COD values of up to 220 g l^{-1}, and a corresponding biochemical oxygen demand (BOD) of up to 100 g l^{-1} [51], thus causing serious environmental issues. The OMWW organic fraction consists of sugars, polyalcohols, pectins, lipids, and significant amounts of aromatic

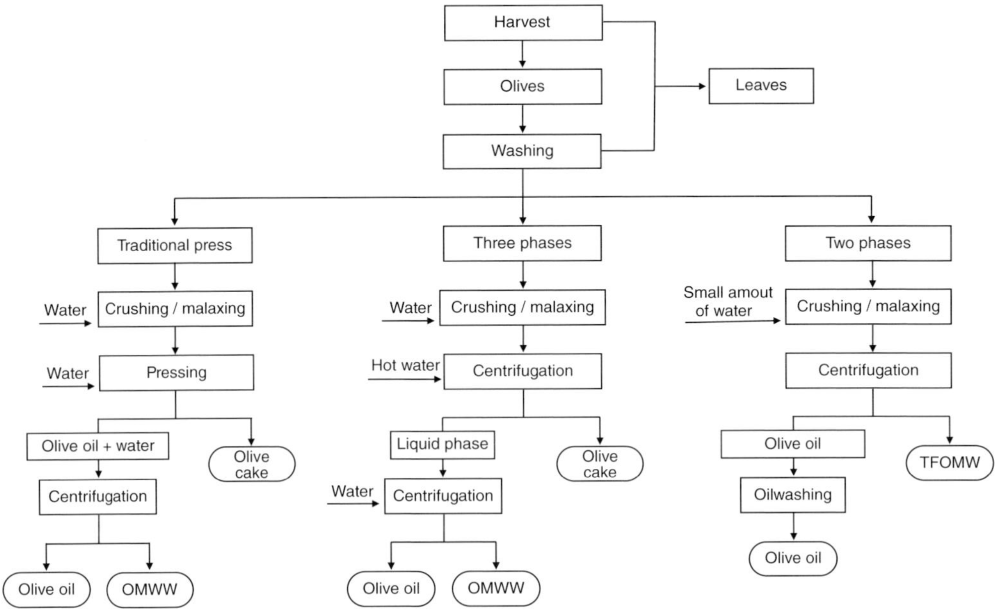

Figure 8.1 Main processes for olive oil extraction.

compounds such as tannins and phenols, responsible for antimicrobial activities and phytotoxicity [38, 52, 53]. To reduce water consumption during olive oil extraction, consequent generation of OMWW and washing of phenols, the two-phase decanter centrifugation was developed in the 1990s, thus reducing 75% of olive mill wastes [47].

The two-phase system separates the olive paste into olive oil and a semisolid byproduct known as two-phase olive mill waste (TPOMW) or “alperujo,” without requiring the addition of huge amounts of water. This residue is a combination of olive cake and OMWW, which can be reprocessed to further extract the remaining oil by either solvent extraction or a second centrifugation. TPOMW has a high moisture content (56.6–74.5%) and an organic fraction or “orujo” (dry weight) with lignin (45.8%), hemicellulose (37.7%), cellulose (20.8%), carbohydrates soluble in water (10.1%), fats (13.0%), proteins (7.7%), and phenols (1.5%) as main components [49]. Table 8.4 shows the chemical characteristics of the different olive oil wastes.

The large amounts of wastes from the olive oil industry represent a serious environmental problem, mainly in the Mediterranean region as they are produced in short periods of time and must be properly disposed of to avoid environmental risks. The safe disposal of this waste is crucial because of its polluting effects on soil and water. Several ways for safe disposal and partial or full exploitation of olive oil waste have so far been developed.

Olive tree leaves constitute a waste associated with the production of olive oil, but it also results from pruning during the spring–summer time.

Table 8.4 Chemical characteristics of OMWW, TPOMW, and solid residues from olive oil production [38].

Parameters	Olive oil by-products		
	OMWW	Solid residue (olive cake)	TPOMW
Pulp (%)	—	12–35	10–15
Olive-stone (%)	—	15–45	12–18
Dry matter (%)	6.33–7.19	87.1–94.4	—
Ash (%)	1	1.7–4	1.42–4
pH	2.24–5.9	—	4.9–6.8
Electrical conductivity ($dS\,m^{-1}$)	5.5–10	—	1.78–5.24
Total carbon (%)	2–3.3	29.03–42.9	25.37
Organic matter (%)	57.2–62.1	85	60.3–98.5
Total organic carbon ($g\,l^{-1}$)	20.19–39.8	—	—
Total suspended solids ($g\,l^{-1}$)	25–30	—	—
Mineral suspended solids ($g\,l^{-1}$)	1.5–1.9	—	—
Volatile suspended solids ($g\,l^{-1}$)	13.5–22.9	—	—
Volatile solids ($g\,l^{-1}$)	41.9	—	—
Mineral solids ($g\,l^{-1}$)	6.7	—	—
Volatile acidity ($g\,l^{-1}$)	0.64	—	—
Inorganic carbon ($g\,l^{-1}$)	0.2	—	—
Total nitrogen (%)	0.63	0.2–0.3	0.25–1.85
P (%)	0.19	0.03–0.06	0.03–0.14
K (%)	0.44–5.24	0.1–0.2	0.63–2.9
Na (%)	0.15	—	0.02–0.1
Ca (%)	0.42–1.15	—	0.23–1.2
Mg (%)	0.11–0.18	—	0.05–0.17
Fe (%)	0.26 ± 0.03	—	0.0526–0.26
Cu (%)	0.0021	—	0.0013–0.0138
Mn (%)	0.0015	—	0.0013–0.0067
Zn (%)	0.0057	—	0.0010–0.0027
Lipids (%)	0.03–4.25	3.5–8.72	3.76–18.0
Total phenols (%)	0.63–5.45	0.2–1.146	0.4–2.43
Total sugars (%)	1.5–12.22	0.99–1.38	0.83–19.3
Total proteins (%)	—	3.43–7.26	2.87–7.2
Chemical oxygen demand ($g\,l^{-1}$)	30–320	—	—
Biological oxygen demand ($g\,l^{-1}$)	35–132	—	—
Cellulose (%)	—	17.37–24.14	14.54
Hemicellulose (%)	—	7.92–11.00	6.63
Lignin (%)	—	0.21–14.18	8.54

8.3.2
Exploitation of Alpechín

Alpechín is a major environmental problem affecting olive oil producer countries, which results every year during VOO production for about 3 months. More than 30 million m^3 of OMWW are produced in the Mediterranean basin during the press season [50], which must be properly managed to avoid the negative environmental impact associated with its disposal. The main alpechín exploitation has so far been as fertilizer.

For many years, direct spreading of alpechín on agricultural lands was its most common use in the Mediterranean region [54], as it constitutes a low cost source for water and nutrients. In addition, spreading OMWW on the ground was a simple and relatively inexpensive method for disposal that contributed to the development of sustainable agriculture [55]. Nevertheless, the chemical properties of OMWW are affected by the ripeness of the olive fruit and the irrigation level, which in turn affect the spread of agriculture with alpechín. Two major factors affect OMWW composition: (i) solute concentrations in the fruit vegetable water and (ii) extraction efficiency of the vegetable water into the wastewater phase [56]. However, the agronomic use of alpechín for direct spreading on agricultural lands is limited by some constraints, such as oil and grease, high salinity, acidity, and phenol concentration. These last compounds are mainly responsible for the phytotoxic and antimicrobial actions of OMWW. Generally, these environmental issues arise from the constraints mentioned above: (i) negative effects on the physical, chemical, and biological properties of the soil; (ii) phytotoxic effects on crops; and (iii) groundwater pollution [57].

The impact of alpechín application on the structure of bacterial communities in soil was analyzed using actinobacteria and ammonia-oxidizing bacteria as models. Two types of soils were irrigated weekly with aqueous solutions of 2% and 4% of alpechín for a period of 49 days, after which changes in the structure of the communities of actinobacteria and ammonia-oxidizing bacteria were observed and considered as induced by alpechín. The alterations attributed to alpechín application are potentially the result of environmental changes including lowered oxidative conditions, strong competition for mineral-N and phenol availability [58].

8.3.3 Overall Use of Either Alperujo or Orujo

Two main overall uses of both alperujo and orujo have been reported: bioremediation and brick fabrication.

The main overall use of the solid waste from the olive oil industry is bioremediation that involves microorganisms to remove pollutants and reduce the COD of organic wastes. They are considered environmentally friendly, reliable, and cost effective. Degradation of the organic material produces a biomass that yields a mixture of gases and a fertilizer free from phenols, and thus, without danger for bacterial communities. When degradation is carried out under aerobic conditions, the main gas produced is CO_2 instead of CH_4, the latter being characteristic of degradation under anaerobic conditions and commonly known as *biogas*. Under anaerobic conditions, the maximum rate of biogas production from orujo has been reported around of 0.70 l of biogas per liter of digester volume per day during 20 days with a yield of 0.08 l biogas per gram of COD added to the digester. The methane content of the biogas was around of 80%, the remainder being principally carbon dioxide [59].

Among nonbiological methods recently developed for overall exploitation of alperujo or orujo, the most effective has been to enter them into the composition of bricks. Studies on this application show that, in addition to helping to reduce pollution, these raw materials decrease the amount of energy required for brick fabrication. The studies on this use led to replacement of a part of clay by alperujo. Different amounts (3, 6, and 12 wt%) of clay have been substituted by alperujo, and the properties of the resulting ceramic units were compared to those of conventional products. Addition of 12 wt% alperujo showed advantages such as [60] (i) lower density (1710 kg m^{-3} vs 1850 kg m^{-3} of the reference value). The lower density is caused by greater water absorption in experimental bricks than in traditional bricks. This allows fabricating blocks of the same size but with lower weight than traditional bricks, thus facilitating handling and reducing transportation costs [61]. (ii) Higher thermal insulation effectiveness (18% reduction in the bulk of fired clay thermal conductivity). (iii) The strength attained (approximately 14 N mm^{-2}) is enough high for this type of units. Finally, (iv) the reduction of the traditional firing temperature of 920–880 °C when alperujo is added to the raw material, which results in heating requirements reduced within the range 2.4–7.3%, depending on the final product [60]. In all the studies, the resulting units are equivalent to traditional bricks in mechanical properties and can save water and energy in the production process. This application can alleviate environmental impact from the two-phase olive oil extraction industry and, at the same time, results in economic savings for the brick-manufacturing industry [62].

8.3.4
Partial Use of Either Alpechín or Alperujo

Alpechín has as favorable characteristic a very significant phenol content (it is estimated that 53% of olive phenols remain in alpechín after oil separation [63]). The most abundant phenol in alpechín is hydroxytyrosol, representing around 70% of total phenols, followed by tyrosol, *p*-coumaric acid, oleuropein, and caffeic acid [64, 65]. In addition, the presence of elements such as Pb (6.7–10 μg l^{-1}), Cd (0.03–10 μg l^{-1}), Fe (0.45–20 mg l^{-1}), Zn (1.7–4.98 mg l^{-1}), Cu (0.49–2.96 mg l^{-1}), Mn (0.46–20 mg l^{-1}), Mg (0.03–0.17 g l^{-1}), Ca (0.03–0.29 g l^{-1}), K (0.73–6.1 g l^{-}), Cl (0.76–1 g l^{-}), and Na (0.03–0.13 g l^{-}) has been reported [47]. The favorable characteristics of this waste make it a potential application for preparation of functional beverages after treatment by the appropriate methods (e.g., ultrafiltration) to maximize recovery of the bioactive constituents and for removal of undesirable suspended solids [47].

In recent years, the use of membranes for filtration of wastewater has generated many expectations, mainly by allowing, in addition to water purification, recovery of compounds of interest contained in it without the use of solvents. Filtration membranes have also been used for alpechín, and the best results were provided by an integrated system constituted by in-series high-to-low porosity membranes, thus allowing fractionation in mainly three phases: (i) substances with high molecular weight, which can be subjected to anaerobic digestion for biogas production; (ii) bioactive compounds such as phenols, suitable for cosmetics, food, and pharmaceutical industries; and (iii) purified water, which can be reused in the olive oil extraction process [66]. Generally, the ultrafiltration process starts with a coagulation–flocculation treatment to remove as much as possible of suspended solids. Systems based on ultrafiltration–nanofiltration [66] or ultrafiltration–osmotic distillation [67] allow recovery of 96% or 78% of total phenols in alpechín, respectively.

In addition, alperujo is a great source of natural antioxidants, as it is estimated that 45% of total phenols in olive drupes remains in alperujo after oil separation [63]. Extraction of phenols from alperujo can be based on contact of the semisolid residue with the extractant (usually water or, better, an ethanol–water mixture if the extract is for edible uses); or a methanol–water mixture if only analytical information is the aim of extraction. Acceleration of the extraction step can be achieved by assisting it with auxiliary energies such as ultrasound [68], microwaves [69], or high temperature + pressure [70]. The extracts thus obtained have been used for enrichment of refined edible oils that improve their stability and healthy properties in this way [69]. At present, the use of these extracts to enrich solid foods is under study [71].

8.3.5 Olive Leaf Exploitation

Traditionally, olive leaves have been used mainly in composting, but in recent years, investigation on this material has demonstrated their high content of phenols, higher than in other parts of the olive tree. As an example, taking oleuropein as a phenol model, the leaf contains100 times more than olive oil and 10 times more than alperujo. Olive phenols show a high antioxidant and free radical scavenging power, which have led to use olive leaf extracts as a natural source of antioxidants to improve the quality and stability of refined edible oils [69, 72]. In addition, the cosmetics industry takes advantage of this raw material to extract phenols for enrichment of its products [71, 73]. Most olive phenol extracts in the market (both liquid and solid extracts) come from olive leaves [74].

8.3.6 Foreseeable/Desirable Future Uses of Olive Tree–Olive Oil Wastes

Alpechín and alperujo contain a large amount of olive phenols which possess several beneficial health effects and make these wastes attractive as sources of natural antioxidants for the manufacture of added-value products for human consumption. The use of olive phenols to enrich edible oils, elaborate functional beverages, or a number of cosmetics products has started to be successfully implemented.

VOO contains a significant amount of natural antioxidants that have demonstrated multiple beneficial effects on health; therefore, this oil does not need addition of artificial antioxidants to inhibit or delay oxidation [45]. On the other hand, other edible oils, all them refined, such a soy oil, high-oleic sunflower oil, sunflower oil, or rapeseed oil, or even refined olive oil, are deficient in natural antioxidants. To reduce the oxidative reactions and improve stability of these oils, synthetic antioxidants are typically added to them. However, the reported deleterious effects on human health of synthetic additives such as butylated hydroxyanisole, butylated hydroxytoluene, propyl gallate, or tert-butyhydroquinone have decreased their use and promoted the general consumer's rejection of synthetic food additives. As a result, enrichment of edible oils with natural antioxidants to inhibit or suppress oil oxidation becomes of great interest. These substances are cheap and drastically delay oil deterioration under deep-frying [44]. Because of the antioxidant and nutraceutical properties of olive phenols, there is a growing interest in the use of these compounds to enrich low-priced edible oils, thus obtaining a healthy added-value product with oxidative stability similar to VOO [69].

In addition to improvement of edible oils, the use of alpechín or alperujo antioxidants to produce functional beverages is an attractive alternative of easy implementation. Generally, a functional or supplemented food is consumed as part of a normal diet and contains biologically active components that provide health benefits and reduce illness risk. There is no legislation in the European Union regarding safety of functional foods as such, but aspects of food security such as amount and frequency of consumption, possible interactions with other dietary constituents, impact on metabolic pathways, and potential adverse effects such as allergy and/or intolerance [75] should be taken into account.

Thanks to the number of reported biological activities of phenols, the inclusion in beverages of phenol extracts from alpechín or alperujo may have a significant impact on the health of the population by reducing the incidence of cardiovascular and chronic degenerative diseases [76]. The potential application of alpechín or alperujo for preparation of functional beverages is discussed mainly in terms of bioavailability, defined as the amount of a food constituent in soluble form and released from the food matrix that is able to cross the intestinal barrier. Phenols in olive fruit products have through centuries been demonstrated to be highly bioavailable and safe as compared to synthetic antioxidants [47].

Two main studies are mandatory before incorporation of alpechín or alperujo phenols into beverages or other foods: (i) development of methods to obtain them by separation from other compounds present in the given raw material and (ii) enough knowledge on the stability of these compounds during processing and storage of the final product, with the aim of reducing adverse reactions between

Table 8.5 Main phenols in olives and olive oil wastes.

Compound	Olives	Alpechin	Alperujo	Leaves	References
Cinnamic acid	√	—	—	—	[53]
Tyrosol	√	√	√	√	[53, 64, 69]
p-Hydroxy-benzoic acid	√	—	—	—	[53]
p-Hydroxy-phenyl-acetic acid	√	—	—	—	[53]
p-Hydroxy-phenyl-propanoic acid	√	—	—	—	[53]
Vanillic acid	√	√	√	—	[53, 64, 69]
Hydroxy-tyrosol	√	√	√	√	[53, 64, 69]
Protocatechuic acid	√	—	—	—	[53]
3,4-Dihydroxy-phenyl-acetic acid	√	—	—	—	[53]
p-Coumaric acid	√	√	—	√	[53, 64]
Ferulic acid	√	√	√		[53, 64, 69]
Caffeic acid	√	√	—	—	[53, 64]
Oleanolic acid	√	—	—	—	[53]
Oleuropein	—	√	√	√	[53, 64, 69]
Verbascoside	—	√	—	—	[64]
Vainillin	—	—	√	—	[69]
Luteolin	—	—	√	√	[44, 69]
Apigenin	—	—	√	√	[44, 69]
3,4,DHPEA-EDA[a)]	—	—	√	√	[44, 69]
p-HPEA-EDA[b)]	—	—	√	√	[44, 69]
p-HPEA-FA[c)]	—	—	√	√	[44, 69]

a) Decarboxymethyl elenolic acid linked to hydroxytyrosol.
b) Dialdehydic form of decarboxymethyl elenolic acid linked to tyrosol.
c) Aglycon of the dialdehydic form of ligstroside.

food components and phenols. Table 8.5 lists the main phenols in olive drupes and olive oil wastes.

A very different alternative use of alpechín is as dye for wool. Protein fibers in wool have shown a high affinity for alpechín dyes, which generally bind dark shades with good fastnesses. The main influential factors on dyeability of wool by alpechín are pH, temperature, and time of dyeing. The best results were obtained at pH 2, 100 °C and 90 min dyeing. In addition, the large amounts of natural dyes in alpechín allow its use for dyeing wool at least twice and the wash, rubbing, and perspiration fastness of the dyed wool fabrics were good; however, light fastness was medium. Metal mordants when used in conjunction with alpechín were found to enhance the dyeability and fastness properties [77]. The use of alpechín to dye wool is very promising as it contributes to solve the environmental problem of olive mill wastes and avoids allergenic problems caused by synthetic colorants. This natural dye is renewable and available free of cost in large quantities in three-phase olive oil–producing countries. The dye process is economical and environmental pollution can be minimum if employed with an ecofriendly mordanting metal. This exploitation also applies to alperujo as studies on extraction of colorants from this raw material are successfully being developed [78].

Olive tree leaves have traditionally been used mainly as a feed resource for ruminants. Different studies have been devoted to improve this use in terms of both animal performance and product quality [79]. Olive leaves provide half of the energy and amino acid requirements of sheep and goats, if appropriately supplemented. It is recommended to feed fresh olive leaves rather than dried or ensiled; nevertheless, their use may be restricted because they contain significant amounts of Cu, which is toxic at high concentrations [80].

In addition, orujo has been used as a feed resource both for ruminants and pigs after preservation in different ways such as dried, ensiled, or incorporated into multi-nutrient blocks. This last way,the most successfully developed so far, should also been extended to alperujo. The use of these olive oil by-products, which are rich in oil, seems to increase the content of monounsaturated fatty acids in milk and lower the content of saturated fatty acids [79].

Ultrasound application has been proposed in recent years to improve olive oil extraction by reducing the duration of malaxation [81]. The authors of this research emphasize the growing interest in the use of ultrasound in food processing to achieve higher product yields, shorter processing times, reduced operating and maintenance costs, besides improved taste, texture, flavor, and color. Therefore, they suggest that olive oil industry could take advantage of the use of ultrasound in the extraction process, mainly because it provides a quick heating of olive paste and sonication can be easily transferred to an industrial scale. This proposal should be taken with enormous caution as previous studies had revealed that the flavor and composition of edible oils is deteriorated by ultrasound treatment [82], probably owing to oxidation under ultrasound application attributed to cavitation that in turn is responsible for the formation of oxy radical species [83]. Actually, ultrasound energy has given place to a fast method for determination of the oxidative stability of VOO, in which direct application of an ultrasound probe on the sample allowed oxidation of VOO 110 times faster than the traditional Rancimat method [84]. Oxidation of olive oil is also produced by microwave irradiation, but this occurs 40 times slower than under ultrasound application [85]. The use of ultrasound to facilitate extraction of VOO foreseeably produces undesirable changes in the composition of fatty acids and, also in addition, degradation of organoleptic properties such as color and/or flavor.

8.4 Exploitation of By-products from Vineyards and Wine Production

8.4.1 Generalities

Winemaking is one the most important economic activities in Mediterranean countries, with a production of 15 million tons of wine per year by cultivation

Table 8.6 Total production of grapes and wine in Mediterranean countries [1].

Mediterranean country	Grapes		Total wine production (tons)
	Production (tons)	Harvested area (ha)	
France	6 588 904	764 124	6 533 646
Italy	7 115 500	725 353	4 673 400
Spain	5 809 315	963 095	3 339 700
Greece	856 600	103 200	295 000
Croatia	204 373	32 485	48 875
Algeria	402 592	72 042	47 500
Morocco	381 861	44 905	33 300
Turkey	4 296 351	472 545	29 000
Slovenia	121 396	16 352	24 000
Tunisia	114 000	29 471	23 200
Albania	195 200	9 077	18 000
Montenegro	32 815	9 000	15 000
Lebanon	89 000	10 000	14 200
Cyprus	24 656	8 336	12 000
Israel	89 476	7 890	5 000
Egypt	1 320 801	64 835	4 400
Bosnia and Herzegovina	21 601	5 100	3 354
Malta	4 478	1 600	2 450
Syrian Arab Republic	337 961	46 295	72
Libya	35 115	8 350	a)
Total Mediterranean production	28 041 995	3 394 055	15 122 097

a) Datum no provided by FAO.

of around 3.4 million ha to harvest about of 28 millions tons of grapes per year (Table 8.6) [1].

The grape (*Vitis vinifera*) is the raw material for winemaking. Wine is a beverage produced by fermentation of grape juice by yeasts, mainly *Saccharomyces cerevisiae*. This yeast converts simple sugars to ethanol and carbon dioxide. The alcoholic medium that results from fermentation contains 12–15% of ethanol, a percentage above which the yeast cannot survive [86].

The winemaking process (Figure 8.2) begins with grape harvest. The selected grapes are washed to remove leaves and other undesirable material. The fruit is then crushed or pressed to release the juice and begin the maceration process that facilitates extraction of nutrients, flavorants, and other constituents from the pulp, skins, and seeds. Maceration of white wines is kept to a minimum and seldom lasts more than a few hours; the step is prolonged for red wines and occurs simultaneously with alcoholic fermentation. Rosé wines are typically made from red grapes exposed to a short prefermentative maceration. Fermentation may start spontaneously, caused by endemic yeasts in grapes; however, the standard practice is to inoculate the juice or must with a yeast strain of known characteristics.

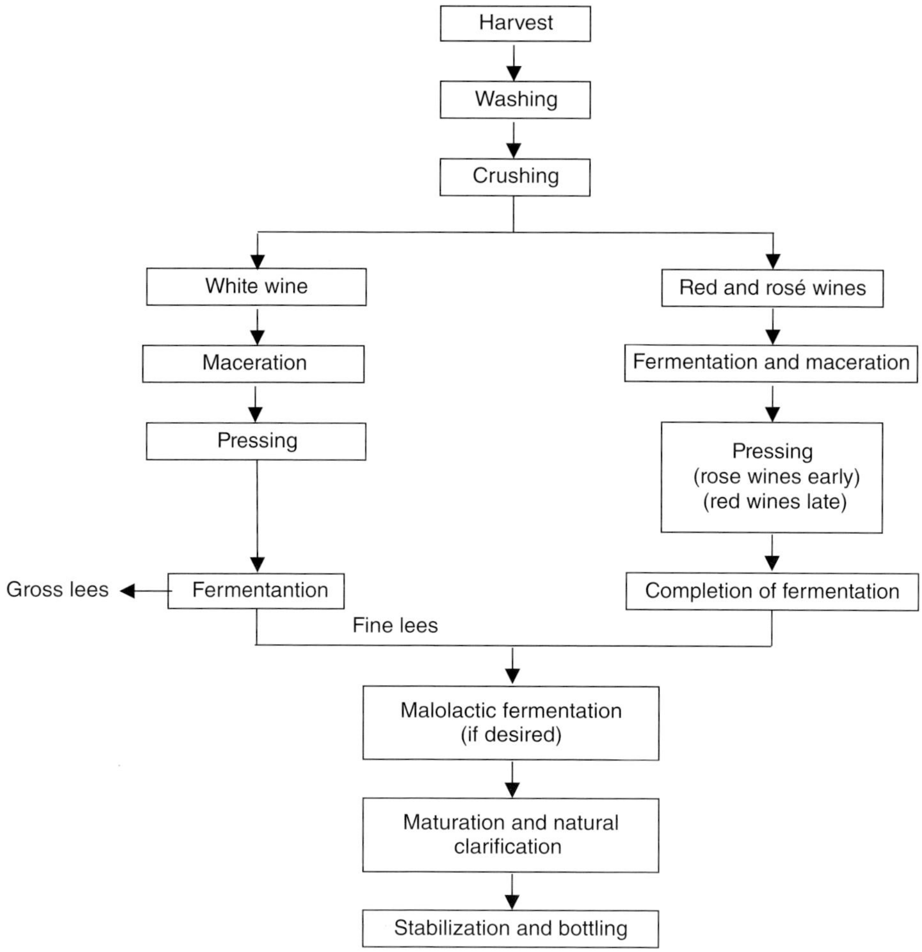

Figure 8.2 Main processes for wine production.

Yeasts not only produce ethanol but also generate the general bouquet and flavor attributes that typify wines. After completion of alcoholic fermentation, the wine may be treated to foster a malolactic fermentation that converts malic acid into the weaker lactic acid, thus reducing wine acidity. Thereafter, the wine is stored for maturation, during which the excess of carbon dioxide escapes, yeasty odors dissipate, and suspended material precipitates. Changes in aroma and development of an aged bouquet may begin during maturation. After several weeks or months, the wine is racked, a process that separates the wine from solids that settle out during clarification. The sediment consists primarily of yeast and bacterial cells, remained grape cells, and precipitated tannins, proteins, and potassium tartrate crystals; products that in longer contact with wine can endow it with undesirable characteristics. Before bottling, wine must be clarified to remove traces of dissolved proteins and other material – otherwise, they could generate turbidity, especially on exposure to heat. In addition, clarification may be used to soften the wine's taste by removing excess tannins. Wines are commonly chilled and filtered to further enhance clarification and stability; after which the wine is ready for bottling or further aging [86].

Most of the residues generated in a winery (80–85%) are organic wastes. Grape pomace, produced during grape press, is constituted by skin and seeds. The rest of residues are lees, which are generated in the fermentation and clarification processes; stalks, branches, and leaves of the grapevine, and wastewater sludge from wastewater treatment [87].

8.4.2
Types and Characteristics of Vineyard Residues

Vineyards residues are vine shoots and leaves.

Vine shoots are the branches of the grape bushes. Most research about vine shoot exploitation has been focused on production of paper pulp, ethanol, lactic acid, methanol, fuels, biomass, biosurfactants, and activated carbon for wine treatment, extraction of volatile compounds, phenols, and ferulic and coumaric acids, among others [88]. The composition of vine shoots is characterized by a lignin content around 20% (dry weight), which can be hydrolyzed to release phenols such as low-molecular-mass alcohols, aldehydes, ketones, or acids; thus making vine shoots a source for phenol production [89].

Leaves constitute a residue to which scant attention has so far been paid, usually remaining in the vineyard or being collected for compost preparation. However, vine leaves contain large amounts of compounds such as anthocyanins and flavonols that make leaves an attractive residue to obtain natural colorants, which are in high demand at present [90].

8.4.3
Present and Potential Exploitation of Vineyard Residues

Vine shoots have for many year been used to produce fertilizers by composting [91] or anaerobic digestion [92]. In addition, this material has been used to produce syngas by gasification process, which reaches maximum efficiency at 800 °C in a short time (1–2 min). A power of 6 MW could be obtained from the gases produced in the process, with a global yield of 20%. This is an acceptable efficiency taking into account the large amount of vine shoots produced per year [93].

The present trend to replace the large amounts of wood used to manufacture paper by nonwoody plants and/or lignocellulose agrofood wastes has also reached vine shoots. In general, agricultural residues appear to be effective alternatives to wood raw materials as they provide excellent special paper and constitute the sole source of raw materials for paper in some regions of the world. This is the case with vine shoots, which are highly abundant in some areas such as the Mediterranean basin in a way that their disposal poses major economic and environmental problems [94]. The best papermaking process is by adding ethylene glycol, with an approximate yield of 50% under the following optimal operating conditions: 155 °C, 60 min of reaction, and 60% ethylene glycol [95].

Another use of vine shoots is production of activated carbon. Activation can be accomplished by a physical process (heating of the raw material) or a chemical process (by addition of phosphoric acid) [17]. Comparison of products from both processes shows that chemical synthesis is more effective, because it is a one-step process with higher yields and products with areas above 1000 $m^2 g^{-1}$; therefore, the chemical process seems to be the best for industrial applications [96].

Other, less massive uses of vine shoots could be for obtaining ferulic and *p*-cumaric acids [97], and also as a source of food fiber, especially of polyphenols [98].

Recent research on the composition of extracts from different varieties of vine shoots has been developed, then compared with extracts from different types oak chips, with the final goal of using the former for wine-aging purposes [88]. The comparative profiling analysis of woody flavoring from vine shoots and oak chips allowed to identify those varieties of vine shoots providing extracts with a more similar composition to those obtained from oak chips. Present research involves the use of both oak chips and the best varieties of vine shoots to age wine in a comparative study (M.P. Delgado de la Torre, personal communication, November 21, 2013). Taking into account the present trend to check other aging materials to endow wine with different characteristics [99], the variety of cultivars opens a wide fan of possibilities to vine shoots.

Leaves from *Vitis vinifera* L. have received a dissimilar attention as a function of their age. Some high-standing cosmetics include in their composition extract from vine leaf shoots, with subsequent damage to the plant. Green adult leaves (both fresh and in brine) are used as food in countries such as Greece and Turkey [100]. In addition, extracts from fresh leaves – which are rich in flavonols, mainly quercetin and kaempferol; and anthocyanins, mainly cyanidin, petunidin, peonidin, and malvidin [90] – have been used for treatment of high blood pressure, diarrhea, hemorrhages, varicose veins, inflammation, and diabetes, and because of their protective effect on the liver [101]. These extracts may be used to enrich grape seed oil, thus improving its stability and functional properties. Finally, senescent leaves are an ignored raw material, despite the fact that this stage endows leaves with color from yellow to brown, with intermediate reddish and reddish–brown tonalities, depending on the cultivar. This wide range of color demonstrated also a wide range of colorants that, taking into account the traditional culinary use of leaves, can enter the food market without the need for obtaining the GRAS (Generally Recognized as Safe) status.

On the other hand, leaves from *Vitis vinifera* L. have been used to extract functional compounds to develop dietary ingredients. These extracts are rich in flavonols, mainly quercetin and kaempferol, and anthocyanins, mainly cyanidin, petunidin, peonidin, and malvidin [90]. These extracts may be used to enrich grape seed oil, thus improving its stability and functional properties.

8.4.4 Types and Characteristics of Wine Residues

Typical residues from wine production are grape seeds, grape skin, stalks, and lees. A "residue from residue" results from ethanol distillation, the raw material of which is the wine pomace, constituted by seeds and grape skin.

Grape seeds are a waste generated both in wineries and grape juice industries. This residue contains mainly lipids, proteins, and carbohydrates. The lipid fraction of grape seeds has a high commercial value, only partially exploited so far. Grape seed oil is gaining popularity as a culinary oil that has been studied as a possible source of special lipids. The main characteristic of this oil is its high content in unsaturated fatty acids such as linoleic acid (72–76%, w/w); an acid associated with promotion of cardiovascular health by downregulation of low-density lipoprotein cholesterol and clearance enhancement [102]. In addition, grape seed oil contains tocopherols, also of interest because of their antioxidant properties [103].

Grape skins are major components of wine pomace. This residue may be used in animal feeds, but little more is known about other applications. The lack of other uses is mainly associated with a rather scant fundamental knowledge about the chemical composition and structure of the main components of grape skins. Most studies have dealt with extraction of different classes of compounds such as anthocyanins, hydroxycinnamic acids, flavanols, and flavonol glycosides, and very few with the assessment of basic macromolecular components of grape skins [104, 105].

Stalks, the main solid waste from winemaking, constitute the skeleton of the grape bunch and consist of lignified tissues. Grape stalk could be an interesting raw material for industrial bioconversion processes given that it has a high content of biomolecules such as lignin, cellulose, and hemicellulose [106]. As compared to other lignocellulosic feedstocks, grape stalk also contains high amounts of condensed tannins (proanthocyanidins), which are chemically reactive molecules able to react with electrophilic and nucleophilic reagents, with proteins, or to produce auto-condensation reactions. Condensed polyphenols could also potentially be converted to priced chemicals and materials [107].

Wine lees – the solid residue remaining at the bottom of reservoirs after wine fermentation – have received less attention than other wine residues. Despite their variable composition, wine lees are mainly composed of microorganisms (particularly yeasts) and, in a less proportion, by tartaric acid and inorganic matter. Wine lees from red wine production possess an intense red color that suggests they could have an added value as a source of colorants [78].

8.4.5 Present and Potential Exploitation of Wine Residues: Overall and Partial Exploitation

Traditionally, wastes from winemaking (seeds, skins, stalks, and lees) have been treated by composting to produce fertilizers for further application in vineyards

Table 8.7 Fatty acid composition of grape seed oil [109, 110].

Fatty acid	Abbreviated formula	Oil contents (%, v/w dry basis)
Myristic	C14:0	0.06–0.16
Palmitic	C16:0	6.17–8.5
Palmitoleic	C16:1	0.13–0.24
Margaric	C17:0	0.07–0.14
Margaroleic	C17:1	0.0–0.04
Stearic	C18:0	4.09–5.91
Oleic	C 18:1n9c	13.7–20.8
trans-Oleic	C 18:1n9t	0.0–0.16
Linoleic	C18:2cc	63.0–73.1
trans-Linoleic	C18:2t	0.0–0.16
Linolenic	C18:3	0.36–0.51
Arachidic	C20:0	0.18–0.27
Gadoleic	C 20:1n9	0.11–0.22
Eicosadienoic	C20:2n6	0.0–0.05
Behenic	C22:0	0.0–0.09
Oleic/linoleic	18:1/18:2	0.22 to 0.01

[108]. However, these wastes can be utilized individually for different applications, improving their fields of application.

Grape seeds comprise about 5% of the fruit weight and correspond to 38–52% of dry matter in the pomace [102]. This is the most valuable residue from the wine industry because of its lipid profile (see Table 8.7); it is rich in linoleic acid, as mentioned before [111], has a high content of vitamin E, a powerful antioxidant [109]. Grape seed oil is gaining popularity as a culinary oil that has been studied as a possible source of special lipids [110]. These characteristics make the grape seed oil an interesting product which could promote valorization of this residue so abundant in the Mediterranean basin. In this and other regions with large wine production, a massive exploitation of grape seeds oil can be biodiesel production by transesterification [112].

In addition, grape seed oil is a potential source of natural antioxidants and other healthy bioactive compounds, mainly phenolic compounds [103]. In fact, grape seed proanthocyanidins have shown an antioxidant power 20 times greater than vitamin E and 50 times greater than vitamin C [113]. These characteristics make grape seed extracts exploitable as additives in the food industry, in which the extracts have been used as antimicrobial additives in soy protein edible films, showing to be effective in inhibiting the growth of some pathogens such as *L. monocytogenes, E. coli,* and *S. Typhimurium* [114]. In addition to their use as colorants, flavor modifiers, or antioxidants in the food industry, these characteristics can be exploited in the cosmetics and nutraceutical industries [103].

Grape skins and stalks are not hazardous residues from the wine industry, but the high content of organic matter and their high seasonal production can give rise to pollution problems, mainly regarding COD and BOD of groundwater [106].

Grape skins are the major component of grape pomace contributing roughly to a half of its weight. In addition, this residue has a high content of phenols, mainly polyphenols (Table 8.8), which varies greatly according to the grape cultivar, season, and environmental factors [104]. The potential health benefits of these compounds have been widely reported. Phenols are partially extracted into the

Table 8.8 Phenolic profile of red and white grape skin [105].

Phenol compound	Formula	Grape skin	
		Red	White
Epicatechin	$C_{15}H_{14}O_6$	√	—
Epicatechin-o-gallate	$C_{22}H_{18}O_{10}$	√	√
Epigallocatechin	$C_{15}H_{14}O_7$	√	√
Gallocatechin-o-gallate	$C_{22}H_{18}O_{11}$	—	√
Procyanidin B2	$C_{30}H_{26}O_{12}$	√	—
Myricetin	$C_{15}H_{10}O_8$	—	√
Quercetin	$C_{15}H_{10}O_7$	√	√
Myricetin-3-o-glucoside	$C_{21}H_{20}O_3$	√	√
Quercetin-3-o-glucoside	$C_{21}H_{20}O_{12}$	—	√
Astilbin	$C_{23}H_{22}O_{33}$	√	—
Caffeic acid	$C_9H_8O_4$	√	√
o-Coumaric acid	$C_9H_8O_3$	√	√
Ferulic acid	$C_{10}H_{10}O_4$	√	√
Gallic acid	$C_7H_6O_5$	√	√
4-Hydroxybenzoic acid	$C_7H_6O_3$	√	√
Protocatechuic acid	$C_7H_6O_4$	√	√
trans-Resveratrol	$C_{14}H_{12}O_3$	√	√
Astringinin	$C_{14}H_{12}O_4$		√
Piceid	$C_{20}H_{22}O_8$	√	—
trans-Pterostilbene	$C_{16}H_{16}O_3$	√	—
5-Acetoxymethylfurfural	$C_8H_8O_4$	—	√
2-Furancarboxylic acid ethyl esther	$C_7H_8O_3$	—	√
Ethyl ferulate	$C_{12}H_{14}ClO_4$	√	√
Ethyl protocatechuate	$C_9H_{10}O_4$	√	√
Pyrocatechol	$C_6H_6O_2$	√	√
Pyrogallol	$C_6H_6O_3$	√	√
4-Methylpyrocatechol	$C_7H_8O_2$	√	√
4-Vinylguaiacol	$C_9H_{10}O_2$	√	√
2-Phenylacetaldehyde	C_8H_8O	√	√
4-Hydroxybenzaldehyde	$C_7H_6O_2$	√	√
Coniferaldehyde	$C_{10}H_{10}O_3$	√	√
p-Hydroxybenzalacetone	$C_{10}H_{10}O_2$	—	√
Vanillin	$C_8H_8O_3$	√	√
4-Ethoxyphenol	$C_8H_{10}O_2$	—	√

must during the winemaking process. In red wines, the crushed grapes are kept in contact with the juice during fermentation for several days to enrich it with these compounds (mainly with anthocyanins). Nevertheless, the extraction estimated for anthocyanins during the contact time is 30–40% [115]; therefore, skins from the remaining solid waste of pressed grapes represent an inexpensive source of these phenols. In addition, cellulose and hemicelluloses (approximately 20.8% and 12.5%, respectively, of dry skin) can be exploited for production of "green paper" [116] in the case of cellulose; while hemicellulose, mainly pectin, can be used in the food and pharmaceutical industries as a food thickener [117] or as encapsulant in emulsions [118].

Grape stalks represent approximately 4 kg of waste per 100 l of produced wine. This residue is mainly composed of cellulose (30.3%), hemicellulose (21.0%), lignin (17.4%), tannins (15.9%), and proteins (6.1%) [119]. In addition to the cellulose and hemicellulose applications described above, condensed tannins have promising properties as adhesives. In fact, they are chemically and economically interesting for preparation of adhesives and resins usually destined to the wood industry, and could be successfully used as substitutes for phenol in the production of phenol–formaldehyde resins [120]. In addition, grape stalk residues may be used for decontamination of metal-containing effluents, as their capability for absorbing Cu(II) and Ni(II) from aqueous solutions by an ion exchange mechanism has been demonstrated [121]. Similarly to vine shoots, grape stalks can be used to obtain activated carbon by chemical activation with phosphoric acid [17].

The contribution of wine lees to the quality of wines during the time they remain in contact is undeniable; however, more research in this field is needed to clarify the influence of each lees component on wine characteristics. Wine–lees contact seems to be positive concerning the removal of most undesirable compounds from wine, but negative in the case of biogenic amines. On the other hand, the interaction of wine lees with phenolic compounds, lipids, and mannoproteins released by them as well as the influence of all of these on wine quality is not clear [122]. Wine lees have received less attention than other wine residues. They are mainly composed of yeasts and, in a less proportion, by tartaric acid and inorganic matter. In red wine production, the intense red color of wine lees, characteristic of a high content of anthocyanins, suggests that this residue could have an added value as source of colorants [77].

8.5 Exploitation of By-products from the Citrus Juice Industry

8.5.1 Generalities

Citrus fruits are well known for their refreshing fragrance, thirst-quenching ability, and adequate provision of vitamin C as per recommended dietary allowance [123]. In addition to ascorbic acid, citrus contain several other phytochemicals, for example, carotenoids (lycopene and β-carotene), limonoids, flavanones (naringins and rutinoside), and vitamin-B complex and related nutrients (thiamine, riboflavin, nicotinic acid/niacin, pantothenic acid, pyridoxine, folic acid, biotin, choline, and inositol), which play a key role as nutraceuticals [124].

Citrus is the most abundant crop in the world. Its worldwide production is over 88 millions tons per year and one-third of the crop is processed. Oranges, lemons, grapefruits, and mandarins represent approximately 98% of the entire industrialized crop, oranges being the most relevant with approximately 82% of the total [125]. In Mediterranean countries, an area of around a million hectares is dedicated to produce 23 million tons of citrus per year, which represent, approximately, a quarter of the worldwide annual production (Table 8.9) [1].

Citrus fruits are processed mainly to obtain juice and the remaining materials (including peel, segment membranes, and other by-products) are considered as citrus residues (CRs) (Figure 8.3) [126]. Peels, the main fraction of CRs, represent roughly half of the fruit mass and have been widely studied because they contain numerous biologically active compounds, carbohydrates, and polymers that can be exploited in the pharmaceutical and food industries or for energy production [127].

Table 8.9 Total production of citrus fruit in Mediterranean countries [1].

Mediterranean country	Total citrus fruits	
	Production (tons)	Harvested area (ha)
Spain	5 773 619	317 605
Italy	3 840 388	169 906
Egypt	3 730 685	166 207
Morocco	1 642 244	105 410
Turkey	3 613 766	100 397
Algeria	1 107 329	55 677
Greece	938 866	50 336
Syrian Arab Republic	1 163 718	41 673
Tunisia	448 863	33 421
Israel	476 665	16 197
Lebanon	220 000	9 050
Libya	77372	7 601
Cyprus	118 226	2 910
France	41 584	2 418
Croatia	42 916	2 336
Montenegro	8 912	1 000
Albania	14 800	700
Bosnia and Herzegovina	161	261
Malta	2 380	193
Slovenia	a)	a)
Total Mediterranean production	23 262 494	1 083 298

a) Datum no provided by FAO.

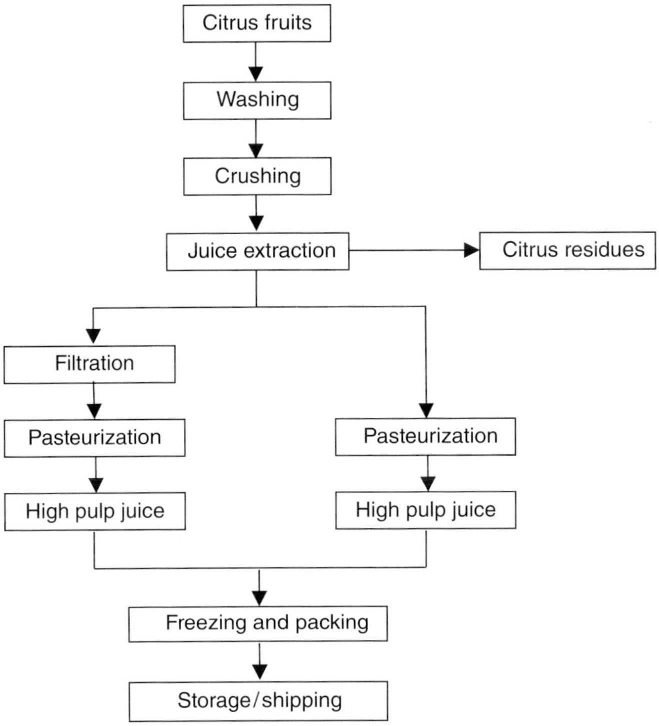

Figure 8.3 Main processes for citrus fruit juice extraction.

8.5.2
Uses and Potential Applications of Bioactive Compounds from Citrus Residues

The present interest in the use of bioactive compounds from natural sources makes CRs especially attractive as they can provide a large amount of these compounds, mainly essential oils, phenols, and pectin.

Citrus essential oils are the most widely used essential oils in the world. They are obtained as by-products of citrus processing, mainly from the peel (Table 8.10). These volatile compounds, constituted by secondary metabolites, are characterized by a strong odor. The antimicrobial and insecticidal properties of citrus essential oils have been exploited in pharmaceutical, sanitary, cosmetics, agricultural, and food industries [129]. In particular, they are used as flavor agents in many foods (including alcoholic and nonalcoholic beverages, candy, and gelatins); and in pharmaceuticals to mask unpleasant tastes of drugs. Citrus essential oils are widely used as odorants in cosmetics [130]. The traditional way to extract essential oils is either by hydrodistillation or cold-pressing of citrus peels. A major component of peel essential oils is *d*-limonene [128] (a strong inhibitor of microorganisms [131]); therefore, removal of essential oils before development of most biotechnological processes involving citrus peel is mandatory.

Table 8.10 Main essential oils in citrus fruits (%) [128].

Chemical name	Formula	Orange (*Camellia Sinensis*)	Lemon (*C. Limon*)	Grapefruit (*C. Paradisi*)	Mandarin (*C. Resticulata*)
Limonene	$C_{10}H_{16}$	91.38	56.04	88.65	86.59
Linalool	$C_{10}H_{18}O$	1.21	12.56	0.95	1.65
β-Myrcene	$C_{10}H_{16}$	2.93	2.14	4.18	2.92
γ-Terpinene	$C_{10}H_{16}$	0.05	6.99	0.05	4.08
2-β-Pinene	$C_{10}H_{16}$	0.06	7.44	0.05	0.46
1R-α-Pinene	$C_{10}H_{16}$	0.87	1.70	1.20	1.34
α-Terpineol	$C_{10}H_{18}O$	0.25	1.77	0.34	0.56
Linalyl acetate	$C_{12}H_{20}O_2$	0.00	2.95	0.00	0.08
E-Citral	$C_{10}H_{16}O$	0.29	1.37	0.51	0.04
Sabinene	$C_{10}H_{16}$	0.49	0.53	0.50	0.31
Caryophyllene	$C_{15}H_{24}$	0.10	0.56	0.74	0.04
β-Citral	$C_{10}H_{16}O$	0.22	1.07	0.40	0.03
trans-β-Ocimene	$C_{10}H_{16}$	0.06	0.96	0.27	0.21
Linalool oxide	$C_{10}H_{18}O_2$	0.11	0.22	0.51	0.05
4-Terpineol	$C_{10}H_{18}O$	0.10	0.54	0.14	0.20
Germacrene D	$C_{15}H_{24}$	0.10	0.49	0.20	0.06
Geraniol acetate	$C_{12}H_{20}O_2$	0.05	0.50	0.15	0.05
L-Phellandrene	$C_{10}H_{16}$	0.00	0.37	0.16	0.21
1-Octanol	$C_8H_{18}O$	0.81	0.08	0.01	0.11
Citronellal	$C_{10}H_{18}O$	0.12	0.10	0.14	0.18
β-Citronellol	$C_{10}H_{20}O$	0.12	0.10	0.05	0.34
Geraniol	$C_{10}H_{18}O$	0.07	0.23	0.15	0.03
Nerol	$C_{10}H_{20}O$	0.06	0.40	0.04	0.07
Neryl acetate	$C_{12}H_{20}O_2$	0.01	0.34	0.02	0.02
trans-Carveol	$C_{10}H_{16}O$	0.03	0.04	0.15	0.03
α-Humulene	$C_{15}H_{24}$	0.01	0.09	0.12	0.01
Nonanal	$C_9H_{18}O$	0.24	0.03	0.03	0.07
L-Carvone	$C_{10}H_{14}O$	0.05	0.05	0.08	0.10
γ-Elemene	$C_{15}H_{24}$	0.08	0.07	0.07	0.01
Limonene oxide	$C_{10}H_{16}O$	0.04	0.01	0.05	0.05
trans-β-Farnesene	$C_{15}H_{24}$	0.01	0.08	0.02	0.01
Perilla alcohol	$C_{10}H_{16}O$	0.01	0.02	0.04	0.01
cis-Carveol	$C_{10}H_{16}O$	0.03	0.03	0.01	0.03
Nerolidol	$C_{15}H_{26}O$	0.01	0.01	0.03	0.01
Camphene	$C_{10}H_{16}$	0.01	0.05	0.00	0.01
Citronellyl acetate	$C_{12}H_{22}O_2$	0.01	0.05	0.00	0.01
1-Hexanol	$C_6H_{14}O$	0.02	0.01	0.00	0.01

Citrus flavonoids are among the most studied bioactive compounds in citrus peel because many epidemiological and intervention studies have associated consumption of these compounds with lower risks of different types of cancer and cardiovascular diseases. This behavior has been associated to the antioxidant, anti-inflammatory, and radical-scavenging activity of flavonoids [132]. In most cases, CRs are subjected to a drying process before solid–liquid extraction of metabolites, mainly obtained by maceration, Soxhlet extraction, ultrasound-assisted extraction (USAE), microwave-assisted extraction (MAE), supercritical fluid extraction (SFE), or superheated liquid extraction (SHLE). The use of any of these techniques requires previous optimization of the methodology, mainly when auxiliary energies are used, as they can lead to degradation of the target compounds. Regardless of the extraction technique, most studies conclude that the best extractants for these compounds are pure ethanol or methanol, or a mixture of any of them with water. The profile and flavonoid content varies depending on the citrus variety. The major phenolic compounds in citrus are listed in Table 8.11.

Citrus peel extracts have been used as natural antioxidant additives to replace synthetic antioxidants (Butylated hydroxyanisole (BHA) or Butylated hydroxytoluene (BHT)) in refined corn oil [133]. The study shows that the addition of

Table 8.11 Main phenols in citrus fruits [124, 127, 133].

Class/compound	Orange (*C. Sinensis*)	Lemon (*C. Limon*)	Grapefruit (*C. Paradisi*)	Mandarin (*C. Resticulata*)
Flavonoid				
Scopoletin	—	—	√	—
Dihydroquercetin	—	—	—	√
Dihydrokaempferol	—	—	√	√
Quercetin	√	√	√	√
Dihydroisorhamnetin	√	√	√	√
Luteolin	√	√	√	√
Naringenin	√	√	√	√
Apigenin	√	√	√	√
Isorhamnetin	√	√	—	√
Diosmetin	√	√	—	√
Eriodictyol	√	√	—	√
Hesperetin	√	—	√	√
Homoeriodictyol	—	√	—	—
Hesperetin	√	√	√	√
Isorhamnetin	√	√	—	√
Chrysoeriol	—	√	—	√
Tamarixetin	—	—	—	√
Kaempferol	√	—	√	√
Isosakuranetin	√	√	√	√
Phenolic acid				
Sinapic	√	√	√	√
Ferulic	√	√	√	√
p-Hydroxybenzoic	√	√	√	√
Vainillic	√	√	√	√
p-Coumaric	√	√	√	√
Caffeic	√	√	√	√
Gallic	√	√	√	√

citrus antioxidants is better than that of synthetic antioxidants to delay rancidity in corn oil. After 6 months storage at 45 °C, inhibition of corn oil peroxidation was similar in the case of addition of synthetic antioxidants (0.2 mg ml^{-1} oil) or citrus peel extract (1.6 or 2.0 mg ml^{-1} oil). For this reason, natural antioxidant extracts of citrus peel would be preferred over synthetic antioxidants to avoid the adverse health effects of the latter. An aspect not discussed by the authors is the high amount of added extracts – a part of which are not antioxidants – that can influence the taste of the enriched oil.

Pectin from citrus peel has been used in the food industry as a gelling agent, and also in the pharmaceutical industry as an ingredient for preparation of antidiarrheal and detoxifying drugs. Recently, pectin has been used for preparation of suspensions, showing potential applications in controlled release drug delivery [134, 135]. Some fragments of pectin have also shown to exert both *in vitro* and *in vivo* positive effects on reducing metastatic stages; therefore, pectin is also considered an anticancer agent [136].

8.5.3
Potential Exploitation of Citrus Residues for Energy Production

Production of biofuels from CRs has focused on obtaining bioethanol and/or biogas. In both cases, the residue is pretreated to remove *d*-limonene, which, as stated before, is toxic to the microorganisms involved in biomass conversion. Traditionally, hydrodistillation has been used to remove *d*-limonene, but recent steam explosion pretreatment has shown to lower the content of essential oils below 0.025%; reducing significantly the hydrolytic enzyme requirements as a result [137].

Bioethanol production using *Saccharomyces cerevisiae* as biocatalyst makes it necessary to lower the concentration of *d*-limonene to 0.01% w/v [126]. Once removal of *d*-limonene has lowered its concentration to the required limit, the residue is subjected to hydrolysis to transform glucose polymers into simpler, fermentable sugars. The hydrolysis step can be carried out by either addition of sulfuric acid (chemical hydrolysis) or enzymes (enzymatic hydrolysis). The hydrolyzate is subjected to alcoholic fermentation for an estimated optimal fermentation time of around 72 h; after which 60 l of ethanol per 1000 kg CRs are obtained [137].

Another application of CRs is as low-priced substrates to produce methane under anaerobic conditions, also after *d*-limonene extraction. The highest methane yield, around 0.27–0.29 l CH_4 per g CRs, occurs under thermophilic conditions (55 °C), with biodegradability between 84% and 90% [138]. Anaerobic digestion can easily be integrated in the citrus juice industry to treat the organic residues, thus providing a biogas which can be used for electricity or thermal energy. In addition to recovery of *d*-limonene in the initial stage of biofuel production, many studies report the recovery of pectin as a by-product at the end of the process [126]. An integrated process for production of ethanol, biogas, pectin, and limonene from CRs resulted in 39.64 l of ethanol,

45 m^3 methane, 8.91 limonene, and 38.8 kg pectin per ton CRs with 20% dry weight [126].

Fast pyrolysis (carrier gas at 700 °C and 400 m s^{-1}) has been used for bio-oil production. This yield represents up to 60% in weight of the raw material used, and the remainder can be used for charcoal production. Around 25% of CRs remaining after the pyrolysis process can be used for charcoal production [139].

8.5.4
Other Overall and Partial Uses of Citrus Residues

As part of the worldwide trend toward searching for new raw materials production, the recovery of polysaccharides from CRs as polymer material is being investigated [139]; on the other hand, CRs have not been used for hydrolysis and/or fermentation processes. The contents of cellulose and hemicellulose in CRs (ranges 12.7–13.6% and 5.3–6.1%, respectively) allow partial use of this raw material as component of paper pulp at percentages lower than 10% [140]. The effects of this addition are nil on breaking length, positive on bursting strength, and negative on tearing resistance. Brightness was negatively affected at proportions above 10% because citrus peel contains colored pigments [141]. The addition of citrus peels to paper pulp has no effect on breaking length, but it has a positive effect on bursting strength and negative on tearing resistance. Brightness was negatively affected at proportions above 10%, because citrus peel particles contain colored pigments [141].

Citrus peel has demonstrated a great ability to adsorb metallic ions from simulated industrial effluent individually contaminated with Ni(II), Co(II), or Cu(II) solutions. Under dynamic adsorption conditions, the maximum adsorption capacities varied with the working pH, the highest capacity occurring at pH 4.8 and 298 K. The citrus peel saturation took place for 1.85; 1.35; and 1.30 mmol g^{-1} of Ni(II), Co(II), and Cu(II), respectively. For this reason, citrus peel can be considered a natural adsorbent available from the citrus juice industry, which can be used without prior treatment, has a low cost, and a great potential in ecosystem cleanup [142].

Finally, attempts have been made to use dried peel as natural feed additive and even as medicinal supplement for animals [143]. Nevertheless, the low nutritional value of this material and/or the diseases caused to the animals (mycotoxicosis, rumen parakeratosis) proved that this use is not appropriate [144].

Acknowledgments

The Spanish Ministerio de Economía y Competitividad (MINECO) and FEDER program are thanked for financial support through project CTQ2012-37428. C.A.L.E. Thanks are also due to the Spanish International School for Ph.D. in Agrofood (ceiA3) and Mexican National Council for Science and Technology (CONACYT) for grants to support this research in both countries.

List of Abbreviations

AOP	advanced oxidation process
BOD	biochemical oxygen demand
COD	chemical oxygen demand
CR	citrus residue
EVOO	extra virgin olive oil
MAE	microwave-assisted extraction
OMWW	olive mill waste water
SFE	supercritical fluid extraction
SHLE	superheated liquid extraction
TPOMW	two-phase olive mill waste
USAE	ultrasound-assisted extraction
UV	ultraviolet
VOO	virgin olive oil

References

1. Food and Agriculture Organization of the United Nations *http://faostat.fao.org/* (accessed 20 September 2013).
2. Estruch, R. and Salas-Salvadó, J. (2013) Towards an even healthier mediterranean diet. *Nutr. Metab. Cardiovasc. Dis.*, **23** (12), 1163–1166.
3. United Nations Educational, Scientific and Cultural Organization *http://www.unesco.org/* (accessed 27 January 2014).
4. Peralbo-Molina, A. and Luque de Castro, M.D. (2013) Potential of residues from the Mediterranean agriculture and agrifood industry. *Trends Food Sci. Technol.*, **32** (1), 16–24.
5. Brown, D., Ramos-Padrón, E., Gieg, L., and Voordouw, G. (2013) Effect of calcium ions and anaerobic microbial activity on sedimentation of oil sands tailings. *Int. Biodeterior. Biodegrad.*, **81**, 9–16.
6. Chiu, Y.S. and Keh, H.J. (2014) Sedimentation velocity and potential in a concentrated suspension of charged soft spheres. *Colloids Surf., A: Physicochem. Eng. Asp.*, **440**, 185–196.
7. Inan, H., Dimoglo, A., Simsek, H., and Karpuzcu, M. (2004) Olive oil mill wastewater treatment by means of electro-coagulation. *Sep. Purif. Technol.*, **36** (1), 23–31.
8. Berk, Z. (2009) in *Food Process Engineering and Technology* (ed. Z. Berk), Academic Press, San Diego, CA, pp. 195–216.
9. Berk, Z. (2009) in *Food Process Engineering and Technology* (ed. Z. Berk), Academic Press, San Diego, CA, pp. 217–232.
10. Amuda, O.S. and Alade, A. (2006) Coagulation/flocculation process in the treatment of abattoir wastewater. *Desalination*, **196** (1-3), 22–31.
11. Amuda, O.S. and Amoo, I.A. (2007) Coagulation/flocculation process and sludge conditioning in beverage industrial wastewater treatment. *J. Hazard. Mater.*, **141** (3), 778–783.
12. Khoufi, S., Feki, F., and Sayadi, S. (2007) Detoxification of olive mill wastewater by electrocoagulation and sedimentation processes. *J. Hazard. Mater.*, **142** (1-2), 58–67.
13. Aguilar, M.I., Sáez, J., Lloréns, M., Soler, A., and Ortuño, J.F. (2002) Nutrient removal and sludge production in the coagulation–flocculation process. *Water Res.*, **36** (11), 2910–2919.
14. Berk, Z. (2009) in *Food Process Engineering and Technology* (ed. Z. Berk),

Academic Press, San Diego, CA, pp. 279–294.

15. Chand, R., Narimura, K., Kawakita, H., Ohto, K., Watari, T., and Inoue, K. (2009) Grape waste as a biosorbent for removing Cr(VI) from aqueous solution. *J. Hazard. Mater.*, **163** (1), 245–250.
16. Annadurai, G., Juang, R.S., and Lee, D.J. (2002) Use of cellulose-based wastes for adsorption of dyes from aqueous solutions. *J. Hazard. Mater.*, **92** (3), 263–274.
17. Deiana, A.C., Sardella, M.F., Silva, H., Amaya, A., and Tancredi, N. (2009) Use of grape stalk, a waste of the viticulture industry, to obtain activated carbon. *J. Hazard. Mater.*, **172** (1), 13–19.
18. Gogate, P.R. and Pandit, A.B. (2004) A review of imperative technologies for wastewater treatment I: oxidation technologies at ambient conditions. *Adv. Environ. Res.*, **8** (3-4), 501–551.
19. Bhatti, Z.A., Mahmood, Q., Raja, I.A., Malik, A.H., Rashid, N., and Wub, D. (2011) Integrated chemical treatment of municipal wastewater using waste hydrogen peroxide and ultraviolet light. *Phys. Chem. Earth*, **36** (9-11), 459–464.
20. Neyens, E. and Baeyens, J. (2003) A review of classic Fenton's peroxidation as an advanced oxidation technique. *J. Hazard. Mater.*, **98** (1-3), 33–50.
21. Karci, A. (2013) Degradation of chlorophenols and alkylphenol ethoxylates, two representative textile chemicals, in water by advanced oxidation processes: the state of the art on transformation products and toxicity. *Chemosphere*. doi: 10.1016/j.chemosphere.2013.10.034
22. Oller, I., Malato, S., and Sánchez-Pérez, J.A. (2011) Combination of advanced oxidation processes and biological treatments for wastewater decontamination–A review. *Sci. Total Environ.*, **109** (20), 4141–4166.
23. Arvanitoyannis, I.S. and Kassaveti, A. (2008) in *Waste Management for the Food Industries* (ed. I.S. Arvanitoyannis), Academic Press, Amsterdam, pp. 453–568.
24. Waste-to-Energy Research and Technology Council *http://www.wtert.eu* (accessed 28 October 2013).
25. Caputo, A.C., Scacchia, F., and Pelagagge, P.M. (2003) Disposal of by-products in olive oil industry: waste-to-energy solutions. *Appl. Therm. Eng.*, **23** (2), 197–214.
26. Encinar, M.J., González, J.F., Martínez, G., and Román, S. (2009) Catalytic pyrolysis of exhausted olive oil waste. *J. Anal. Appl. Pyrolysis*, **85** (1-2), 197–203.
27. Tillman, D.A. (1991) *The Combustion of Solid Fuels and Wastes*, Academic Press, San Diego, CA.
28. Belgiorno, V., De Feo, G., Della Rocca, C., and Napoli, R.M.A. (2003) Energy from gasification of solid wastes. *Waste Manag.*, **23** (1), 1–15.
29. Sutton, D., Kelleher, B., and Ross, J.R.H. (2001) Review of literature on catalysts for biomass gasification. *Fuel Process Technol.*, **73** (3), 155–173.
30. Rapagnà, S., Jand, N., and Foscolo, P.U. (1998) Catalytic gasification of biomass to produce hydrogen rich gas. *Int. J. Hydrogen Energy*, **23** (7), 551–557.
31. Bilitewski, B., Härdtle, G., and Marek, K. (1997) *Waste Management*, Springer-Verlag, Berlin.
32. Roig, A., Cayuela, M.L., and Sánchez-Monedero, M.A. (2006) An overview on olive mill wastes and their valorisation methods. *Waste Manag.*, **26** (9), 960–969.
33. Bajaj, M. and Winter, J. (2013) Biogas and biohydrogen production potential of high strength automobile industry wastewater during anaerobic degradation. *J. Environ. Manage.*, **128**, 522–529.
34. Azbar, N., Keskin, T., and Yuruyen, A. (2008) Enhancement of biogas production from olive mill effluent (OME) by co-digestion. *Biomass Bioenergy*, **32** (12), 1195–1201.
35. Himanen, M. and Hänninen, K. (2011) Composting of bio-waste, aerobic and anaerobic sludges – Effect of feedstock on the process and quality of compost. *Bioresour. Technol.*, **102** (3), 2842–2852.

36. Buendía, I.M., Fernández, F.J., Villaseñor, J., and Rodríguez, L. (2008) Biodegradability of meat industry wastes under anaerobic and aerobic conditions. *Water Res.*, **42** (14), 3767–3774.
37. Muktadirul Bari Chowdhury, A.K.M., Akratos, C.S., Vayenas, D.V., and Pavlou, S. (2013) Olive mill waste composting: a review. *Int. Biodeterior. Biodegrad.*, **85**, 108–119.
38. Dermeche, S., Nadour, M., Larroche, C., Moulti–Mati, F., and Michaud, P. (2013) Olive mill wastes: biochemical characterizations and valorization strategies. *Process Biochem.*, **48** (10), 1532–1552.
39. Authority, EFSA (2011) Scientific opinion on the substantiation of health claims related to polyphenols in olive and protection of LDL particles from oxidative damage (ID 1333, 1638, 1639, 1696, 2865), maintenance of normal blood HDL cholesterol concentrations (ID 1639), maintenance of normal blood pressure (ID 3781), "anti-inflammatory properties" (ID 1882), "contributes to the upper respiratory tract health" (ID 3468), "can help to maintain a normal function of gastrointestinal tract" (3779), and "contributes to body defences against external agents" (ID 3467) pursuant to Article 13(1) of Regulation (EC) No 1924/2006. *EFSA J.*, **9**, 25.
40. European Commission (2013) Commission implementing regulation (EU) no 299/2013 of March 2013 amending regulation (EEC) no. 2568/91 on the characteristics of olive oil and olive-residue oil and on the relevant methods of analysis. *Off. J. Eur. Union*, **L90**, 52–70.
41. Sandasi, M., Kamatou, G.P., and Viljoen, A.M. (2012) An untargeted metabolomic approach in the chemotaxonomic assessment of two Salvia species as a potential source of alpha-bisabolol. *Phytochemistry*, **84**, 94–101.
42. García-González, D.L., Aparicio-Ruiz, R., and Aparicio, R. (2008) Virgin olive oil – Chemical implications on quality and health. *Eur. J. Lipid Sci. Technol.*, **110** (7), 602–607.
43. El Riachy, M., Priego-Capote, F., Rallo, L., Luque de Castro, M.D., and Léon, L. (2012) Phenolic profile of virgin olive oil from advanced breeding selections. *Span. J. Agric. Res.*, **10** (2), 443–453.
44. Sánchez de Medina, V., Priego-Capote, F., Jiménez-Ot, C., and Luque de Castro, M.D. (2011) Quality and stability of edible oils enriched with hydrophilic antioxidants from the olive tree: the role of enrichment extracts and lipid composition. *J. Agric. Food Chem.*, **59** (21), 11432–11441.
45. Baccouri, O., Guerfel, M., Baccouri, B., Cerretani, L., Bendini, A., Lercker, G., Zarrouk, M., and Daoud, D. (2008) Chemical composition and oxidative stability of Tunisian monovarietal virgin olive oils with regard to fruit ripening. *Food Chem.*, **109** (4), 743–754.
46. Werth, M.T., Halouska, S., Shortridge, M.D., Zhang, B., and Powers, R. (2010) Analysis of metabolomic PCA data using tree diagrams. *Anal. Biochem.*, **399** (1), 58–63.
47. Zbakh, H. and El Abbassi, A. (2012) Potential use of olive mill wastewater in the preparation of functional beverages: a review. *J. Funct. Foods*, **4** (1), 53–65.
48. Di Giovacchino, L., Sestili, S., and Di Vincenzo, D. (2002) Influence of olive processing on virgin olive oil quality. *Eur. J. Lipid Sci. Technol.*, **104** (9-10), 587–601.
49. Alburquerque, J. (2004) Agrochemical characterisation of "alperujo", a solid by-product of the two-phase centrifugation method for olive oil extraction. *Bioresour. Technol.*, **91** (2), 195–200.
50. Barbera, A.C., Maucieri, C., Cavallaro, V., Ioppolo, A., and Spagna, G. (2013) Effects of spreading olive mill wastewater on soil properties and crops, a review. *Agric. Water Manage.*, **119**, 43–53.
51. Paraskeva, P. and Diamadopoulos, E. (2006) Technologies for olive mill wastewater (OMW) treatment: a review. *J. Chem. Technol. Biotechnol.*, **81** (9), 1475–1485.
52. He, J., Alister-Briggs, M., de Lyster, T., and Jones, G.P. (2012) Stability and antioxidant potential of purified olive

mill wastewater extracts. *Food Chem.*, **131** (4), 1312–1321.

53. Boskou, G., Salta, F.N., Chrysostomou, S., Mylona, A., Chiou, A., and Andrikopoulos, N.K. (2006) Antioxidant capacity and phenolic profile of table olives from the Greek market. *Food Chem.*, **94** (4), 558–564.
54. Paredes, C., Cegarra, J., Roig, A., Sánchez–Monedero, M.A., and Bernal, M.P. (1999) Characterization of olive mill wastewater (alpechín) and its sludge for agricultural purposes. *Bioresour. Technol.*, **67** (2), 111–115.
55. Komilis, D.P., Karatzas, E., and Halvadakis, C.P. (2005) The effect of olive mill wastewater on seed germination after various pretreatment techniques. *J. Environ. Manage.*, **74** (4), 339–348.
56. Aviani, I., Raviv, M., Hadar, Y., Saadi, I., Dag, A., Ben-Gal, A., Yermiyahu, U., Zipori, I., and Laor, Y. (2012) Effects of harvest date, irrigation level, cultivar type and fruit water content on olive mill wastewater generated by a laboratory scale 'Abencor' milling system. *Bioresour. Technol.*, **107**, 87–96.
57. Saadi, I., Laor, Y., Raviv, M., and Medina, S. (2007) Land spreading of olive mill wastewater: effects on soil microbial activity and potential phytotoxicity. *Chemosphere*, **66** (1), 75–83.
58. Ntougias, S., Iskidou, E., Rousidou, C., Papadopoulou, K.K., Zervakis, G.I., and Ehaliotis, C. (2010) Olive mill wastewater affects the structure of soil bacterial communities. *Appl. Soil Ecol.*, **45** (2), 101–111.
59. Tekin, A.R. and Dalgiç, A.C. (2000) Biogas production from olive pomace. *Resour. Conserv. Recycl.*, **30** (4), 301–313.
60. de la Casa, J.A., Romero, I., Jiménez, J., and Castro, E. (2012) Fired clay masonry units production incorporating two-phase olive mill waste (alperujo). *Ceram. Int.*, **38** (6), 5027–5037.
61. Mekki, H., Anderson, M., Benzina, M., and Ammar, E. (2008) Valorization of olive mill wastewater by its incorporation in building bricks. *J. Hazard. Mater.*, **158** (2-3), 308–315.
62. de la Casa, J.A., Lorite, M., Jiménez, J., and Castro, E. (2009) Valorisation of wastewater from two-phase olive oil extraction in fired clay brick production. *J. Hazard. Mater.*, **169** (1-3), 271–278.
63. Rodis, P.S., Karathanos, V.T., and Mantzavinou, A. (2002) Partitioning of olive oil antioxidants between oil and water phases. *J. Agric. Food Chem.*, **50** (3), 596–601.
64. Azaizeh, H., Halahlih, F., Najami, N., Brunner, D., Faulstich, M., and Tafesh, A. (2012) Antioxidant activity of phenolic fractions in olive mill wastewater. *Food Chem.*, **134** (4), 2226–2234.
65. El-Abbassi, A., Kiai, H., and Hafidi, A. (2012) Phenolic profile and antioxidant activities of olive mill wastewater. *Food Chem.*, **132** (1), 406–412.
66. Cassano, A., Conidi, C., Giorno, L., and Drioli, E. (2013) Fractionation of olive mill wastewaters by membrane separation techniques. *J. Hazard. Mater.*, **248–249**, 185–193.
67. García-Castello, E., Cassano, A., Criscuoli, A., Conidi, C., and Drioli, E. (2010) Recovery and concentration of polyphenols from olive mill wastewaters by integrated membrane system. *Water Res.*, **44** (13), 3883–3892.
68. Priego-Capote, F., Ruiz-Jiménez, J., and Luque de Castro, M.D. (2004) Fast separation and determination of phenolic compounds by capillary electrophoresis–diode array detection. *J. Chromatogr. A*, **1045** (1-2), 239–246.
69. Sánchez de Medina, V., Priego-Capote, F., and Luque de Castro, M.D. (2012) Characterization of refined edible oils enriched with phenolic extracts from olive leaves and pomace. *J. Agric. Food Chem.*, **60**, 5866–5873.
70. Peralbo-Molina, A., Priego-Capote, F., and Luque de Castro, M.D. (2012) Tentative identification of phenolic compounds in olive pomace extracts using liquid chromatography-tandem mass spectrometry with a quadrupole-quadrupole-time-of-flight mass detector. *J. Agric. Food Chem.*, **60** (46), 11542–11550.
71. Tergum *http://www.tergum.es* (accessed 12 December 2013).

72. Japón-Luján, R. and Luque de Castro, M.D. (2006) Superheated liquid extraction of oleuropein and related biophenols from olive leaves. *J. Chromatogr. A*, **1136** (2), 185–191.
73. SK-II *http://www.sk-ii.com/en* (accessed 20 January 2014).
74. Ferrer Healt Tech *http://www.exquim.com* (accessed 11 December 2013).
75. Yapo, B.M. (2009) Biochemical characteristics and gelling capacity of pectin from yellow passion fruit rind as affected by acid extractant nature. *J. Agric. Food Chem.*, **57** (4), 1572–1578.
76. Kountouri, A.M., Mylona, A., Kaliora, A.C., and Andrikopoulos, N.K. (2007) Bioavailability of the phenolic compounds of the fruits (drupes) of *Olea europaea* (olives): impact on plasma antioxidant status in humans. *Phytomedicine*, **14** (10), 659–667.
77. Meksi, N., Haddar, W., Hammami, S., and Mhenni, M.F. (2012) Olive mill wastewater: a potential source of natural dyes for textile dyeing. *Ind. Crops Prod.*, **40**, 103–109.
78. Delgado de la Torre, M.P., Ferreiro-Vera, C., Priego-Capote, F., and Luque de Castro, M.D. (2013) Anthocyanidins, proanthocyanidins, and anthocyanins profiling in wine lees by solid-phase extraction–liquid chromatography coupled to electrospray ionization tandem mass spectrometry with data-dependent methods. *J. Agric. Food Chem.* doi: 10.1021/jf404194q
79. Molina-Alcaide, E. and Yáñez-Ruiz, D.R. (2008) Potential use of olive by-products in ruminant feeding: a review. *Anim. Feed Sci. Technol.*, **147** (1-3), 247–264.
80. Minervino, A.H.H., Júnior, R.A.B., Ferreira, R.N.F., Rodrigues, F.A.M.L., Headley, S.A., Mori, C.S., and Ortolani, E.L. (2009) Clinical observations of cattle and buffalos with experimentally induced chronic copper poisoning. *Res. Vet. Sci.*, **87** (3), 473–478.
81. Clodoveo, M.L., Durante, V., and La Notte, D. (2013) Working towards the development of innovative ultrasound equipment for the extraction of virgin olive oil. *Ultrason. Sonochem.*, **20** (5), 1261–1270.
82. Chemat, F. (2004) Deterioration of edible oils during food processing by ultrasound. *Ultrason. Sonochem.*, **11** (1), 13–15.
83. Chemat, F., Grondin, I., Costes, P., Moutoussamy, L., Shum Cheong Sing, A., and Smadja, J. (2004) High power ultrasound effects on lipid oxidation of refined sunflower oil. *Ultrason. Sonochem.*, **11** (5), 281–285.
84. Cañizares-Macías, M.P., García-Mesa, J.A., and Luque de Castro, M.D. (2004) Fast ultrasound-assisted method for the determination of the oxidative stability of virgin olive oil. *Anal. Chim. Acta*, **502** (2), 161–166.
85. Canizares-Macías, M.P., García-Mesa, J.A., and Luque de Castro, M.D. (2004) Determination of the oxidative stability of olive oil, using focused-microwave energy to accelerate the oxidation process. *Anal. Bioanal. Chem.*, **378** (2), 479–483.
86. Jackson, R.S. (2008) in *Wine Science*, 3rd edn (ed. R.S. Jackson), Academic Press, San Diego, CA, pp. 332–417.
87. Ruggieri, L., Cadena, E., Martínez-Blanco, J., Gasol, C.M., Rieradevall, J., Gabarrell, X., Gea, T., Sort, X., and Sánchez, A. (2009) Recovery of organic wastes in the Spanish wine industry. Technical, economic and environmental analyses of the composting process. *J. Cleaner Prod.*, **17** (9), 830–838.
88. Delgado de la Torre, M.P., Priego-Capote, F., and Luque de Castro, M.D. (2012) Evaluation of the composition of vine shoots and oak chips for oenological purposes by superheated liquid extraction and high-resolution liquid chromatography-time-of-flight/mass spectrometry analysis. *J. Agric. Food Chem.*, **60** (13), 3409–3417.
89. Delgado de la Torre, M.P., Priego-Capote, F., and Luque de Castro, M.D. (2014) Comparative profiling analysis of woody flavouring from vine-shoots and oak chips. *J. Agric. Food Chem.*, **94** (3), 504–514.
90. Monagas, M., Hernández-Ledesma, B., Gómez-Cordovez, C., and Bartolomé,

B. (2005) Commercial dietary ingredients from *Vitis vinifera L.* leaves and grape skins: antioxidant and chemical characterization. *J. Agric. Food Chem.*, **54** (2), 319–327.

91. Lobo, M.C. (1988) The effect of compost from vine shoots on the growth of Barley. *Biol. Waste.*, **25** (4), 281–290.
92. Jiménez-Gómez, S., Cartagena-Causapé, M.C., and Arce-Martínez, A. (1993) Distribution of nutrients in anaerobic digestion of vine shoots. *Bioresour. Technol.*, **45** (2), 93–97.
93. Gañán, J., Al-Kassir Abdulla, A., Cuerda-Correa, E.M., and Macías-García, A. (2006) Energetic exploitation of vine shoot by gasification processes: a preliminary study. *Fuel Process. Technol.*, **87** (10), 891–897.
94. Jiménez, L., Angulo, V., Ramos, E., De la Torre, M.J., and Ferrer, J.L. (2006) Comparison of various pulping processes for producing pulp from vine shoots. *Ind. Crops Prod.*, **23** (2), 122–130.
95. Jiménez, L., Angulo, V., Rodríguez, A., Sánchez, R., and Ferrer, J.L. (2009) Pulp and paper from vine shoots: neural fuzzy modeling of ethylene glycol pulping. *Bioresour. Technol.*, **100** (2), 756–762.
96. Corcho-Corral, B., Olivares-Marín, M., Fernández-González, C., Gómez-Serrano, V., and Macías-García, A. (2006) Preparation and textural characterisation of activated carbon from vine shoots (*Vitis vinifera*) by H_3PO_4–Chemical activation. *Appl. Surf. Sci.*, **252** (17), 5961–5966.
97. Max, B., Torrado, A.M., Moldes, A.B., Converti, A., and Domínguez, J.M. (2009) Ferulic acid and *p*-coumaric acid solubilization by alkaline hydrolysis of the solid residue obtained after acid prehydrolysis of vine shoot prunings: effect of the hydroxide and pH. *Biochem. Eng. J.*, **43** (2), 129–134.
98. Luque-Rodríguez, J.M., Pérez-Juan, P., and Luque de Castro, M.D. (2006) Extraction of polyphenols from vine shoots of *Vitis vinifera* by superheated ethanol–water mixtures. *J. Agric. Food Chem.*, **54** (23), 8775–8781.
99. Pérez-Juan, P.M. and Luque de Castro, M.D. (2014) in *Processing and Impact on Active Components in Food* (ed. V. Preedy), Elsevier Inc., pp. 469–479.
100. Orhan, D.D., Orhan, N., Ergun, E., and Ergun, F. (2007) Hepatoprotective effect of *Vitis vinifera L.* leaves on carbon tetrachloride-induced acute liver damage in rats. *J. Ethnopharmacol.*, **112** (1), 145–151.
101. Dani, C., Oliboni, L.S., Agostini, F., Funchal, C., Serafini, L., Henriques, J.A., and Salvador, M. (2010) Phenolic content of grapevine leaves (*Vitis labrusca var.* Bordo) and its neuroprotective effect against peroxide damage. *Toxicol. in Vitro*, **24** (1), 148–153.
102. Maier, T., Schieber, A., Kammerer, D.R., and Carle, R. (2009) Residues of grape (*Vitis vinifera L.*) seed oil production as a valuable source of phenolic antioxidants. *Food Chem.*, **112** (3), 551–559.
103. Peralbo-Molina, A., Priego-Capote, F., and Luque de Castro, M.D. (2013) Characterization of grape seed residues from the ethanol-distillation industry. *Anal. Methods*, **5** (8), 1922–1929.
104. Mendes, J.A.S., Prozil, S.O., Evtuguin, D.V., and Cruz-Lopes, L.P. (2013) Towards comprehensive utilization of winemaking residues: characterization of grape skins from red grape pomaces of variety *Touriga Nacional*. *Ind. Crops Prod.*, **43**, 25–32.
105. Peralbo-Molina, A., Priego-Capote, F., and Luque de Castro, M.D. (2012) Comparison of extraction methods for exploitation of grape skin residues from ethanol distillation. *Talanta*, **101**, 292–298.
106. Spigno, G., Pizzorno, T., and De Faveri, D.M. (2008) Cellulose and hemicelluloses recovery from grape stalks. *Bioresour. Technol.*, **99** (10), 4329–4337.
107. Ping, L., Brosse, N., Sannigrahi, P., and Ragauskas, A. (2011) Evaluation of grape stalks as a bioresource. *Ind. Crops Prod.*, **33** (1), 200–204.
108. Bertran, E., Sort, X., Soliva, M., and Trillas, I. (2004) Composting winery waste: sludges and grape stalks. *Bioresour. Technol.*, **95** (2), 203–208.

109. Fernandes, L., Casal, S., Cruz, R., Pereira, J.A., and Ramalhosa, E. (2013) Seed oils of ten traditional Portuguese grape varieties with interesting chemical and antioxidant properties. *Food Res. Int.*, **50** (1), 161–166.
110. Beveridge, T.H.J., Girard, B., Kopp, T., and Drover, J.C.G. (2005) Yield and composition of grape seed oils extracted by supercritical carbon dioxide and petroleum ether: varietal effects. *J. Agric. Food Chem.*, **53** (5), 1799–1804.
111. Wijendran, V. and Hayes, K.C. (2004) Dietary n-6 and n-3 fatty acid balance and cardiovascular health. *Annu. Rev. Nutr.*, **24**, 597–615.
112. Fernández, C.M., Ramos, M.J., Pérez, A., and Rodríguez, J.F. (2010) Production of biodiesel from winery waste: extraction, refining and transesterification of grape seed oil. *Bioresour. Technol.*, **101** (18), 7030–7035.
113. Shi, J., Yu, J., Pohorly, J.E., and Kakuda, Y. (2003) Polyphenolics in grape seeds-biochemistry and functionality. *J. Med. Food*, **6** (4), 291–299.
114. Sivarooban, T., Hettiarachchy, N.S., and Johnson, M.G. (2008) Physical and antimicrobial properties of grape seed extract, nisin, and EDTA incorporated soy protein edible films. *Food Res. Int.*, **41** (8), 781–785.
115. Luque-Rodríguez, J.M., Pérez-Juan, P., and Luque de Castro, M.D. (2007) Dynamic superheated liquid extraction of anthocyanins and other phenolics from red grape skins of winemaking residues. *Bioresour. Technol.*, **98** (14), 2705–2713.
116. Arbatan, T., Zhang, L., Fang, X.-Y., and Shen, W. (2012) Cellulose nanofibers as binder for fabrication of superhydrophobic paper. *Chem. Eng. J.*, **210**, 74–79.
117. Chen, J., Liang, R.-h., Liu, W., Luo, S.-j., Liu, C.-m., Wu, S., and Wang, Z. (2014) Extraction of pectin from Premna microphylla turcz leaves and its physicochemical properties. *Carbohydr. Polym.*, **102**, 376–384.
118. Tsai, R.-Y., Chen, P.-W., Kuo, T.-Y., Lin, C.-M., Wang, D.-M., Hsien, T.-Y., and Hsieh, H.J. (2014) Chitosan/pectin/gum Arabic polyelectrolyte complex: process-dependent appearance, microstructure analysis and its application. *Carbohydr. Polym.*, **101**, 752–759.
119. Prozil, S.O., Evtuguin, D.V., and Lopes, L.P.C. (2012) Chemical composition of grape stalks of *Vitis vinifera L.* from red grape pomaces. *Ind. Crops Prod.*, **35** (1), 178–184.
120. Ping, L., Brosse, L., Chrusciel, L., Navarrete, P., and Pizzi, A. (2011) Extraction of condensed tannins from grape pomace for use as wood adhesives. *Ind. Crops Prod.*, **33** (1), 253–257.
121. Villaescusa, I., Fiol, N., Martínez, M., Miralles, N., Poch, J., and Serarols, J. (2004) Removal of copper and nickel ions from aqueous solutions by grape stalks wastes. *Water Res.*, **38** (4), 992–1002.
122. Pérez-Serradilla, J.A. and Luque de Castro, M.D. (2008) Role of lees in wine production: a review. *Food Chem.*, **111** (2), 447–456.
123. Ladaniya, M.S. (ed.) (2008) *Citrus Fruit*, Academic Press, San Diego, CA.
124. Abeysinghe, D., Li, X., Sun, C.D., Zhang, W.-S., Zhou, C.H., and Chen, K.S. (2007) Bioactive compounds and antioxidant capacities in different edible tissues of citrus fruit of four species. *Food Chem.*, **104** (4), 1338–1344.
125. Marín, F.R., Soler-Rivas, C., Benavente-García, O., Castillo, J., and Pérez-Álvarez, J.A. (2007) By-products from different citrus processes as a source of customized functional fibres. *Food Chem.*, **100** (2), 736–741.
126. Pourbafrani, M., Forgacs, G., Horvath, I.S., Niklasson, C., and Taherzadeh, M.J. (2010) Production of biofuels, limonene and pectin from citrus wastes. *Bioresour. Technol.*, **101**, 4246–4250.
127. Hayat, K., Zhang, X., Chen, H., Xia, S., Jia, C., and Zhong, F. (2010) Liberation and separation of phenolic compounds from citrus mandarin peels by microwave heating and its effect on antioxidant activity. *Sep. Purif. Technol.*, **73**, 371–376.
128. Hosni, K., Zahed, N., Chrif, R., Abid, I., Medfei, W., Kallel, M., Brahim, N.B., and Sebei, H. (2010) Composition of

peel essential oils from four selected Tunisian Citrus species: evidence for the genotypic influence. *Food Chem.*, **123** (4), 1098–1104.

129. Bakkali, F., Averbeck, S., Averbeck, D., and Idaomar, M. (2008) Biological effects of essential oils– a review. *Food Chem. Toxicol.*, **46** (2), 446–475.

130. Bousbia, N., Vian, M.A., Ferhat, M.A., Meklati, B.Y., and Chemat, F. (2009) A new process for extraction of essential oil from Citrus peels: microwave hydrodiffusion and gravity. *J. Food Eng.*, **90** (3), 409–413.

131. Lohrasbi, M., Pourbafrani, M., Niklasson, C., and Taherzadeh, M.J. (2010) Process design and economic analysis of a citrus waste biorefinery with biofuels and limonene as products. *Bioresour. Technol.*, **101**, 7382–7388.

132. Tripoli, E., La Guardia, M., Giammanco, S., Di Majo, D., and Giammanco, M. (2007) Citrus flavonoids: molecular structure, biological activity and nutritional properties: a review. *Food Chem.*, **104** (2), 466–479.

133. Zia ur, R. (2006) Citrus peel extract – A natural source of antioxidant. *Food Chem.*, **99** (3), 450–454.

134. Liu, L., Cao, J., Huang, J., Cai, Y., and Yao, J. (2010) Extraction of pectins with different degrees of esterification from mulberry branch bark. *Bioresour. Technol.*, **101**, 3268–3273.

135. Piriyaprasarth, S. and Sriamornsak, P. (2011) Flocculating and suspending properties of commercial citrus pectin and pectin extracted from pomelo (Citrus maxima) peel. *Carbohydr. Polym.*, **83** (2), 561–568.

136. Maxwell, E.G., Belshaw, N.J., Waldron, K.W., and Morris, V.J. (2012) Pectin – An emerging new bioactive food polysaccharide. *Trends Food Sci. Technol.*, **24**, 64–73.

137. Boluda-Aguilar, M. and López-Gómez, A. (2013) Production of bioethanol by fermentation of lemon (Citrus limon L.) peel wastes pretreated with steam explosion. *Ind. Crop Prod.*, **41**, 188–197.

138. Martin, M.A., Siles, J.A., Chica, A.F., and Martín, A. (2010) Biomethanization of orange peel waste. *Bioresour. Technol.*, **101** (23), 8993–8999.

139. Rezzadori, K., Benedetti, S., and Amante, E.R. (2012) Proposals for the residues recovery: orange waste as raw material for new products. *Food Bioprod. Process.*, **90** (4), 606–614.

140. Bicu, I. and Mustata, F. (2011) Cellulose extraction from orange peel using sulfite digestion reagents. *Bioresour. Technol.*, **102** (21), 10013–10019.

141. Ververis, C., Georghiou, K., Danielidis, D., Hatzinikolaou, D.G., Santas, P., Santas, R., and Corleti, V. (2007) Cellulose, hemicelluloses, lignin and ash content of some organic materials and their suitability for use as paper pulp supplements. *Bioresour. Technol.*, **98**, 296–301.

142. Pavan, F.A., Lima, I.S., Lima, C.E., Airoldi, C., and Gushikem, Y. (2006) Use of Ponkan mandarin peels as biosorbent for toxic metals uptake from aqueous solutions. *J. Hazard. Mater.*, **137** (1), 527–533.

143. Bampidis, V.A. and Robinson, P.H. (2006) Citrus by-products as ruminant feeds: a review. *Anim. Feed Sci. Technol.*, **128** (3-4), 175–217.

144. Duoss-Jennings, H.A., Schmidt, T.B., Callaway, T.R., Carroll, J.A., Martin, J.M., Shields-Menard, S.A., Broadway, P.R., and Donaldson, J.R. (2013) Effect of citrus byproducts on survival of O157:H7 and non-O157 Escherichia coli serogroups within in vitro bovine ruminal microbial fermentations. *Int. J. Microbiol.* doi: 10.1155/2013/398320

9
Selective Extraction from Food Plants and Residues by Pulsed Electric Field

Eugene Vorobiev and Nikolai Lebovka

9.1
Introduction

The main principles of green extraction of natural products were recently formulated. They assume the use of renewable plant resources, alternative solvents (water or agro-solvents), reduction in energy consumption and unit operations, production of high-quality and high-purity extracts (non-denatured and biodegradable), and extract coproducts instead of wastes [1]. The concept of green extraction is rather new and is based on implementation of innovations and emerging technologies. Extraction by solvents (diffusion) and force fields (pressing, filtration, and centrifugation) is widely used for production of fruit juices, wines, sugar, vegetable oils, and starch, as well as for extraction of different molecules of agricultural origin (carbohydrates or polysaccharides, colorants, antioxidants, essential oils, proteins, aromas, flavors, etc.). Extraction is often linked with the use of environmentally polluting chemical or biological agents. For example, organic solvents (hexane, heptane) are used for the extraction of oil from oilseeds; enzymes are used for ripening and pressing of apples; and mineral agents, such as $CaCO_3$, are used for adsorption of impurities from sugar beet juices. Among solvents considered to be "green," water should be noted first of all, and renewable solvents (biosolvents, such as ethanol or isopropanol) should be also mentioned [2]. Unfortunately, "green" solvents, particularly room-temperature water, are often inappropriate for efficient extraction from food plants. The classical treatments (grinding, heating) and different alternative treatments currently used in industry to make extraction easier degrade and disrupt the tissue structure (membranes and cellular walls) in an uncontrollable way.

Recent scientific and practical efforts show full correspondence of pulsed electric field (PEF) techniques with green extraction concepts [3]. The use of PEF became very popular in food engineering as it allows critical acceleration of the solid–liquid extraction. Application of traditional methods (thermal, chemical, etc.) for this purpose is not consistent with the concept of green extraction and is commonly accompanied by degradation, necessity to use unsafe organic solvents,

Green Extraction of Natural Products: Theory and Practice, First Edition.
Edited by Farid Chemat and Jochen Strube.

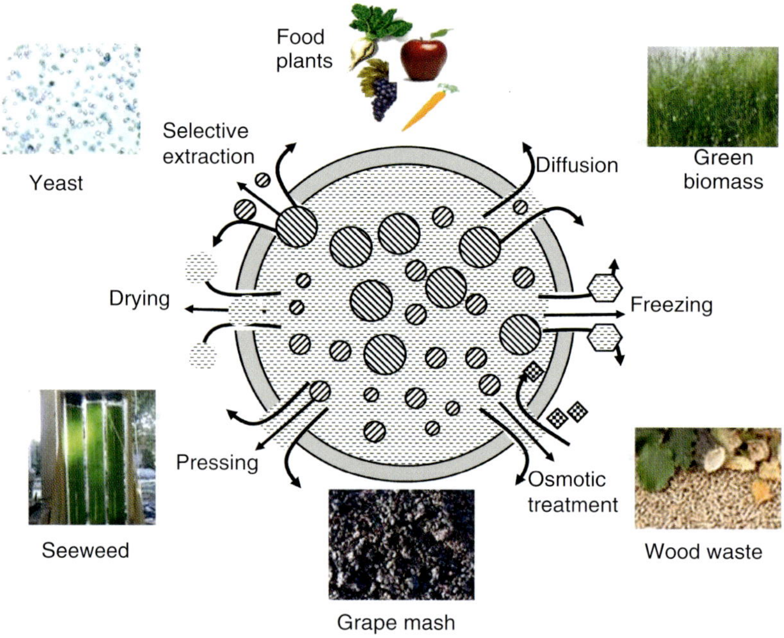

Figure 9.1 Different PEF applications for food and biomass transformation.

enzymes, detergents, temperature elevation, or unreasonably high energy consumption. The supplementary attractiveness of PEF-assisted methods is related to their potential for inactivation of pathogenic and spoilage microorganisms [4, 5].

Different food and biomass transformation processes can be enhanced by PEF as presented in Figure 9.1. There are undoubtedly many interesting potential applications of PEF for selective and green extraction from some food plants and residues. Some of them were recently reported in the literature [6–10]. Combinations of PEF with other methods (e.g., with ultrasound or supercritical fluids) can have synergetic effects on extraction processes. Different technologies of agro-industrial extraction become more selective, "green," and less energy consuming if a PEF is applied.

This chapter provides a review of the modern state and recent developments of PEF-assisted selective extraction techniques with examples of its applications to materials of different origin.

9.2
Basics of PEF-Assisted Extraction

PEF-assisted extraction focuses on a combination of conventional and electrically based techniques. Electrical processing can be incorporated immediately into solvent extraction or can be applied as a pretreatment step before extraction [6–8]. Electrical treatment can result in phenomena of different origin: heating of material, "electroporation," or "electropermeabilization," and electrostimulation of cells.

PEF is a nonthermal treatment of very short duration (from several nanoseconds to several milliseconds) and large pulse amplitude from 100–300 V cm^{-1} up to 300 kV cm^{-1}. It differs from other treatments by limiting damage to cellular membranes and demonstrating that it can be nondestructive to the tissue matrix. The biological membrane becomes electrically pierced under the effect of PEF and loses its semipermeability, either temporarily or permanently [11]. The electric field induces a potential difference across a membrane u_m, and electroporation occurs when u_m exceeds certain threshold value, typically, 0.5–1.5 V. The value of u_m is directly proportional to the cell size, so the larger cells get damaged before the smaller ones.

In practice, the extent of electroporation depends also on the properties of the material (electrical conductivity of constituents, humidity, presence of air, etc.) and details of pulse protocol. An important electroporation of water-rich plant tissues with large cells (≈50–100 μm) can be observed with PEF treatment at $E = 500–1000$ V cm^{-1} and treatment time within 10^{-4} to 10^{-1} s [7]. detailed studies have shown that electroporation can also be essential during ohmic heating even at moderate electric fields (MEFs), when the electric field strength E is under 100 V cm^{-1} [12]. Usually, different mass transfer effects can accompany electroporation [13]. Electrical permeabilization of biological membranes can be partly reversible or irreversible. Recovery of the membrane integrity within several seconds after pulse termination is possible if the tissue is treated at low values of E [14]. Very promising results were obtained in the United States and Europe on PEF application for the nonthermal inactivation of microorganisms and pasteurization of liquid foods [4, 15, 16]. Note that electroporation and killing of microbes (<5 μm) requires PEF treatment at larger fields ($E = 20–50$ kV cm^{-1}) and smaller treatment times (10^{-5} to 10^{-4} s) [4].

It was also shown that electroporation allows introduction of small and/or large molecules into cells or their extraction from cells, insertion of proteins into the membrane, and fusion of cells [11]. As a result, electroporation, because of its efficiency, has rapidly found applications in many fields of biochemistry, molecular biology, and medicine. Probably, one of the most important functional properties of electroporation is its ability to create conditions for the transport of small or large molecules through the cellular membrane and enhance extraction processes [8]. The cell network can act as a barrier for the passage of some undesired compounds, which is a big advantage for the purposes of extraction and allows to improve selectivity. Furthermore, plant materials, treated by PEF and exhausted of solutes, seem to be less altered than thermally treated materials, and can be used in some new auxiliary applications as part of their biorefinement.

Examples of different PET-assisted extraction processes are discussed in the following sections.

9.3 Application of PEF for Different Food Plants and Residues

Among very promising potential applications of PEF are the development of technologies of "cold" extraction from fresh food plants, for example, sucrose extraction from sugar beetroot, betanin extraction from red beet, inulin extraction from chicory, beta-carotene extraction from carrot, and phenolics extraction from grapesn [8].

9.3.1 Sugar Beets

The conventional sugar extraction and purification technology, practically, did not change during the last century. It utilizes a power-consuming hot-water diffusion of sugar from sliced cossettes at 70–75 °C and a very complex multistaged juice purification using big quantity of lime [17].

PEF application has a large potential for replacement or modification of the conventional thermal technology. For example, electroporation of cell membranes of sugar beetroot considerably enhances the diffusivity of sucrose even at ambient temperature and preserves the cell walls from thermal degradation. Better conservation of the tissue structure under cold conditions during electrical treatment would allow more pectins to remain in the cellular matrix. Recently, PEF-assisted pressing and aqueous extraction from sugar beets were intensively studied on the laboratory scale [18–31]. Different aspects were investigated, and it was shown that up to 82% of the overall yield could be achieved by two-stage pressing with an intermediate PEF application. The efficiency of so-called, cold pressing of PEF-treated sugar beet cossettes was demonstrated [18]. The cold juices, expressed from the sugar beet gratings after the intermediate PEF treatment, had higher purity (95–98%) as compared to those before PEF application (90–93%). In addition, the quantity of pectin was noticeably lower and the color of juice was systematically three to four times less intense than the color of factory juices [23].

Alternative studies have shown the principle of possibility of sugar extraction by cold or moderately heated water [22]. Solute diffusivities inside sugar beet tissue were nearly the same for sugar diffusion from untreated tissue at 60 °C and from PEF-pretreated tissue at 30 °C. The purest juice was obtained after cold diffusion. However, juice purity was higher for slices pretreated by PEF than for untreated slices even after thermal diffusion at 70 °C [30].

Recently, the scale-up study of PEF-assisted aqueous extraction of sugar using a pilot countercurrent extractor with 14 extraction sections was reported [31]. The sugar beet cossettes were prepared using industrial knives and PEF treatment of the cossettes was done with electric field strength E varying from 100 to 600 V cm^{-1}. The total time of PEF treatment was $t_{PEF} = 50$ ms. The effects of the main extraction parameters (temperature of extracting water and juice to beet mass ratio (draft)) on the extraction kinetics, as well as on the juice and cossette

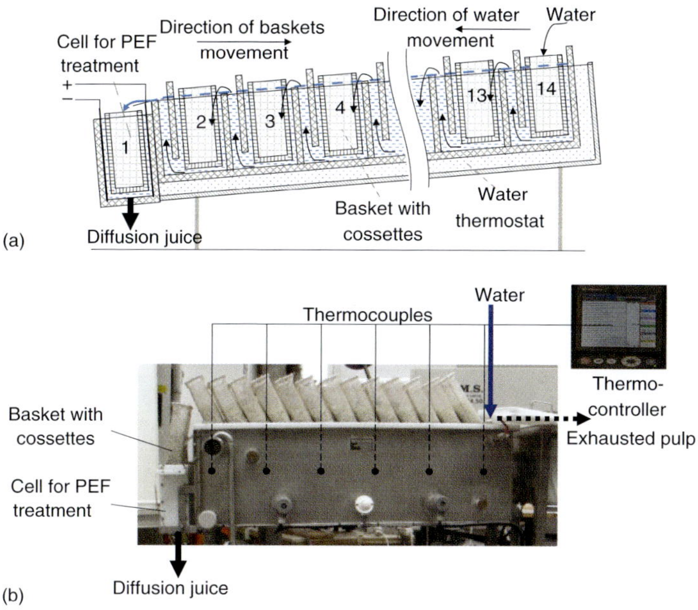

Figure 9.2 (a) Scheme of experimental equipment: schematic representation of extractor. (b) Photo of extractor. (Reproduced with permission from Ref. [31] © 2011 Elsevier).

(pulp) characteristics, were investigated. The temperature of extraction from PEF-treated cossettes was varied from 30 to 70 °C; the draft was varied from 120% to 90%. The scheme of experimental apparatus is presented in Figure 9.2.

Figure 9.3 compares the extraction data, obtained for the PEF-treated cossettes at different diffusion temperatures. In the case of PEF treatment, the sucrose extraction kinetics depends less on the temperature. However, the following general tendency can be clearly observed: the temperature increase results in better juice concentration in the first sections and exhaustion of cossettes in the last sections. For example, two last sections BE were useless for exhaustion of PEF-treated cossettes at 50 °C. However, all 14 sections were necessary for effective exhaustion of PEF-treated cossettes at 30 °C. Therefore, a combination of mild heating (for example, at 50 °C) and PEF treatment can be useful for shortening the diffusion time.

The effects of an electric field of field strength E varied from 1 to 7 kV cm^{-1} and pulse width of 2–5 μs on the kinetics of sucrose extraction from the sugar beet at different temperatures (20–70 °C) was studied [28]. The sucrose yield increased with field strength, time of extraction, and temperature, and the effect of the field strength was higher if the temperature was lower.

It was shown recently that sugar juice extracted without heating is less colored and more pure in comparison to the juice derived using industrial thermal diffusion [32–35]. It is expected that PEF treatment will simplify (or even eliminate) the very complicated and polluting carbonic purification process, existing today

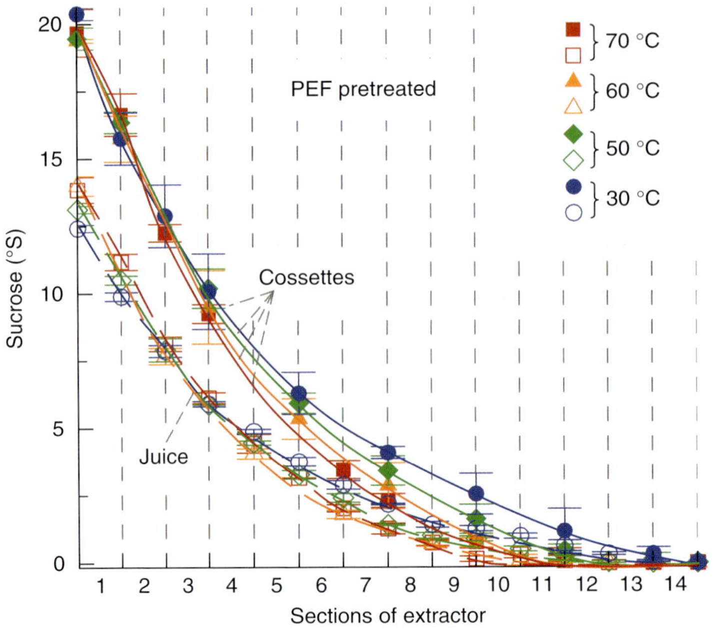

Figure 9.3 Sucrose concentration in the juices and cossettes, measured in extractor sections in the case of PEF-treated cossettes, at different diffusion temperatures. (Reproduced with permission from Ref. [31] © 2011 Elsevier).

in sugar production. Sugar technology of the future will use less energy and will be less polluting. Moreover, pulps exhausted in sugar will be richer in pectin and more valuable. The purity of the juice obtained by cold and moderate thermal extraction was found to be higher than the purity of juice obtained by conventional hot water diffusion at 70 °C.

Decrease of draft from 100% to 90% permitted increase of the extracted juice concentration, but it made the cossettes less exhausted. The pulp obtained by cold extraction of PEF-treated cossettes had dryness >30%, which was noticeably higher than the dryness of pulp obtained by conventional hot-water extraction technique. The optimal parameters of PEF protocol for cold and mild heat extraction (30–60 °C), were between $E = 600\,\mathrm{V\,cm^{-1}}$ (at 30 °C) and $E = 260\,\mathrm{V\,cm^{-1}}$ (at 60 °C). The value of draft could be decreased from 120% to 100%, or even to 90%, for the extraction duration of 70 min if PEF treatment was combined with mild heating. The estimated energy gain for cold extraction with temperature reduction from 70 to 30 °C (i.e., by $\Delta T = 40$ °C) was ≈46.7 kWh $\mathrm{t^{-1}}$, and it was noticeably higher than the power consumption required for PEF treatment (≈5.4 kWh $\mathrm{t^{-1}}$). The PEF power consumption could even e reduced by further optimization of PEF parameters and minimization of the liquid-to-solid ratio during PEF treatment [31].

A detailed comparison of the qualitative characteristics (concentration of soluble solids, purity, nature of impurities, coloration, and filterability) of sugar beet

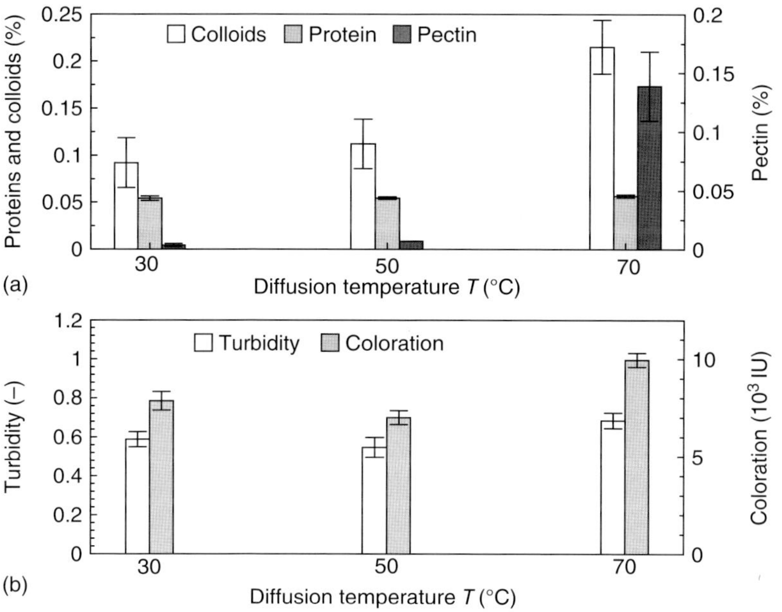

Figure 9.4 (a) Concentration of total colloidal compounds, proteins and pectin in extracted juices versus the temperature of extraction. (b) Coloration and turbidity of extracted juices versus the temperature of extraction. (Reproduced with permission from Ref. [34] © 2011 Elsevier).

juices, extracted at 30 and 50 °C after PEF treatment and juices obtained by classical "hot" extraction at 70 °C was done [33–35]. It was shown that PEF-assisted "cold" extraction results in lower concentration of colloidal impurities (especially pectins), lower coloration, and better filterability of juice. Concentration of various colorants and their intermediates decreased significantly with decrease of the extraction temperature from 70 to 30 °C (Figure 9.4).

Increase of the extraction temperature can result in decomposition of the cell walls and substantially accelerate extraction of different oligo- and polymolecular compounds, which are the main constituents of cellular walls and membranes. The temperature increase also accelerates various chemical reactions between extracted components of the cellular juice (e.g., color development due to Maillard reaction). Temperature-enhanced extraction of high-molecular-weight compounds and formation of undesirable products of chemical reactions result in decrease in the quality of the extracted juice [33].

9.3.2
Red Beets

The water-soluble betalains are the main red–purple pigments extracted from red beets (Beta vulgaris L.). Very good extractability of betalains (release of about 90% of their total red pigment) after PEF treatment at field strength of 1 kV cm^{-1}

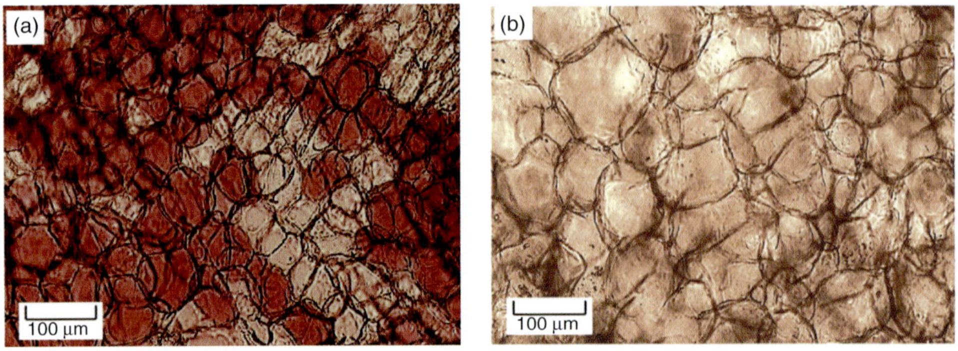

Figure 9.5 Microscopic images of the red beet tissue after aqueous extraction: untreated (a) and PEF pre-treated (b) samples. (Reproduced with permission from Ref. [37] © 2011 Elsevier).

was demonstrated [36, 37]. Application of the PEF treatment at 7 kV cm^{-1} enabled increase of the maximum yield of betanine by a factor of 4.2 compared to samples not subjected to PEF treatment and achievement of almost complete betanine release [36]. The combination of PEF at 7 kV cm^{-1} and pressing at 14 kg cm^{-2} permitted 18-fold shortening of extraction time. It can be noted that effective electroporation of the red beet tissue at ambient temperature can be attained even at lower electric fields of 400–600 V cm^{-1} [38]. A discussion of the PEF and temperature effects ($T = 30-80\,°C$) on degradation of colorants and the kinetics of their extraction from the red beet was presented in [37]. PEF treatment was done using trains of monopolar rectangular 100 μs pulses at electric field strength $E = 375-1500$ V cm^{-1} with total treatment time $t_{PEF} = 0-0.2$ s. The degradation of colorants and their extraction were characterized by means of absorbance and electrical conductivity measurements. PEF treatment was found effective for acceleration of the extraction of betalains and shortening the time of extraction. Electroporation was shown to be responsible for intensification of the release of colorants through aqueous extraction.

Optical microscopy images of red beet tissues after 3 h of aqueous extraction are presented for untreated (Figure 9.5a) and PEF-treated (Figure 9.5a) samples. Typically, nontreated tissue patterns were a mixture of intensively colored and almost colorless pale cells (Figure 9.5a). The intensively colored cells corresponded to undamaged cells and the colorless pale cells to damaged cells with lost vacuolar sap and diluted pigment [37]. The PEF pretreated tissues demonstrated a homogenous distribution of cells with equivalently pale color (Figure 9.5b). It shows that PEF-induced changes in the red beet tissue are related to electroporation of cells.

The extraction B and colorant degradation D indexes were defined for the PEF and thermal treatments as

$$B = (\sigma - \sigma_i)/(\sigma_f - \sigma_i)$$

and

$$D = (A - A_i)/(A_f - A_i)$$

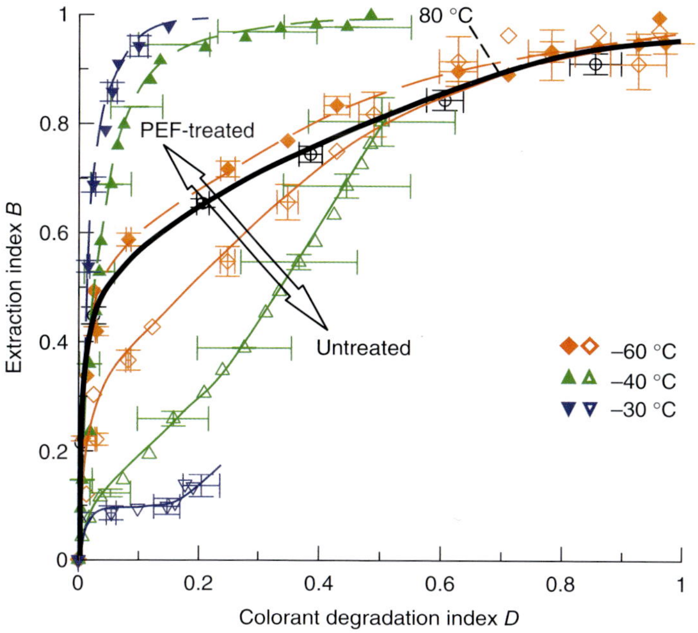

Figure 9.6 Dependencies of extraction index B versus colorant degradation index of the red beet juice D at different temperatures for untreated and PEF-treated samples [37]. (Reproduced with permission from Ref. [37] © 2011 Elsevier).

where σ is the electrical conductivity of the extract, A is the red beet juice absorbance measured at a wavelength $\lambda = 536$ nm (it corresponds to the maximum absorbance for betacyanins), and the subscripts i and f refer to the initial ($t = 0$) and final ($t = 3$ h) extracts, respectively. Figure 9.6 shows B versus D dependencies under the same extraction conditions.

The character of $B(D)$ dependencies was similar at $T = 80$ °C but rather different at lower temperatures $T = 30–60$ °C. At lower temperatures, the $B(D)$ curves of untreated samples were always below the $B(D)$ curves of PEF-treated samples. It reflects the positive effect of PEF treatment in the improvement of extraction and decrease of degradation. For example, PEF treatment at "cold" extraction at $T = 30$ °C allowed reaching a high yield of colorants ($B = 0.95$) at lower level of colorant destruction ($D = 0.10$). This is important because 1 h of thermal extraction at 80 °C resulted in nearly complete damage of betalains. Therefore, increase of temperature caused acceleration of both extraction and colorant degradation processes on the aqueous extraction of colorants from the red beet. PEF treatment removed the membrane barriers with smaller activation energy and permitted very efficient "cold" extraction with high level of colorant release and small level of colorant degradation. So, PEF treatment seems to be very promising and opens encouraging possibilities for very efficient "cold" diffusion with high level of colorant extraction and low level of colorant degradation (5–10%) [37].

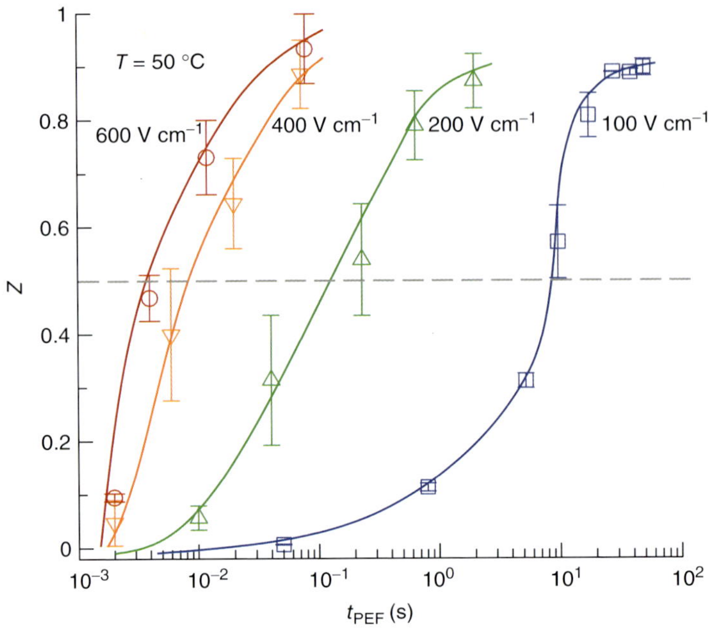

Figure 9.7 PEF treatment of chicory tissue. Disintegration index Z versus PEF treatment time t_{PEF} at different field strengths E and the temperature $T = 50\,°C$. (Reproduced with permission from Ref. [39] © 2010 Elsevier).

9.3.3 Chicory Roots

Chicory (*Cichorium intybus*) roots contain many useful components, such as sucrose, proteins, and inulin (becoming an important food industry ingredient, which permits replacement of up to 100% of fat in the dairy and meat products, thus allowing to create high-quality dietary foods). Inulin can be also used as a dietary fiber or sugar substitute in tablets.

The effects of PEF treatment on the efficiency of soluble solid extraction from chicory tissues was discussed in [39]. The degree of tissue damage estimated from the electrical conductivity disintegration index Z [7] is presented in Figure 9.7.

The damage degree Z increased during the PEF treatment. This effect was more pronounced at higher electric field strengths (Figure 9.7). PEF treatment of chicory tissue with field strength between 100 and 600 V cm^{-1} enhanced the membrane permeabilization and allowed reaching a high level of tissue disintegration.

The Arrhenius plots of the effective diffusion coefficient D for untreated and PEF-pretreated chicory slices are presented in Figure 9.8. The highest values of D were observed for both untreated and PEF-pretreated slices at high temperatures (60–80 °C). At lower temperatures, diffusion was noticeably higher for PEF-pretreated than for untreated slices.

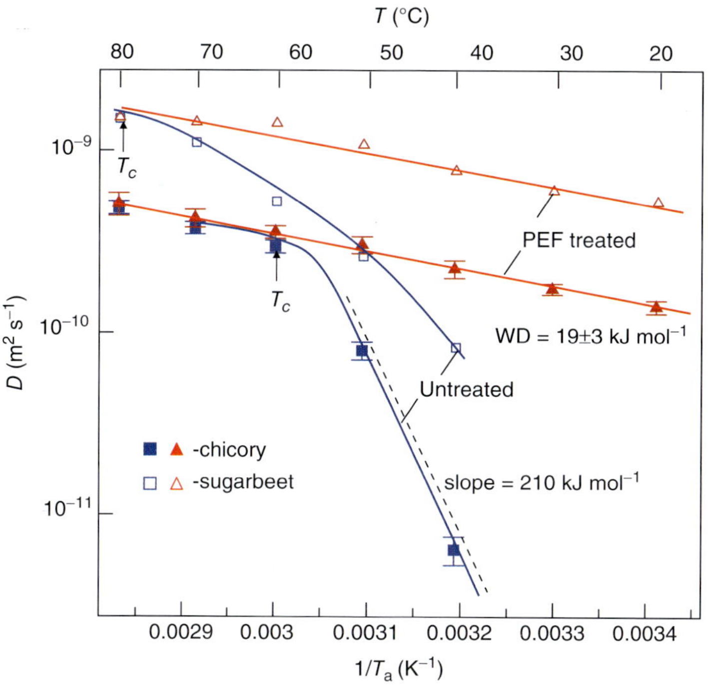

Figure 9.8 Arrhenius plots of the effective solute diffusion coefficient D for the untreated and PEF-pretreated chicory and sugar beet slices [39]. (Reproduced with permission from Ref. [39] © 2010 Elsevier).

As shown in Figure 9.8, solute diffusivity is more rapid in sugar beet than in chicory and sugar beet crossover temperature (80 °C) is higher in comparison to that observed for chicory (60 °C). The activation energy of the usual thermal damage is rather high (more than 200 kJ mol^{-1}, Figure 9.8); however, it could be noticeably reduced to about 20 kJ mol^{-1} by PEF treatment. The temperature contribution to electroporation efficiency is important, reflecting the synergetic effect of simultaneous thermal and PEF application.

9.3.4 Apples

The models of dielectric breakage were developed and tested using experimental data obtained for PEF-treated apple tissue [40]. The effects of apple tissue anisotropy and orientation with respect to the applied electric field on electropermeabilization were reported in [41]. It was shown that elongated cells (taken from the inner region of the apple parenchyma) responded to the electric field in a different manner, while no field orientation dependence was observed for round cells (taken from the outer region of parenchyma). The textural relaxation data support higher apple damage efficiency at longer pulse duration. The effect of the synergy of PEF and thermal treatments on the textural properties of apple tissue and apple juice expression was demonstrated in [42]. It was shown that mild thermal treatment allows increase of the damage efficiency of PEF treatment, and apple tissue preheated at 50 °C and treated by PEF at $E \approx 500\,\mathrm{V\,cm^{-1}}$ exhibits a noticeable enhancement of juice extraction by pressing.

A potential application of electrical pulsed energy concerns production of the apple juice without using enzymes, which will actually serve to increase the expressed yield [43]. The impact of PEF treatment in additional yield of apple juice extraction was compared for monopolar- and bipolar-shaped pulses [44]. PEF treatment was done using electric field strength E of $100-1300\,\mathrm{V\,cm^{-1}}$, pulse duration $20-300\,\mu\mathrm{s}$, number of pulses n within $10-200$, and frequency of 200 Hz. The monopolar pulses have demonstrated smaller efficiency of juice extraction than bipolar ones; however, the total soluble dry matters were similar for monopolar and bipolar pulses.

Many studies were devoted to the investigation of the influence of PEF treatment on juice expression from apples. PEF application improved noticeably the juice yield and soluble matter content in the juice. It was demonstrated that juice characteristics and yield are directly related to the size of slices [42]. As an example, the yield of juice from Golden Delicious apple slices sized $2 \times 3.5 \times 55$ mm increased after PEF treatment ($E = 400\,\mathrm{V\,cm^{-1}}$, $t_{\mathrm{PEF}} = 0.1$ s) by 28%, while the increase in juice yield was only 5% when the apple slices were finer ($1 \times 1.9 \times 55$ mm) [45, 46]. Juice yield Y increased significantly after PEF treatment of large apple mash (Y = 71.4%) at $E = 450\,\mathrm{V\,cm^{-1}}$ as compared with the check sample with small apple mash (45.6%) [43]. PEF treatment did not alter the acid–sweet balance and pH; however, a decrease in the yield of native polyphenols was observed after PEF treatment (control: 9.6%; treated: 5.9% for small mash). The PEF pretreatment was accompanied by a noticeable improvement of the apple juice clarity, an increase of the total soluble matter and the content of polyphenols, and intensification of the antioxidant capacities of juice [47].

9.3.5 Grapes

One more PEF application concerns the pressing of white wine grapes combined with their PEF treatment. The juice extracted from grapes after PEF application was less turbid (clearer) than the untreated juice [42]. Consequently, the treated juice would not require filtration using polluting filter aids (such as diatomite). Another possibility is PEF-assisted extraction of aromatic compounds and phenolic bays from red grapes, which would be quicker and more selective if PEF is applied [48, 49].

PEF-induced electroporation of wine grapes was shown to be a promising alternative process leading to prudent extraction of colorants and valuable constituents [42, 49]. The PEF treatment ($E = 750\,\mathrm{V\,cm^{-1}}$, $t_{\mathrm{PEF}} = 0.3$ s) resulted in increase of the final juice yield up to 73–78% as compared to 49–54% for the untreated grapes (Muscadelle, Sauvignon, and Semillon) [42]. PEF treatment enhanced compression kinetics and extraction of polyphenols from Chardonnay grapes [45]. PEF-assisted extraction of polyphenols from grape skins by water was investigated experimentally [50]. A microscopic observation has clearly shown that PEF treatment invoked extraction of red color pigments from anthocyanoplasts into the cell. It was shown that increase in pulse width resulted in increase in the polyphenol yield and decrease in the energy consumption. PEF application is promising for reduction of the maceration time during vinification and production of wines with better characteristics. Some aspects and potential applications of PEF technology in the winemaking industry were reported in [49].

9.3.6 Other Fruits and Vegetables

Potato was used as a model system for testing electroporation effects and studying reversible electroporation, transient viscoelastic behavior, different stress-induced effects, metabolic responses, and electrostimulated effects in plant tissues [40, 42, 45, 51, 52]. The effects of PEF treatment on the textural and compressive properties of potato were studied in detail [45]. It was also shown that application of only PEF treatment was not sufficiently effective for complete elimination of the textural strength; however, mild thermal pretreatment at 45–55 °C allowed to increase the PEF efficiency [9]. Potato was used also in the investigation of temperature and PEF protocol effects on characteristic damage time, dehydration, freezing, and drying [40, 42, 51, 52]. Attempts of extraction-oriented practical application of PEF were made. As an example, PEF application for facilitation of starch extraction from potato and enhancement of the extractability of an anthocyanin-rich pigment were reported [53].

PEF treatment was also applied to other tissues, such as red bell pepper and paprika, fennel, alfalfa, and red cabbage (see, e.g., [7, 39]). These studies have demonstrated improvement in juice yield and extraction of dry matter, carotenoids, vitamins, sucrose, proteins, and other compounds. A principal possibility of PEF-assisted selective extraction of water-soluble components (soluble sugars) and production of a "sugar-free" concentrate, rich in vitamins and carotenoids, was demonstrated for carrot [45]. The high extraction yield (up to 96.7%) by lycopene from tomato residual was achieved using PEF treatment with electric field strength 30 kV cm^{-1}, liquid-stuff ratio 9 ml g^{-1}, temperature 30 °C, and ethyl acetate as an extraction solvent [54]. Moreover, the PEF-assisted extraction method was proved to be a fast method for lycopene extraction from the tomato residual. It was applied to mashed red cabbage in water in a batch treatment chamber. PEF treatment at electric field strength of 2.5 kV cm^{-1}, pulse duration $t_i = 15\,\mu s$, and number of pulses $n = 50$ enhanced total anthocyanin extraction by 2.15 times [55].

9.3.7
Egg Yolk

Egg yolk is a rich source of lecithin (a mixture of phospholipids: phosphatidylcholine, 29–46%, phosphatidylethanolamine, 21–34%, phosphatidylinositol, 13–21%). However, it is too expensive to extract pure lecithin on an industrial scale. Usually, lecithin is extracted from plant seeds, for example, soybean oilseeds [56]. It was shown that PEF-assisted organic solvent extraction in 95% ethanol gives good opportunity to produce pure phosphatidylcholine from the egg yolk [57]. Yolk powder was used as a raw material and the phosphatidylcholine content of extract was increased to 90%.

9.3.8
Bio-suspensions and Yeasts

PEF also allows weakening of the cellular structure and increase in the effectiveness of conventional or alternative treatments. For example, PEF pretreatment can weaken the structure of yeast used by a bakery, so, it could be combined with high-pressure homogenization (HPH) for the more efficient extraction of proteins [58]. Heating makes easier the electroporation of biological membranes; therefore, a combination of PEF treatment and mild temperature (lower than 50 °C) heating and use of microwaves could be an interesting method of processing.

It was demonstrated that PEF and HVED (high-voltage electrical discharge) treatments always resulted in incomplete damage of yeast cells, although efficiency of HVED was higher than that of PEF [59]. Figure 9.9 shows the degree of extraction of ionic components Z versus the number of passes N_p for HPH treatment (Figure 9.9a) and versus the number of pulses N for HVED and PEF treatments (Figure 9.9b). HPH treatment at $p = 80$ MPa and $N_p = 20$ resulted in almost complete damage of yeast with the highest release of ionic substances, $Z \approx 1$. These data are in correspondence with data reported in [58].

High selectivity in the extraction of the ionic substances, proteins, and nucleic acids was demonstrated. For example, treatment at $E = 40\,\text{kV cm}^{-1}$ allowed extraction of ≈80 and ≈70% of ionic substances, ≈4 and ≈1% of proteins and ≈30 and ≈16% of nucleic acids in cases of HVED and PEF treatments, respectively.

PEF was applied to beer waste brewing yeasts in order to extract trehalose from the brewing yeasts [60]. Treatment at electric field strength ≈20 kV cm^{-1} and liquid–solid ratio of 30 : 1 allowed to reach the rate of trehalose extraction of up to 2.635%. PEF-assisted extraction efficiency was also the highest as compared to the other two extraction techniques. It was 15.96 times higher than that of microwave extraction and 34.08 times higher than that of the ultrasound technique. PEF treatment at electric field strength of 50 kV cm^{-1} with pulse duration of 0.0067 s enhanced the RNA yield waste of Brewer's yeast by water and ethanol as extraction solvents [61]. The RNA yield in ethanol extraction was 1.69 times higher than that of water extraction.

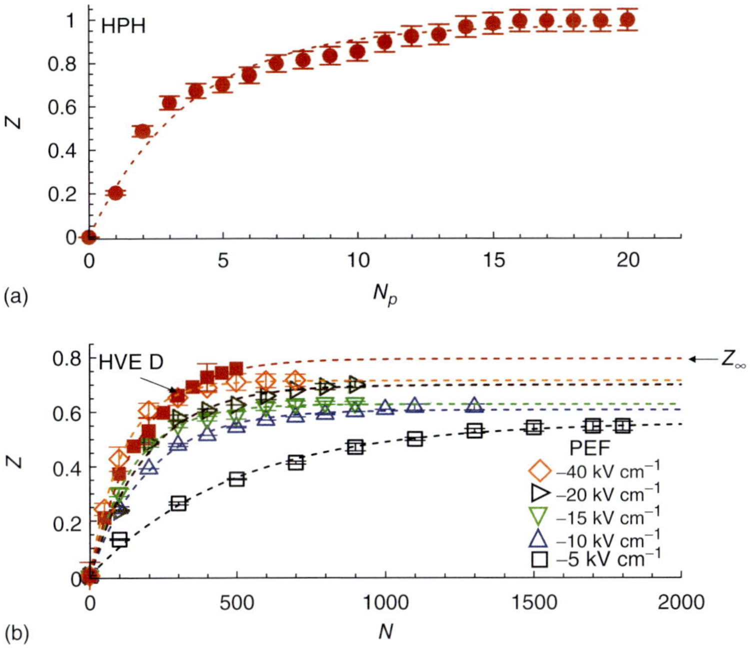

Figure 9.9 Degree of release of the ionic components Z versus the number of passes N_p for HPH treatment ($p = 80$ MPa) (a) and versus the number of pulses N for HVED ($U = 40$ kV) and PEF ($E = 5–40$ kV cm^{-1}) treatments (b). (Reproduced with permission from Ref. [59] © 2011 Springer Science and Business Media).

The possibility of PEF-enhanced extraction of intracellular proteins without killing the cells was recently demonstrated [62]. Chinese hamster ovary cells, cultured in a microfluidic channel were used as a source of proteins and PEF treatment at the electric field strength $E = 0.4-1.1$ kV cm^{-1} with pulse duration $t_i = 10-50$ ms was applied. Tuning of the E and t_i values allowed control of the average amount of protein release and the percentage of viable cells.

9.3.9
Microalgae

Microalgae have high content of lipids, proteins, polyunsaturated fatty acids, carotenoids, valuable pigments, and vitamins, and can be used in the food, feed, cosmetics, pharmaceutical, and biofuel industries. Traditional methods for extraction of these components use environmentally toxic solvents. Recently, several groups have developed PEF-assisted techniques for extraction of valuable components from microalgae. PEF treatment in a flow cell at electric field strength E of 23–43 kV cm^{-1} and pulse duration of 1 μs was applied for concentration of the aqueous suspension of microalgae *Auxenochlorella protothecoides* (36–167 g dry weight per kilogram suspension) [63]. A noticeable disintegration of cells in suspensions with biomass content of 100 g dry weight per kg suspension required energy consumption of the order of 1 MJ/kg, which was equivalent to 4.8% of the upper heating value for algae. It was demonstrated that PEF-assisted extraction was highly selective and allowed release of the soluble intracellular matter, while extraction of lipids required application of solvents. Thus, the opportunity to use PEF treatment at the first step of extraction and solvents at the second step of extraction was demonstrated [63]. Continuous PEF treatment was studied and a continuous flow technology was proposed for extraction of the entire cytoplasmic proteins from microalgae (*Nannochloropsis salina* and *Chlorella vulgaris*) [64]. Continuous flow permits the PEF treatment of large volumes. Effective extraction was observed in the case when PEF was followed by a 24-h incubation period in a salty buffer. PEF treatment was applied to enhance lipid extraction from *Ankistrodesmus falcatus* wet biomass using the green solvent ethyl acetate. Application of PEF allowed significant enhancement of the rate of lipid recovery. It was noted that the increase in lipid recovery was due to the electroporation and not due to temperature effects [65].

Extraction of intracellular components from microalgae *Nannochloropsis sp.* with application of different cell disruption techniques, including PEF ($20\,\text{kV cm}^{-1}$, 1–4 ms, $13.3–53.1\,\text{kJ kg}^{-1}$), HVED ($40\,\text{kV cm}^{-1}$, 1–4 ms, $13.3–53.1\,\text{kJ kg}^{-1}$), ultrasonication (USN) (200 W, 1–8 min, $12–96\,\text{kJ kg}^{-1}$), and HPH (150 MPa, 1–10 passes, $150–1500\,\text{kJ kg}^{-1}$) was studied [47].

The data evidence that electrically based disruption techniques (PEF and HVED) allow selective extraction of water soluble ionic components and microelements, small molecular weight compounds, and water soluble proteins. Figure 9.10 compares the degree of extraction of the water-soluble proteins Z_p(%) from 1% microalgae suspensions treated by different methods. The treatment

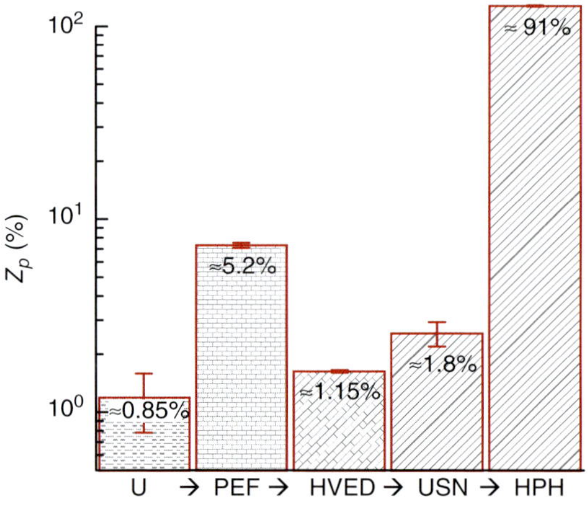

Figure 9.10 Degree of extraction of water soluble proteins Z_p(%) from 1% microalgae suspensions: untreated (U) and pre-treated by PEF ($20\,\text{kV cm}^{-1}$, 4 ms), HVED ($40\,\text{kV cm}^{-1}$, 4 ms), USN (200 W, 4 min), and HPH (150 MPa, 6 passes) The treatment was done sequentially: U → PEF → HVED → USN → HPH [47].

was done sequentially, U → PEF → HVED → USN → HPH. The total quantity of extracted water-soluble proteins was $Z_p \approx 0.7\%$ (w/w dry-weight biomass biomass). Note that the effect of PEF treatment (≈5.2) was rather significant as compared with supplementary contributions of HVED (≈1.15%) or USN (≈1.8%), however, the most fundamental contribution (≈91%) was given by the HPH treatment.

However, the electrically based disruption techniques were ineffective for delivery of pigments (e.g., chlorophylls or carotenoids) and their extraction required subsequent application of more potent methods of disruption.

PEF treatment was also proposed as an effective tool for control of predators in industrial-scale microalgae cultures [66]. It was shown PEF can be used for selective elimination of such contaminant as protozoa that can highly jeopardize the productivity of the culture.

9.3.10 Rhizomes

PEF treatment was applied to *Podophyllum peltatum* in order to enhance extraction of podophyllotoxin, which is valuable for the treatment of cancer and venereal warts [67]. Conventional extraction of this chemical is inefficient and involves mechanical fragmentation of dehydrated rhizomes followed by solvent extraction at warm temperatures. The dried rhizomes of *P. peltatum* were soaked in deionized water and then were PEF treated using electric field strength of 17.7 and 19.4 kV cm^{-1} rectangular pulses with pulse duration $t_i = 2$ μs and treatment time t_{PEF} (pulse duration multiplied by the number of pulses, excluding duration of pauses between the pulses) of 0.504–0.806 s. The color of the sample changed after PEF treatment from a sand-yellow to a deep red. The data demonstrated significant increase (up to 47%) of podophyllotoxin concentration as compare with control samples.

9.3.11 Bones

The PEF technique was used to assist hydrolysis of fishbones and improve efficiency of fishbone calcium extraction [68]. The highest extraction rate (up to 84.2%) was reached under the optimized condition at electric field strength E of 25 kV cm^{-1}. PEF-assisted extraction appeared to be more efficient and less time consuming in comparison with the traditionally applied ultrasonic technique.

PEF treatment allowed effective extraction of pure collagen protein without impurities of miscellaneous other proteins from bovine bone. The highest yield of dissoluble collagen (16.21 mg ml^{-1}) was achieved using electric field strength of ≈22 kV cm^{-1}. PEF treatment was also applied as a fast method for preparation of chitosan from the shrimp shells [69]. The optimum electric field strength of ≈20.5 kV cm allowed higher degree of chitosan deacetylation (up to 92.32%) as compared with traditional methods.

9.3.12
Eggshell

PEF treatment was shown to be an effective tool for production of eggshell calcium malate (ESCM). It was applied to an aqueous suspension of eggshell powder in the presence of 1–3% of citric acid. The treatment was carried out in a continuous flowing chamber by PEF with the strength $E = 4–20\,kV\,cm^{-1}$, and pulse duration $t_i = 4–24\,\mu s$ [70, 71]. PEF – assisted technology allowed obtaining of dissoluble ESCM content up to $7.12\,mg\,ml^{-1}$ under the following best processing conditions: 2.0% citric acid, $E = 15\,kV\,cm^{-1}$, and $t_i = 20\,\mu s$.

9.3.13
Leaves

Thin slices of the fresh tea leaves were subjected to PEF treatment in order to extract polyphenols [72]. The data have shown that PEF treatment accelerate kinetics of extraction. Different PEF protocols with the same number of pulses $n = 30$, pulse duration t_i of 0.05 s, and electric field strength E varied within 0.4 and $1.1\,kV\,cm^{-1}$ were tested. The values of electric field strength E and pause between the pulses Δt were important factors affecting extraction yield. The maximum extraction yield value 27% was reached at $E = 0.9\,kV\,cm^{-1}$ and $\Delta t = 0.5\,s$ and at $E = 0.9\,kV\,cm^{-1}$ and $\Delta t = 3\,s$.

It was demonstrated that PEF treatment (electric field strength of $42.13\,kV\,cm^{-1}$, pulse duration t_i of $5.46\,\mu s$ and liquid–solid ratio of $30.12\,ml\,g^{-1}$) influenced extraction of trehalose and exerted great effect on activation of neutral trehalase in the beer yeast cells [73].

9.3.14
Herbs

The PEF extraction method was developed for enhancement of the yield of an alkaloid (Guanfu base A, GFA) from the Chinese medicinal herb *Aconitum coreanum*. An electric field of $20\,kV\,cm^{-1}$, 1 : 12 solid-to-solvent ratio, and 90% ethanol–water solution were applied [74]. The extraction method demonstrated the highest yield of GFA ($3.94\,mg\,g^{-1}$) with the shortest extraction time (0.5–1 min), and the lowest energy costs as compare to other extraction methods (cold maceration extraction, percolation extraction, heat reflux extraction (HRE), and ultrasonic-assisted extraction (UE)).

The effect of PEF on extraction of major components from saffron (*Crocus sativus*) was studied [75]. The extracted components included crocin (color), safranal (flavor), and picrocrocin (taste). PEF treatment with electric field strength E of 5 kV, pulse number n of 100 pulses, and pulse duration of $35\,\mu s$ caused a noticeable increase in extraction from saffron stigma and saffron pomace.

PEF treatment was applied for reaching the optimized extraction of polysaccharides from corn silk [76]. Corn silk is a traditional Chinese herbal medicine, which is rich in antioxidants, polyphenols, vitamins (vitamin K, C), and minerals.

The yield of polysaccharides under optimal extraction conditions (electric field strength E of 30 kV cm^{-1}, pulse duration $t_i = 6$ μs, and liquid-to-raw material ratio 50 : 1) was ≈7%.

9.3.15 Ginseng

The effects of PEF on the drying of ginseng (at 55 °C) and hot water (95 °C) extraction of dried ginseng was studied [77]. PEF was applied to fresh ginseng using 1 and 2 kV cm^{-1} electric field strengths, 30 μs pulse duration, 25 and 200 Hz frequencies, and pulse number of 175. Such PEF treatment resulted in reduction (≈38%) of the drying time. It also increased the soluble solid content and significantly reduced the sugar content of the extract as compared to samples that were not treated by PEF.

9.3.16 Peels

The pressing extraction of total polyphenols and flavonoids (naringin and hesperin) from orange peel using PEF with electric field strength $E = 1$ kV cm^{-1} was investigated [78]. These compounds can serve as antioxidant additives in the food and pharmaceutical industries. The total polyphenol extraction yield and the antioxidant activity of the extract increased up to 159% and 192%, respectively, if PEF was applied. PEF-assisted pressing allowed reduction in the extraction time and did not require the use of organic solvents.

9.3.17 Mushrooms

High-intensity PEF-assisted extraction technique was used to optimize conditions of extraction of exopolysaccharides (EPSs) from Tibetan spiritual mushroom broth [79]. The results show that the optimal conditions of such extraction technique are the following: 40 kV cm^{-1} electric field intensity, number of pulses $n = 8$, and pH 7. The effect of these factors on EPS extraction increases in the following order: electric field intensity > pH > number of pulses. The influence of the electric field intensity on EPS extraction is significant ($\alpha = 0.05$). Under optimal conditions, EPS extraction increased by 84.3% compared to that of the control group.

The efficiency of extraction and stability of extracts from the mushroom (*Agaricus bisporus*) were compared for different methods of extraction [80]. The traditional hot water extraction (WE) or ethanol extraction gave relatively high contents of proteins, total polyphenols, and polysaccharides. However, the extracts were cloudy and their colloid stability was low. The extracts obtained by pressure extraction (PE) and PE + PEF methods were clear and their colloid stability was high. The PE + PEF method gave higher nucleic acid/proteins ratio as compared with the PE method.

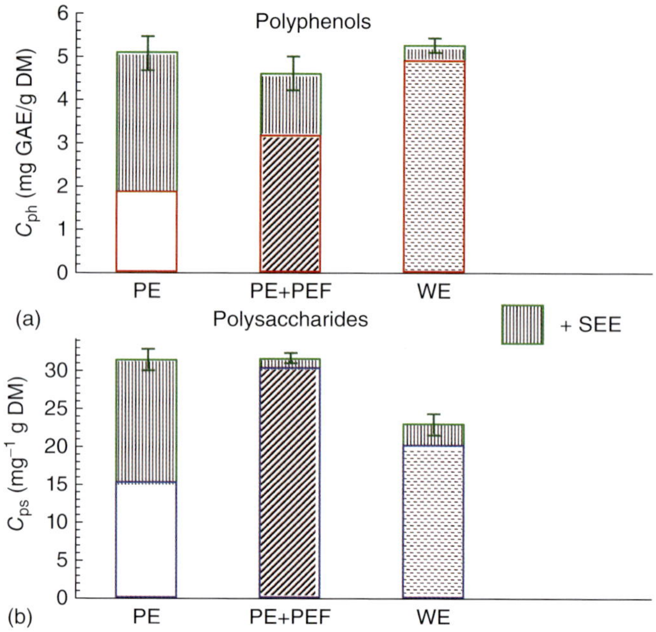

Figure 9.11 The effects of supplementary ethanol extraction (SEE, $T = 298$ K, 24 h) from cakes (after PE or PE + PEF application) or from exhausted slices (after application of WE, $T = 343$ K, $t = 2$ h) on concentration of polyphenols, C_{ph}, (a) and polysaccharides, C_{ps}, (b) in the extracts. (Reproduced with permission from Ref. [80] © 2013 Springer Science and Business Media).

The experiments were also done for checking the possibility of supplementary ethanol extraction (SEE) of total polyphenols and polysaccharides from solid residues obtained using methods of PE, PE + PEF (cakes) and WE (exhausted slices). SEE application to exhausted slices resulting from WE was ineffective and gave small supplementary yield of the total polyphenols (Figure 9.11a) and polysaccharides (Figure 9.11b). However, SEE noticeably increased the content of total polyphenols in the extracts obtained by PE and PE + PEF methods up to a level that was compatible with that in the extracts obtained by WE method. SEE noticeably also increased the content of polysaccharides in extracts, obtained by the PE method. These observations reflect the differences in the selectivity of PE and PE + PEF methods. In general, PE + PEF allowed production of mushroom extracts with high content of proteins and polysaccharides.

9.3.18
Juices and Juice-Based Beverages

A detailed review on PEF treatment application for juices and juice-based beverages can be found in [81]. High-intensity-field PEF treatment (by 4 μs bipolar pulses at $E = 35$ kV cm^{-1}) was shown to be a feasible technology for obtaining the fruit juice–soymilk beverages with extended shelf life and a profile of antioxidant compounds similar to that of freshly made juices [82].

9.4 Conclusions

PEF techniques can be useful as effective tools for selective extraction from food plants and residues. PEF treatment is in full correspondence with the green extraction concept [3] and it allows preservation of integrity of cell walls, color, flavor, vitamin C, and important nutrients of products. Many promising examples of the PEF-assisted extraction of juices, sugars, colors, polyphenolic substances, and oils from solid foods (sugar beets, apples, grapes, etc.) were demonstrated. PEF-assisted processes allow effective extraction of podophyllotoxin from dried rhizomes, calcium from fishbone, calcium malate from eggshell, pure collagen protein from bovine bone, polyphenols from leaves and grape skins, alkaloids from medicinal herbs, valuable components from ginseng, lecithin from egg yolk, polyphenols, EPSs from mushroom broth, proteins, total polyphenols, and polysaccharides from mushrooms and flavonoids from orange peel. Application of PEF reduces time of extraction and does not require the use of organic solvents. PEF treatment also proposes unique possibilities for extraction of valuable proteins, cytoplasmic enzymes, and polysaccharides from cells in suspensions and opens new prospects for production of valuable proteins in cultures of recombinant host cells (*S. cerevisiae, E. coli*). PEF has also been applied to extract lipids, proteins, and other valuable components from microalgae. The opportunity of using PEF treatment at the first step and solvent extraction at the second step was demonstrated. The first important steps for practical implementations of PEF-assisted extraction technologies on the preindustrial pilot scale have already been taken.

Acknowledgments

The authors thank Dr. N. S. Pivovarova for her help with the manuscript preparation.

References

1. Chemat, F., Vian, M.A., and Cravotto, G. (2012) Green extraction of natural products: concept and principles. *Int. J. Mol. Sci.*, **13** (7), 8615–8627.
2. Ahluwalia, V.K. and Verma, R.S. (2009) *Green Solvents in Organic Chemistry*, Alpha Science Publishers, New Delhi.
3. Vorobiev, E. and Lebovka, N. (2010) Enhanced extraction from solid foods and biosuspensions by pulsed electrical energy. *Food Eng. Rev.*, **2** (2), 95–108.
4. Barbosa-Canovas, G.V., Pierson, M.D., Zhang, Q.H., and Schaffner, D.W. (2000) Pulsed electric fields. *J. Food Sci.*, **65**, 65–79.
5. Barbosa-Canovas, G.V., Pothakamury, U.R., Palou, E., and Swanson, B.G.

(1998) *Nonthermal Preservation of Foods*, Marcel Dekker, Inc., New York.
6. Raso, J. and Heinz, V. (eds) (2006) *Pulsed Electric Field Technology for the Food Industry. Fundamentals and Applications*, Springer, New York.
7. Vorobiev, E.I. and Lebovka, N.I. (eds) (2008) *Electrotechnologies for Extraction from Food Plants and Biomaterials*, Springer, New York.
8. Lebovka, N., Vorobiev, E., and Chemat, F. (eds) (2011) *Enhancing Extraction Processes in the Food Industry*, Series: Contemporary Food Engineering, CRC Press, Taylor & Francis Group, Boca Raton, p. 592.
9. Toepfl, S. and Heinz, V. (2011) *Nonthermal Processing Technologies for Food*, Wiley-Blackwell, pp. 190–200.
10. Donsi, F., Ferrari, G., Fruilo, M., and Pataro, G. (2011) Pulsed electric field assisted vinification. *Procedia Food Sci.*, **1**, 780–785.
11. Pakhomov, A.G., Miklavcic, D., and Markov, M.S. (eds) (2010) *Advanced Electroporation Techniques in Biology and Medicine*, CRC Press, Boca Raton, FL.
12. Lebovka, N.I., Praporscic, I., Ghnimi, S., and Vorobiev, E. (2005) Does electroporation occur during the ohmic heating of food? *J. Food Sci.*, **70** (5), E308–E311.
13. Lebovka, N.I., Bazhal, M.I., and Vorobiev, E. (2001) Pulsed electric field breakage of cellular tissues: visualisation of percolative properties. *Innovative Food Sci. Emerg. Technol.*, **2** (2), 113–125.
14. Knorr, D. and Angersbach, A. (1998) Impact of high-intensity electric field pulses on plant membrane permeabilization. *Trends Food Sci. Technol.*, **9** (5), 185–191.
15. Knorr, D., Engel, K.-H., Vogel, R., Kochte-Clemens, B., and Eisenbrand, G. (2008) Statement on the treatment of food using a pulsed electric field. *Mol. Nutr. Food Res.*, **52** (12), 1539–1542.
16. Knorr, D., Froehling, A., Jaeger, H., Reineke, K., Schlueter, O., and Schoessler, K. (2011) Emerging technologies in food processing. *Annu. Rev. Food Sci. Technol.*, **2**, 203–235.
17. der Poel, P.W., Schiweck, H., and Schwartz, T. (1998) *Sugar Technology: Beet and Cane Sugar Manufacture*, Verlag Dr Albert Bartens KG, Berlin, p. 1118.
18. Bouzrara, H. and Vorobiev, E.I. (2000) Beet juice extraction by pressing and pulsed electric fields. *Int. Sugar J.*, **102** (1216), 194–200.
19. Bouzrara, H. and Vorobiev, E.I. (2001) Non-thermal pressing and washing of fresh sugarbeet cossettes combined with a pulsed electrical field. *Zucker*, **126**, 463–466.
20. Bouzrara, H. and Vorobiev, E.I. (2003) Solid/liquid expression of cellular materials enhanced by pulsed electric field. *Chem. Eng. Process. Process Intensif.*, **42**, 249–257.
21. Eshtiaghi, M.N. and Knorr, D. (1999) Method for treating sugar beet *International patent* WO/9964639.
22. Jemai, A.B. and Vorobiev, E. (2003) Enhanced leaching from sugar beet cossettes by pulsed electric field. *J. Food Eng.*, **59** (4), 405–412.
23. Jemai, A.B. and Vorobiev, E. (2006) Pulsed electric field assisted pressing of sugar beet slices: towards a novel process of cold juice extraction. *Biosyst. Eng.*, **93** (1), 57–68.
24. El-Belghiti, K. and Vorobiev, E.I. (2004) Mass transfer of sugar from beets enhanced by pulsed electric field. *Food Bioprod. Process.*, **82**, 226–230.
25. El-Belghiti, K., Rabhi, Z., and Vorobiev, E. (2005) Kinetic model of sugar diffusion from sugar beet tissue treated by pulsed electric field. *J. Sci. Food Agric.*, **85** (2), 213–218.
26. El-Belghiti, K., Rabhi, Z., and Vorobiev, E. (2005) Effect of centrifugal force on the aqueous extraction of solute from sugar beet tissue pretreated by a pulsed electric field. *J. Food Process Eng.*, **28** (4), 346–358.
27. El-Belghiti, K. (2005) *Effets d'un champ électrique pulsé sur le transfert de matière et sur les caractéris-tiques végétales*, Universite de Technologie de Compiegne, Compiegne.
28. Lopez, N., Puertolas, E., Condon, S., Raso, J., and Alvarez, I. (2009) Enhancement of the solid-liquid extraction of

sucrose from sugar beet (Beta vulgaris) by pulsed electric fields. *LWT- Food Sci. Technol.*, **42**, 1674–1680.
29. Vorobiev, E.I., Jemai, A.B., Bouzrara, H., Lebovka, N.I., and Bazhal, M.I. (2005) in *Novel Food Processing Technologies* (eds G. Barbosa-Canovas, M.S. Tapia, and M.P. Cano), CRC Press, New York, pp. 105–130.
30. Lebovka, N.I., Shynkaryk, M.V., El-Belghiti, K., Benjelloun, H., and Vorobiev, E. (2007) Plasmolysis of sugarbeet: pulsed electric fields and thermal treatment. *J. Food Eng.*, **80** (2), 639–644.
31. Loginova, K.V., Vorobiev, E., Bals, O., and Lebovka, N.I. (2011) Pilot study of countercurrent cold and mild heat extraction of sugar from sugar beets, assisted by pulsed electric fields. *J. Food Eng.*, **102** (4), 340–347.
32. Loginova, K. (2011) *Mise en oeuvre de champs electriques pulses pour la conception d'un procede de diffusion a froid a partir de betteraves a sucre et d'autres tubercules alimentaires (etude multi-echelle)*, Universite de Technologie de Compiegne, Compiegne.
33. Loginov, M., Loginova, K., Lebovka, N., and Vorobiev, E. (2011) Comparison of dead-end ultrafiltration behaviour and filtrate quality of sugar beet juices obtained by conventional and 'cold' PEF-assisted diffusion. *J. Membr. Sci.*, **377** (1-2), 273–283.
34. Loginova, K., Loginov, M., Vorobiev, E., and Lebovka, N.I. (2011) Quality and filtration characteristics of sugar beet juice obtained by 'cold' extraction assisted by pulsed electric field. *J. Food Eng.*, **106** (2), 144–151.
35. Loginova, K., Loginov, M., Vorobiev, E., and Lebovka, N.I. (2012) Better lime purification of sugar beet juice obtained by low temperature aqueous extraction assisted by pulsed electric field. *LWT-Food Sci. Technol.*, **46** (1), 371–374.
36. Lopez, N., Puertolas, E., Condon, S., and Alvarez, I. (2009) Enhancement of the extraction of betanine from red beetroot by pulsed electric fields. *J. Food Eng.*, **90**, 60–66.
37. Loginova, K.V., Lebovka, N.I., and Vorobiev, E. (2011) Pulsed electric field assisted aqueous extraction of colorants from red beet. *J. Food Eng.*, **106** (2), 127–133.
38. Shynkaryk, M.V., Lebovka, N.I., and Vorobiev, E. (2008) Pulsed electric fields and temperature effects on drying and rehydration of red beetroots. *Drying Technol.*, **26**, 695–704.
39. Loginova, K.V., Shynkaryk, M.V., Lebovka, N.I., and Vorobiev, E. (2010) Acceleration of soluble matter extraction from chicory with pulsed electric fields. *J. Food Eng.*, **96** (3), 374–379.
40. Bazhal, M. (2001) *Etude du mécanisme d'électropermeabilisation des tissus végétaux. Application à l'extraction du jus des pommes*, Universite de Technologie de Compiegne, Compiegne.
41. Chalermchat, Y., Malangone, L., and Dejmek, P. (2010) Electropermeabilization of apple tissue: effect of cell size, cell size distribution and cell orientation. *Biosyst. Eng.*, **105** (3), 357–366.
42. Praporscic, I. (2005) *Influence du traitement combine par champ electrique pulse et chauffage modere sur les proprietes physiques et sur le comportement au pressage de produits vegetaux*, Universite de Technologie de Compiegne, Compiegne.
43. Turk, M. (2010) *Vers une amelioration du procede industriel d'extraction des fractions solubles de pommes à l'aide de technologies electriques*, Universite de Technologie de Compiegne, Compiegne.
44. Brito, P.S., Canacsinh, H., Mendes, J.P., Redondo, L.M., and Pereira, M.T. (2012) Comparison between monopolar and bipolar microsecond range pulsed electric fields in enhancement of apple juice extraction. *IEEE Trans. Plasma Sci.*, **40** (10, Pt. 1), 2348–2354.
45. Grimi, N. (2009) *Vers l'intensification du pressage industriel des agroressources par champs electriques pulses: etude multi-echelles*, Universite de Technologie de Compiegne, Compiegne.
46. Grimi, N., Mamouni, B.N.,.F., Lebovka, N., Vorobiev, E., and Vaxelaire, J. (2011) Impact of apple processing modes on extracted juice quality: pressing assisted by pulsed electric fields. *J. Food Eng.*, **103** (1), 52–61.

47. Grimi, N., Dubois, A., Marchal, L., Jubeau, S., Lebovka, N.I., and Vorobiev, E. (2014) Selective extraction from microalgae Nannochloropsis sp. using different methods of cell disruption. *Bioresour. Technol.*, **153**, 254–259.
48. El Darra, N., Grimi, N., Vorobiev, E., Louka, N., and Maroun, R. (2013) Extraction of polyphenols from red grape pomace assisted by pulsed ohmic heating. *Food Bioprocess Technol.*, **6** (5), 1281–1289.
49. Puertolas, E., Lopez, N., Condon, S., Alvarez, I., and Raso, J. (2010) Potential applications of PEF to improve red wine quality. *Trends Food Sci. Technol.*, **21** (5), 247–255.
50. Nakagawa, A., Hatayama, H., Takaki, K., Koide, S., and Kawamura, Y. (2013) Influence of pulse width on polyphenol extraction from agricultural products by pulsed electric field. *IEEJ Trans. Fundam. Mater.*, **133** (2), 32–37.
51. Ben Ammar, J. (2011) *Etude de l'effet des champs electriques pulses sur la congelation des produits vegetaux*, Universite de Technologie de Compiegne, Compiegne.
52. Shynkaryk, M. (2006) *Influence de la permeabilisation membranaire par champ electrique sur la performance de sechage des vegetaux*, Universite de Technologie de Compiegne, Compiegne.
53. Toepfl, S. (2006) *Pulsed Electric Fields (PEF) for Permeabilization of Cell Membranes in Food- and Bioprocessing – Applications, Process and Equipment Design and Cost Analysis*, Institut für Lebensmitteltechnologie und Lebensmittelchemie.
54. Jin, S. and Yin, Y. (2010) High intensity pulsed electric fields assisted extraction of lycopene from tomato residual. *Nongye Gongcheng Xuebao* (*Trans. Chin. Soc. Agric. Eng.*), **26** (9), 368–373.
55. Gachovska, T., Cassada, D., Subbiah, J., Hanna, M., Thippareddi, H., and Snow, D. (2010) Enhanced anthocyanin extraction from red cabbage using pulsed electric field processing. *J. Food Sci.*, **75** (6), E323–E329.
56. Xu, Q., Nakajima, M., Liu, Z., and Shiina, T. (2011) Soybean-based Surfactants and their applications, in *Soybean - Applications and Technology* (ed. T.-B. Ng), InTech.
57. Liu, J., Zhou, Y., Liu, D., Lin, S., and Zhang, Y. (2013) Extracting egg yolk lecithin using PEF-assisted organic solvent. *Nongye Gongcheng Xuebao* (*Trans. Chin. Soc. Agric. Eng.*), **29** (5), 251–258.
58. Shynkaryk, M.V., Lebovka, N.I., Lanoisellé, J.-L., Nonus, M., Bedel-Clotour, C., and Vorobiev, E. (2009) Electrically-assisted extraction of bio-products using high pressure disruption of yeast cells (Saccharomyces cerevisiae). *J. Food Eng.*, **92** (2), 189–195.
59. Liu, D., Lebovka, N.I., and Vorobiev, E. (2013) Impact of electric pulse treatment on selective extraction of intracellular compounds from saccharomyces cerevisiae yeasts. *Food Bioprocess Technol.*, **6** (2), 576–584.
60. Jin, Y., Wang, M., Lin, S., Guo, Y., Liu, J., and Yin, Y. (2011) Optimization of extraction parameters for trehalose from beer waste brewing yeast treated by high-intensity pulsed electric fields (PEF). *Afr. J. Biotechnol.*, **10** (82), 19144–19152.
61. Liu, J.-B., Yu, Y.-D., Wang, M., Lin, S.-Y., Wang, Q., Gao, L.-X., and Sun, P. (2010) Fast dissolution of RNA from waste brewing yeast by high-voltage pulsed electric fields. *Jilin Daxue Xuebao (Gongxueban)/J. Jilin Univ.* (*Eng. Technol. Ed.*), **40** (4), 1171–1176.
62. Zhan, Y., Sun, C., Cao, Z., Bao, N., Xing, J., and Lu, C. (2012) Release of intracellular proteins by electroporation with preserved cell viability. *Anal. Chem.*, **84** (19), 8102–8105.
63. Goettel, M., Eing, C., Gusbeth, C., Straessner, R., and Frey, W. (2013) Pulsed electric field assisted extraction of intracellular valuables from microalgae. *Algal Res.*, **2** (4), 401–408.
64. Coustets, M., Al-Karablieh, N., Thomsen, C., and Teissié, J. (2013) Flow process for electroextraction of total proteins from microalgae. *J. Membr. Biol.*, **246** (10), 751–760.
65. Zbinden, M.D., Sturm, B.S., Nord, R.D., Carey, W.J., Moore, D., Shinogle, H., and Stagg-Williams, S.M. (2013) Pulsed electric field (PEF) as an intensification pretreatment for greener solvent lipid

extraction from microalgae. *Biotechnol. Bioeng.*, **110** (6), 1605–1615.

66. Rego, D., Costa, L., Navalho, J., Paramo, J., Geraldes, V., Redondo, L.M., and Pereira, M.T. (2013) Pulsed electric fields applied to the control of predators in production scale microalgae cultures. 2013 Abstracts IEEE International Conference on Plasma Science (ICOPS), doi: 10.1109/PLASMA.2013.6635078, p. 1.
67. Abdullah, S.H., Zhao, S., Mittal, G.S., and Baik, O.-D. (2012) Extraction of podophyllotoxin from Podophyllum peltatum using pulsed electric field treatment. *Sep. Purif. Technol.*, **93**, 92–97.
68. Zhou, Y., Sui, S., Huang, H., He, G., Wang, S., Yin, Y., and Ma, Z. (2012) Process optimization for extraction of fishbone calcium assisted by high intensity pulsed electric fields. *Nongye Gongcheng Xuebao* (*Trans. Chin. Soc. Agric. Eng.*), **28** (23), 265–270.
69. He, G., Yin, Y., Yan, L., and Yu, Q. (2011) Fast extraction of chitosan from shrimp shell by high intensity pulsed electric fields. *Nongye Gongcheng Xuebao* (*Trans. Chin. Soc. Agric. Eng.*), **27** (6), 344–348.
70. Lin, S., Wang, L., Jones, G., Trang, H., Yin, Y., and Liu, J. (2012) Optimized extraction of calcium malate from eggshell treated by PEF and an absorption assessment in vitro. *Int. J. Biol. Macromol.*, **50** (5), 1327–1333.
71. Yu, Y., Liu, J., and Lin, S. (2013) *4th International Conference on Food Engineering and Biotechnology IPCBEE*, Vol. 50, IACSIT Press, Singapore, pp. 136–141.
72. Zderic, A., Zondervan, E., and Meuldijk, J. (2013) Breakage of cellular tissue by pulsed electric field: extraction of polyphenols from fresh tea leaves. *Chem. Eng. Trans.*, **32**, 1795–1800.
73. Ye, H., Jin, Y., Lin, S., Liu, M., Yang, Y., Zhang, M., Zhao, P., and Jones, G. (2012) Effect of pulsed electric fields on the activity of neutral trehalase from beer yeast and RSM analysis. *Int. J. Biol. Macromol.*, **50** (5), 1315–1321.
74. Bai, Y., Li, C., Zhao, J., Zheng, P., Li, Y., Pan, Y., and Wang, Y. (2013) A high yield method of extracting alkaloid from aconitum coreanum by pulsed electric field. *Chromatographia*, **76** (11-12), 635–642.
75. Pourzaki, A., Mirzaee, H., and Hemmati Kakhki, A. (2013) Using pulsed electric field for improvement of components extraction of saffron (Crocus sativus) stigma and its pomace. *J. Food Process. Preserv.*, **37** (5), 1008–1013.
76. Zhao, W., Yu, Z., Liu, J., Yu, Y., Yin, Y., Lin, S., and Chen, F. (2011) Optimized extraction of polysaccharides from corn silk by pulsed electric field and response surface quadratic design. *J. Sci. Food Agric.*, **91** (12), 2201–2209.
77. Lim, J.H., Shim, J.M., Lee, D.U., Kim, Y.H., and Park, K.-J. (2012) Pulsed electric fields effects on drying of white ginseng and extraction of soluble components. *Korean J. Food Sci. Technol.*, **44** (6), 704–710.
78. Luengo, E., Álvarez, I., and Raso, J. (2013) Improving the pressing extraction of polyphenols of orange peel by pulsed electric fields. *Innovative Food Sci. Emerg. Technol.*, **17**, 79–84.
79. Zhang, T.-H., Wang, S.-J., Liu, D.-R., Yuan, Y., Yu, Y.-L., and Yin, Y.-G. (2011) Optimization of exopolysaccharide extraction process from Tibetan spiritual mushroom by pulsed electric fields. *Jilin Daxue Xuebao (Gongxueban)/J. Jilin Univ.* (*Eng. Technol. Ed.*), **41** (3), 882–886.
80. O. Parniakov, O. Lebovka, O. Van Hecke, and E. Vorobiev (2014) Pulsed electric field assisted pressure extraction and solvent extraction from mushroom (Agaricus Bisporus), *Food Bioprocess Technol.* **7**, (1), 174–183.
81. Oms-Oliu, G., Odriozola-Serrano, I., and Martin-Belloso, O. (2012) in *Phytochemicals – A Global Perspective of Their Role in Nutrition and Health* (ed. V. Rao), InTech, pp. 107–126.
82. Morales-De La Peña, M., Salvia-Trujillo, L., Rojas-Graü, M.A., and Martín-Belloso, O. (2011) Changes on phenolic and carotenoid composition of high intensity pulsed electric field and thermally treated fruit juice-soymilk beverages during refrigerated storage. *Food Chem.*, **129** (3), 982–990.

10
Green Extraction of Artemisinin from *Artemisia annua* L

Alexei A. Lapkin

10.1
Introduction

Artemisia annua is a source of artemisinin (**1**), the main precursor to antimalarial drugs against *P. falciparum* malaria, and as such it is one of the most studied plants. Its traditional use in Chinese pharmacopeia includes the treatment of fever and chills [1, 2]. The active component of the plant was isolated and identified as artemisinin in the 1970s. Today it has become established in the form of artemisinin combination therapies (ACTs) as effective antimalarial treatment against uncomplicated malaria [3–5].

Different methods of extraction of artemisinin from *A. annua* were described in an earlier review [6]. Recent developments have seen the introduction of a semisynthetic route to artemisinic acid (**2**), a biological precursor to (**1**) [7] (Scheme 10.1). Starting from artemisinic acid or dihydroartemisinic acid, a continuous photochemical synthesis was developed [8], thus providing a semisynthetic pathway to artemisinin. Further development of the artemisinin derivatization process, either in batch [9] or in a continuous process technology [10, 11] provide a new route to artemisinin-based drugs, avoiding extraction from a plant. However, as extraction will continue to provide the bulk of artemisinin for manufacture of antimalarial drugs in the foreseeable future, research into better methods of its analysis, extraction, and purification continues.

Here we give an overview of the current state of the art of artemisinin extraction and provide some insights into potential use of novel, clean extraction and purification technologies which are likely to improve the quality (purity) of the obtained molecule as well as improving robustness of the manufacturing process and its cost effectiveness and safety.

10.2
Extraction Technologies for Isolation of Artemisinin from *A. annua*

Extraction of artemisinin can be described as a generic solid–fluid extraction from a heterogeneous matrix. In the plant, artemisinin is localized within

Green Extraction of Natural Products: Theory and Practice, First Edition.
Edited by Farid Chemat and Jochen Strube.

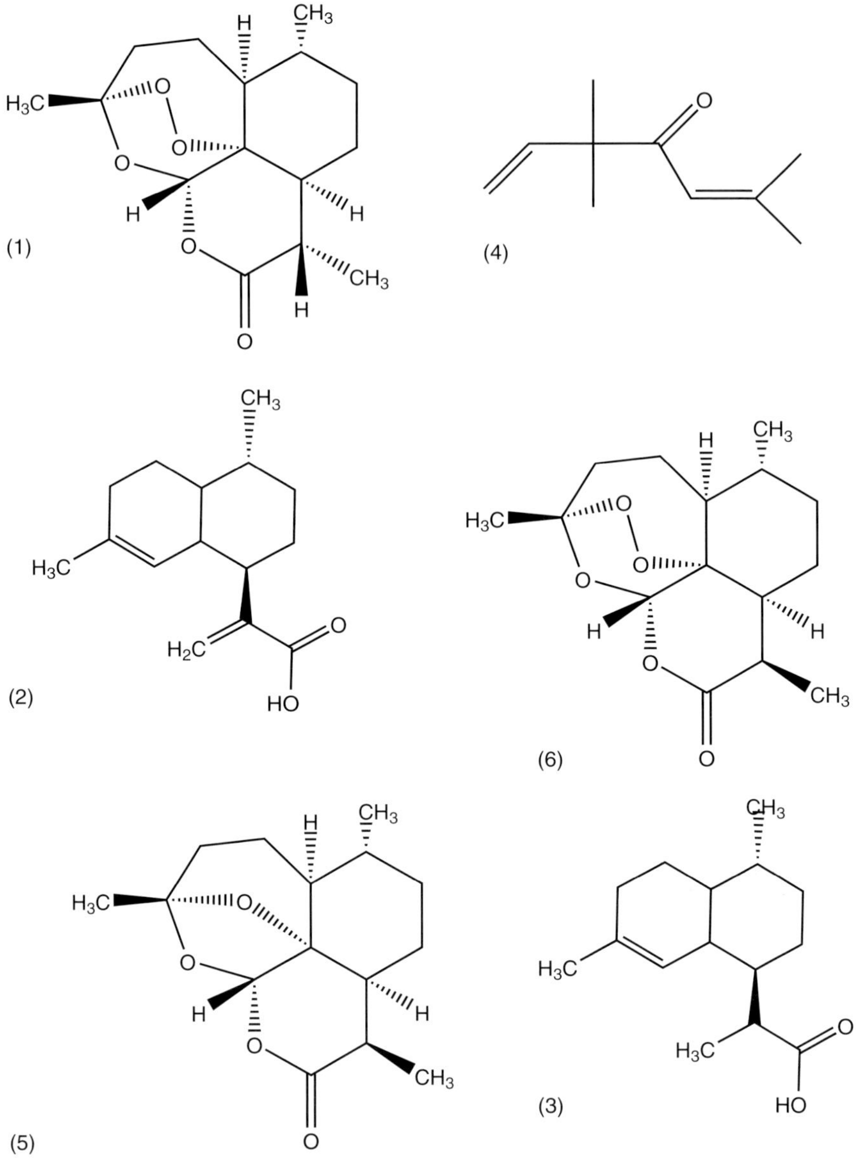

Scheme 10.1 Artemisinin-related metabolites of *Artemisia annua*.

glandular trichomes on the surface of leaves [12, 13] as shown in the SEM image in Figure 10.1. The reason for this secondary metabolite being localized in the specialized structures on the surface rather than inside the leaf structure is believed to be its phytotoxicity. Thus, in the extraction process, the inner structure of the leaf does not need to be accessed, which reduces the requirement for mechanical treatment of the biomass before extraction.

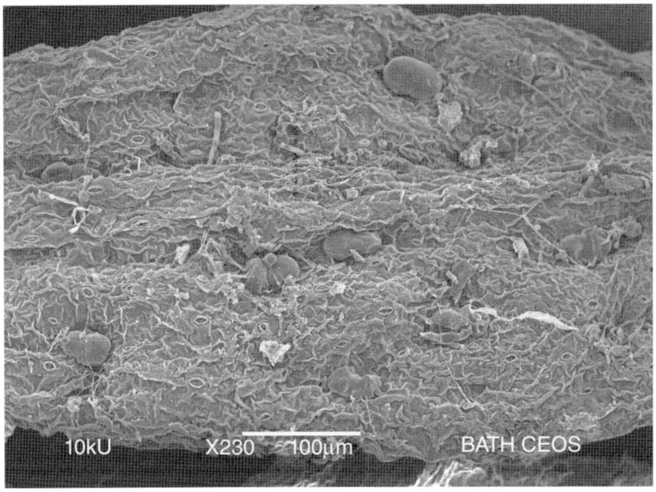

Figure 10.1 An SEM image of *Artemisia annua L.* leaf showing glandular trichomes containing artemisinin. (Image was obtained using JEOL 6480 instrument by Ms Benhilda Mlambo, University of Bath.)

Another generic factor worth discussing before reviewing current approaches to extraction of artemisinin is its electronic structure, which has a direct impact on its solubility in different solvents and on the downstream purification. A detailed analysis of this topic has been presented earlier [14]. Discussion of interactions of artemisinin with solvents could be done from the point of view of different solvation theories. The approach adopted here is from the point of view of a conductor-like screening model or COSMO-RS theory described in detail elsewhere [15, 16].

Specifically in the case of artemisinin, the σ-surface or the screening charge density on the surface of the molecule which will interact with solvent molecules is shown in Figure 10.2. The color coding corresponds to positive and negative charges and relates to a specific scale of charge density. What is important to note are the areas of increased electronic density corresponding to oxygen groups, with the carbonyl of the lactone group presenting the highest charge density, and the areas of lower electronic density corresponding to the saturated ring structure of the molecule. This translates into the surface chemical potential distribution which shows both the area of hydrophobic interactions around zero σ (screening charge density), as well as the areas of polar interactions at nonzero σ values (Figure 10.2). As a consequence, artemisinin interacts with both polar and nonpolar solvents. The choice of extraction solvent will then determine the specificity of extraction and complexity of the downstream separation. Needless to say, hydrophobic solvents such as hexane, toluene, or petroleum ether (PE) would interact with artemisinin through its ring structure. Since this is a rather nonspecific interaction, it is impossible to expect that such extracts would be particularly specific, because dried plant matter contains multiple chemical species that would interact via nonspecific hydrophobic interactions. Discussion

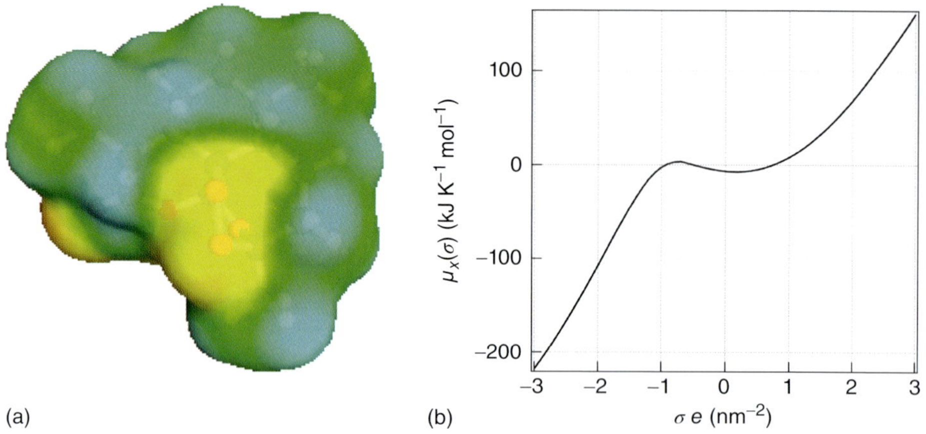

Figure 10.2 Screening charge density of artemisinin calculated in COSMOTherm (a) and probability density of chemical potential as a function of screening charge density (b).

of interactions of various solvents with artemisinin is continued in this chapter in the Section 10.2.2.2 on novel solvents.

10.2.1 Industrial Extraction Processes

Conventional industrial extraction of artemisinin based on hydrocarbon solvents is shown schematically in Figure 10.3. There are several variants of this process, hence the times and conditions are not shown. Typically, the solvent of choice is PE owing to its wide availability and a reasonable solvation of the target molecule from the plant. Alternative commercially used hydrocarbon solvents in this extraction are toluene and hexane/ethyl acetate (95/5 v/v) mixed solvent. The use of EtOAc cosolvent is twofold: it increases solubility of artemisinin and it also improves process safety, as nonconducting hexane tends to accumulate static charge, which in large-scale applications may lead to a discharge and spontaneous explosion. There is a marked difference in solubility of artemisinin in PE and hexane/EtOAc versus toluene [14]. Thus, solubility of pure artemisinin in pure hexane at 20 °C is $0.46\,g\,l^{-1}$, whereas in toluene it is $145.6\,g\,l^{-1}$. Therefore, there is a significant difference in the volume of fresh solvent that is required, as well as in solvent purification and recycling, and downstream purification, which are discussed in detail in the following sections.

However, the analysis presented above is quite simplistic: it has already been shown that other molecules co-extracted along with artemisinin augment its equilibrium solubility [14]. Thus, the real solubility of artemisinin in hexane is significantly higher than the reported value of $0.46\,g\,l^{-1}$ for pure solvent, if measured in a plant extract. This is supported by experimental data, showing that hexane and PE offer better extraction yields of artemisinin, compared to methanol and ethanol, despite the reverse pure substances solubilities [17]. Extraction is also

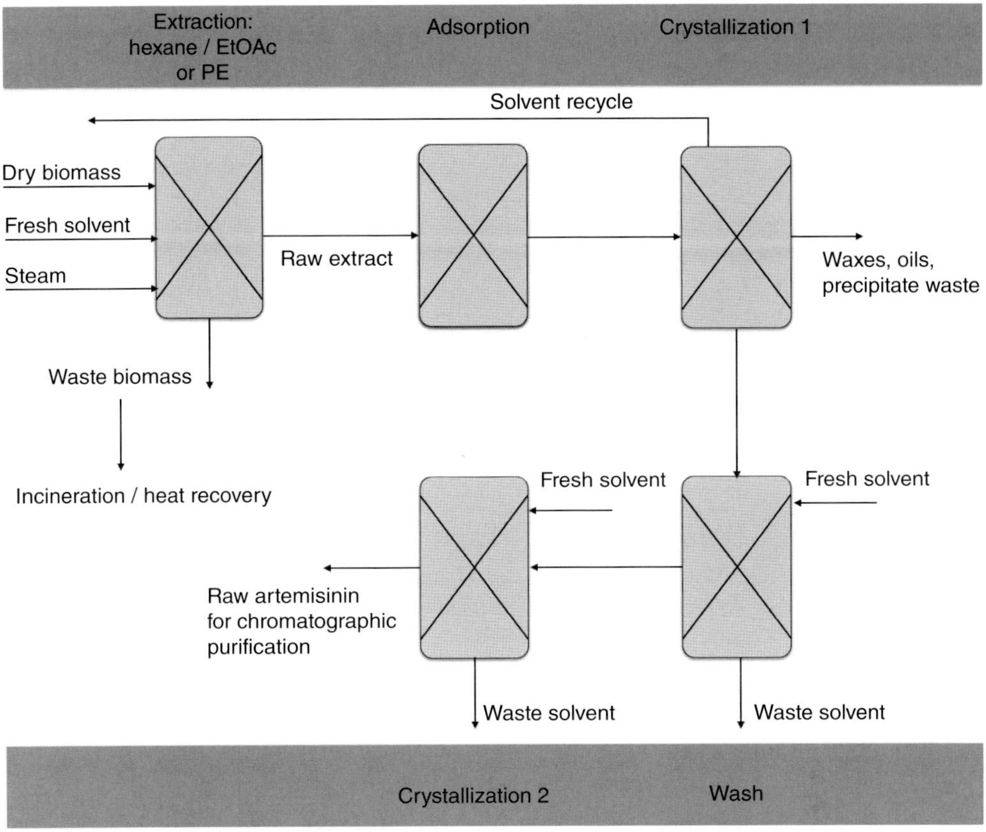

Figure 10.3 A scheme of conventional hexane- or petroleum-ether-based extraction of artemisinin.

highly variable with temperature and duration of extraction, history, and source of the biomass, and the type of the extractor used. It was shown recently that the composition of important co-metabolites that affect solubility of artemisinin and its recoverability from the extracts is strongly dependent on the geographic origin of the biomass [18]. Thus, optimization of extraction is necessarily specific to the origin of the biomass and the location of the extraction facility in addition to the usual consideration of solvent choice and operating conditions.

Typically, industrial artemisinin extraction is done under soaking mode and less frequently under pumped or gravity-driven percolation modes at temperatures in the range of 40 °C up to the solvent's boiling point. If the solvent is used at boiling point, then the extraction could be achieved in a scaled version of a Soxhlet apparatus, which has been done commercially in some Vietnamese extraction facilities. Thermal degradation of artemisinin in such a process must take place, but has not been quantified. The difference in agitation mode translates into the differences in the rates of mass transfer (specifically, the mass transfer coefficient k in Equation 10.1) and hence the duration of the extraction cycle. Cycle times of anything between 2 and 8 h are quite common for various types of installations. Similarly, different numbers of fresh solvent additions are used and also different ratios of solvent to biomass. The solvent-to-biomass ratio impacts on the second part of the mass transfer equation (Equation 10.1), namely, the driving force or the difference between the solubility limit and the actual liquid-phase concentration of the solute. As an extreme example of conventional agitation of large-scale vessels, in the case of the overhead stirring, the ratio of solvent to biomass above 4 : 1 (v/wt) is used to enable stirring of finely cut biomass and solvent slurry. With the overhead stirring of the biomass–solvent slurry, the kinetics of extraction is fast, reaching equilibrium within 1 h at 40 °C. However, overhead stirring is not as scalable as solvent percolation because of the strong dependence of electric motor power on the scale.

The efficiency of conventional extraction of artemisinin (before purification) is reportedly around 70%, with respect to biomass content [6]. For the following reasons, 100% cannot be reached retention of some of the extracted compounds on biomass acting like an adsorbent, incomplete recovery of artemisinin from glandular trichomes due to a significant reduction in the rate of mass transfer closer to saturation of the solvent, poor accessibility of biomass if solvent agitation is inefficient, and degradation of artemisinin during the extraction.

$$\text{Rate of mass transfer} = k \times (\text{Solubility} - \text{Actual liquid phase concentration}) \tag{10.1}$$

Following extraction, the raw extract could be reduced in volume to crystallize raw artemisinin. This has been common practice in Vietnamese extraction facilities. First, crystallization was frequently achieved by evaporating the solvent in open shallow pans in air. This leads to rather low purity of crystallization and low artemisinin yields.

The more commonly used approach is to treat the raw extract with adsorbents before first crystallization. Upon contact of the raw extract with an adsorbent bed, waxes and other large molecules co-extracted along with artemisinin are preferentially removed by an adsorbent, leaving a cleaner extract from which higher purity raw artemisinin with higher crystallization yields are then obtained. Silica gel is reportedly being used in China, whereas a mixed system comprising activated carbon and Celite was described recently [19]. The latter study gave a systematic analysis of the co-metabolites that have the most significant impact on crystallization of artemisinin from raw extracts in the case of the frequently used hexane/EtOAc mixed solvent system. It was found that waxes and pigments have little influence on crystallization, whereas some flavonoids have a significant impact on the amount of artemisinin that is able to crystallize from the raw hexane/EtOAc extracts.

Following the adsorptive treatment, the purified extract is typically evaporated to 10% volume and cooled for the onset of crystallization. The recovered crystals are washed with cold solvent, dried, and recrystallized from ethanol. The final step of purification may be a column chromatography to reach the desired purity for further derivatization into actives, such as artemether or artesunate.

10.2.2 Cleaner and Intensified Processes for Extraction of Artemisinin

There has been a significant effort in developing better extraction protocols that would afford an improved yield of artemisinin. Two parallel and complementing directions were used: (i) increase of mass transfer through use of sonication, percolation at high pressure, and microwave heating and (ii) increase in solubility and specificity of extraction by changing the nature of the solvent. The latter may also require the change of process toward one that simultaneously offers better mass transfer such as in the case of pressurized gaseous solvents, for example, $scCO_2$ (supercritical carbon dioxide). These two approaches affect the two parts of Equation 10.1 independently, although the change of solvent frequently would also change mass transfer through the change of physical properties such as viscosity.

10.2.2.1 Innovative Process Conditions for Extraction

Intensification of solid–liquid extraction through the use of alternative agitation or heating methods is based on the fact that the process of extraction is strongly mass transfer dependent and mass transfer coefficients can be easily affected by the degree of agitation. Thus, in the early studies of enhanced extraction processes applied to *A. annua,* performing extraction with pressurized percolating solvent at up to 35 MPa pressure allowed to reduce extraction time to 20 min [20]. Generally, extraction under pressure also results in decreased solvent use, but this is primarily through shorter extraction times, as pressure does not affect solubility of artemisinin. A more recent study, aimed at analysis of usefulness of waste *A. annua* biomass after distillation of essential oils, confirmed the increased efficiency of pressurized extraction of artemisinin using PE as solvent [21].

Another method of increasing agitation is the use of ultrasound. An earlier study of the use of ultrasound for extraction of essential oils from another *Artemisia* plant showed that ultrasonic extraction resulted in a larger number of compounds being extracted [22]. This suggests that ultrasound disrupts cell membrane, as well as enhancing external mass transfer. In the application of ultrasound to extraction of artemisinin, it was shown that 14– 60% increase in the amount of artemisinin extracted could be attained, the larger effect corresponding to lower extraction temperature due to the decreased efficiency of ultrasonic agitation at elevated temperatures [23]. However, this effect seems to equally affect all other metabolites extracted into hexane, as no significant difference in the relative amount of artemisinin (or its purity) in the extracts was obtained. The general fact that a larger amount of artemisinin was extracted in ultrasound-assisted extraction is most likely due to disruption of glandular trichomes and also due to the decreased re-adsorption of extracted metabolites onto the biomass bed.

Microwave extraction in recent years has become a standard commercial extraction method, and it is rather disappointing to still occasionally see publications repeating old, and now erroneous, statements that microwave extraction or heating cannot be applied on a large scale! A careful study including selection of solvent, extraction technology, and optimization of microwave extraction time showed that unparalleled extraction efficiency of over 90% could be achieved in 12 min extraction time with the use of a good solvent, compared to typical extraction cycle times of 2–12 h even in small-scale experiments, using conventional Soxhlet or soaking extraction [24]. Christen and Veuthey [20] describe earlier results reported in a PhD thesis by M. Kohler on microwave-assisted extraction of artemisinin from three different solvents, confirm that hydrophobic hexane is a better solvent for this process, compared to ethanol or toluene.

In MW extraction, two heating mechanisms are considered: dielectric coupling of microwaves with residual water in plant leaves and MW heating of the solvent. If a solvent is transparent to MWs, then heating would primarily take place in the plant material containing some residual water. However, analysis of the trend of extraction efficiency versus dielectric constant of the solvent had shown that solvent heating plays a more significant role [24].

10.2.2.2 **Alternative Solvents for Extraction of Artemisinin**

Apart from commercially used hydrocarbon solvents, ethanol, $scCO_2$, water, tetrafluoroethane, cyclic carbonates, and ionic liquids were investigated for extraction of artemisinin. Following the classification introduced in an earlier work on comparison of different solvents for extraction of artemisinin [6], we can distinguish solvents according to their physical state under normal conditions. Thus, ethanol, hexane, PE, water, carbonates, and ionic liquids that were reported in extraction of artemisinin from biomass are liquid at 298 K and atmospheric pressure, whereas carbon dioxide and tetrafluoroethane are gases. This implies that the first group of solvents can be used as liquids without pressurization of the extraction apparatus, although, as we already know, forced percolation of a solvent at elevated pressures results in an increased efficiency of extraction in terms of reduced extraction time (see above). The second group of solvents can only be used under pressure, but affords a different mode of solutes recovery: precipitation of the solutes upon depressurization of the solution. These solvents could be used either as liquids or as supercritical fluids, if supercritical conditions are reached. Finally, we may also separate the first group of solvents into high- and low-boiling-point solvents, as this physical property defines the nature of solutes recovery downstream of extraction: low-boiling-point solvents may be stripped, thus triggering crystallization, whereas high-boiling-point solvents cannot and thus require downstream liquid–liquid or supercritical solvent extraction, or anti-solvent-induced crystallization. This classification is shown in Table 10.1 and a detailed discussion of the solvent systems is given below.

Ethanol extraction has been described in a publicly available report [25] and also in peer-reviewed literature [20, 26, 27]. The argument for using ethanol as a solvent is its relatively low cost, local availability as a renewable solvent, and relatively low hazard and toxicity. Multi-objective analysis of different common solvent systems presented earlier [6] did not favor ethanol extraction, as both hazard and cost indicators were worse in comparison to other possible extraction systems, even not accounting for downstream separation.

Table 10.1 Classification of solvents for extraction of artemisinin based on their physical state at normal conditions and boiling point.

Solvents	Physical state at normal conditions	Boiling point	Feasible extraction processes	Downstream separation
Hexane	Liquid	Low	Soaking, percolation at ambient pressure, high-pressure percolation, MW assisted, or ultrasound assisted	Crystallization
Petroleum ether (PE)		Low		
Water		Low		N/A
Ethanol		Low		Crystallization
Cyclic carbonates		High	Soaking	L/L extraction or sc-extraction
Ionic liquids		Very high or N/A	Soaking	L/L extraction, sc-extraction, or antisolvent precipitation
Carbon dioxide	Gas	Very low		Precipitation
HFC R134a		Very low		

One of the most significant disadvantages of ethanol extraction is the more complex downstream separation due to the presence of hydrophilic compounds in the extract, making direct crystallization impossible. The recently developed protocol for recovery of artemisinin from ethanol extracts involves coating of the extract onto diatomite particles, drying, and then elution of the column-packed diatomite with ethyl acetate/hexane-mixed solvents, followed by crystallization [26]. For a full procedure, see the original reference. This procedure is still using hexane/ethyl acetate-mixed solvent, but as dry extract is concentrated, the amount is considerably smaller compared to the primary extraction as in the conventional process.

Extraction with hot water deserves a separate discussion, which slightly breaks the flow of the chapter because of the different application of this extraction process. Solubility of artemisinin in water is negligible and the use of hot-water extraction as a process of manufacture of artemisinin is not a viable option. However, artemisinin tea (hot-water extraction) has been traditionally used as a folk antimalarial remedy [1, 2]. In recent literature, more robust values of solubility of artemisinin and its extraction in hot water under controlled conditions of leaf particle size, temperature, pressure, and duration of extraction appeared [28]. The room temperature aqueous solubility or pure artemisinin was found to be about

51 mg l^{-1} which is an order of magnitude lower than that in pure hexane. A value of 74 mg l^{-1} has been reported elsewhere [29]. At elevated temperatures, the solubility is increased, as also extraction efficiency [28]. A very recent, detailed study of the metabolic profile of hot aqueous extracts of *A. annua* and their bioactivity showed the presence of synergistic interactions between artemisinin and some of the co-metabolites extracted in water, specifically rosmarinic and 3-caffeoylquinic acids [29]. The latter study discovered the differences in potentiation of activity of artemisinin against drug-resistant and nonresistant strains of parasites with arteannuin B as well as a potential contribution of antioxidants to parasite resistance to artemisinin. These findings suggest a way to increase the potentiation of aqueous artemisinin formulations as well as ACT formulations, especially in the case of the drug-resistant strains.

Among the liquid solvents recently reported for extraction of artemisinin, ionic liquids and cyclic carbonates are rather different, as these solvents have very high boiling points, or negligible vapor pressure, making stripping, or distillation of the solvents impossible. Therefore, recovery of solutes must be done either by liquid–liquid extraction, by supercritical fluid stripping, or antisolvent precipitation. Extraction of artemisinin by ionic liquids was first reported in a comparative assessment paper and a publicly available research report [6, 30]. Two ionic liquids were reported, their structures are shown in Figure 10.4.

The characteristic feature of this extraction is its very short extraction time (20 min at ambient temperature). However, in order to recover the dissolved artemisinin, an ionic liquid extract was mixed with water upon which artemisinin precipitated, whereas extracted oils and waxes floated as a light immiscible organic layer. The two ionic liquids that were reported in the original study were completely miscible with water and could be reconstituted upon drying. In this case, water acts as anti-solvent, triggering precipitation of the solute. This hypothetical process offered very favorable economic and environmental indicators [6]. However, at the time, the two ionic liquids were not available commercially at the prices that would allow their use as extraction solvents. Further study of the extraction processes revealed that water-induced precipitation of artemisinin from extracts was not reproducible and frequently resulted in the loss of artemisinin [31]. This study has also shown that water-soluble flavonoids increased aqueous solubility of artemisinin and prevented its precipitation from

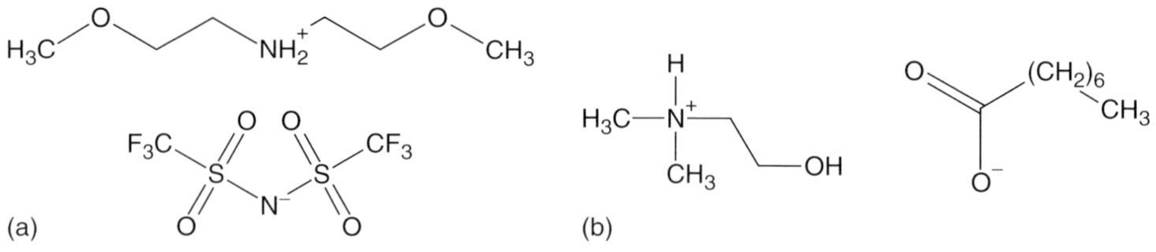

Figure 10.4 Structures of ionic liquid solvents for extraction of artemisinin. (a) Bis(2-methoxyethyl)ammonium bis(trifluoromethylsulfonyl)imide (BMOEA bst), (b) *N,N*-Dimethylethanolammonium octanoate (DMEA oct).

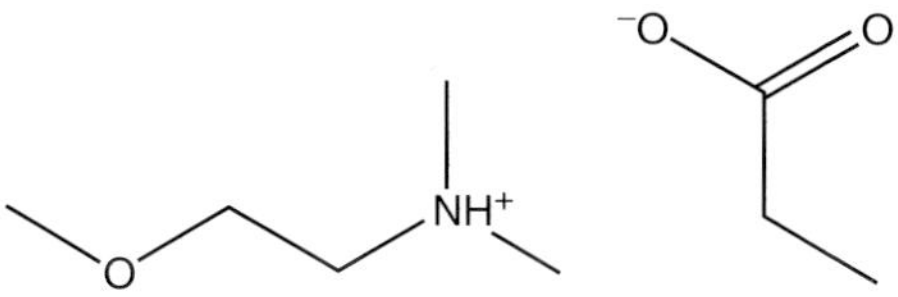

Figure 10.5 Structure of an improved ionic liquid solvent for extraction of artemisinin. N,N-Dimethyl(2-methoxyethyl)ammonium propionate (DMMOEA pro).

dilute aqueous solutions. A new ionic liquid was designed that showed lower solubility toward flavonoids and more stable recovery using a similar aqueous partitioning, the solvent molecule is shown in Figure 10.5. A more detailed study of interaction of ionic liquids with water and artemisinin has also been published more recently [32].

Cyclic carbonates belong to a class of new solvents that could be obtained entirely from renewable resources, thus having a negligible life-cycle impact. From the point of view of clean processing, this is a highly attractive option. A computational study of different new solvents for extraction of artemisinin highlighted high solubility of artemisinin in several carbonate solvents [14]. Thus, for butylene carbonate, the solubility of artemisinin was estimated to be $16\,g\,l^{-1}$ or 35 times higher than that in pure hexane. High solubility of the target compounds would translate into a smaller number of repeat extractions with fresh solvent and a smaller overall solvent inventory. This implies that even if the price of such a solvent is higher, the overall process cost may be competitive with a conventional process, if the reduced material inventory is taken into account. Extraction of artemisinin was confirmed experimentally and clean extracts with very high proportion of artemisinin were obtained, especially in the case of butylene carbonate. Following extraction, the solution of artemisinin and other co-metabolites in butylene carbonate requires a different separation approach compared to conventional volatile solvents. In this case, either liquid–liquid extraction or supercritical fluid extraction could be used, but neither was developed in the original paper [14].

Another class of solvents that also require a different process technology and a different downstream solute recovery technology are $scCO_2$ and tetrafluoroethane (HFC R134a). These solvents are gases under normal conditions and therefore must be pressurized to be used as liquids or as supercritical fluids. The use of $scCO_2$ as a solvent is well known and practiced commercially not only for natural products extraction [33] but also as a reaction solvent [34]. There are commercial small-scale units operating for the extraction of artemisinin and quite a few reports on the use of $scCO_2$ for extraction of artemisinin [35–37]. Thermodynamic data on solubility of artemisinin in $scCO_2$ was recently remeasured and summarized [38]. The solubility of artemisinin in pure $scCO_2$ is rather low. Therefore, methanol or ethanol is used as modifier to increase the solubility and establish a viable process [20, 37, 39, 40]. Such a modification allows attaining high extraction yields. A column chromatography approach similar to that developed for the ethanol extract [26] was also reported for $scCO_2$–ethanol extracts [40]. As an alternative, a direct adsorption of artemisinin onto solids was also considered [41]. However, owing to competitive adsorption of artemisinin and CO_2 onto the solid, this process may require multistep separation and post-extraction column separation seems to be an easier option.

Finally, tetrafluoroethane, and to a lesser extent, iodotrifluoromethane (ITFM) were also suggested as potential extraction solvents for artemisinin (see structures in Figure 10.6). R134a is a well-known refrigerant with large-scale use in car air-conditioning, domestic and commercial refrigeration, as well as in powdered drug formulations as a propellant. It has also been registered as a solvent for extraction of flavors and fragrances in the food industry. The downside of this solvent is its high cost and its very high greenhouse gas potential of 1300 relative to CO_2. For this reason, it is being replaced as a refrigeration media and thus will be less available in the future for solvent applications.

Nevertheless, a considerable effort has gone into developing solid–liquid extraction processes based on R134a solvent. There are several key benefits of this technology:

- R134a is a much better solvent than $scCO_2$ and thus can be used without cosolvent additives.
- It has very low vapor pressure and its pressurization–depressurization cycle is considerably less power intensive compared to that of CO_2.
- The solvent is nonflammable and completely nontoxic; its only hazard is the possibility of asphyxiation.
- The solvent could be easily contained within a closed-loop system.

Small-scale commercial installations based on R134a were developed, notably by Dr Peter Wilde and by Ineos Fluor, and a number of reports of natural product extractions by R134a and by ITFM solvent were published [42–44]. More recently, R134a solvent has received considerably more attention as a viable extraction solvent for food products [45–47]. A number of papers have also appeared on the use of R134a as reaction media for enzyme-catalyzed reactions, for example, Refs. [48, 49].

The use of ITFM solvent is more controversial owing to hazardous degradation products that could be formed (see discussion in the earlier review [6]). The use of a Freon, 1,1,2-trichloro-1,2,2-trifluoroethane, for extraction of *A. annua* and isolation of arteannuin B was reported in 1986 by Nancy Acton and coworkers [50]. The use of Freons has since then been banned, but this early work was the

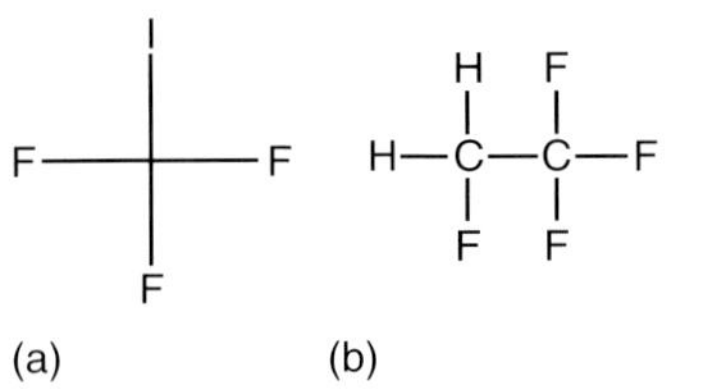

Figure 10.6 Structures of fluorinated solvents tested in extraction of artemisinin. (a) Iodotrifluoromethane (ITFM), (b) Tetrafluoroethane (R134a).

precursor to the use of a Freon replacement, R134a, as a solvent for *A. annua* extraction.

For the industrial use of such solvents as R134a and ITFM, there are two possible process configurations which are shown schematically in Figure 10.7. The first configuration is based on gravity-driven flow and requires minimal instrumentation and automation. Solvent recycle could be achieved either by using a suitable heat exchanger or a compressor. The second configuration is based on pumped flow. In this latter case, the direction of flow through a biomass bed does not have to be downward as shown in the figure, but could be directed upward. The difference will then be in the type of flow established within the biomass bed. Both configurations have been scaled to a 5-l pilot scale and the gravity-driven flow configuration has been used commercially for extraction of flavors. In a recent paper, experimental comparison of the two configurations has been done [18], which has shown that the pumped upflow is the most efficient option. This is reasonable, because pumped flow enables a faster liquid flow rate, whereas upward direction of flow through the bed ensures that the bed is completely wetted.

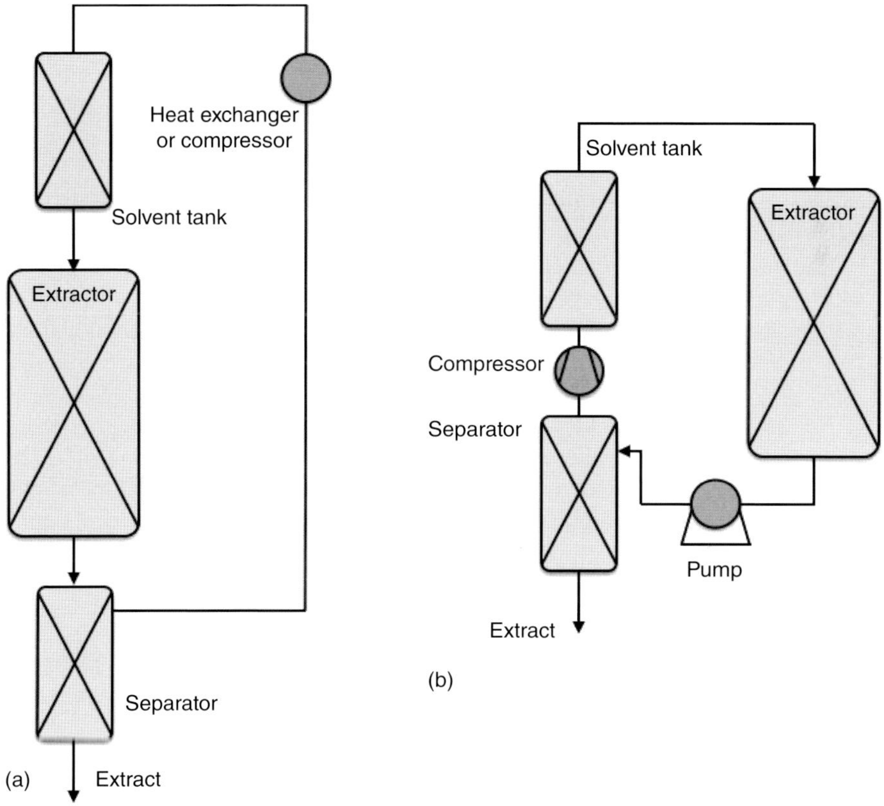

Figure 10.7 A schematic diagram of the two process options for R134a extraction: (a) gravity-driven flow and (b) pumped flow.

10.3
Innovation in Artemisinin Purification

10.3.1
Hybrid Adsorption–Crystallization Separation

All extraction methods that exist for recovery of artemisinin from biomass end up with a raw extract consisting of artemisinin and many other molecules. The initial step of artemisinin recovery from the raw extract has been recently studied in detail for the cases of ethanol [26] and hexane/ethyl acetate extracts [19]. In both cases, the intermediate separation step is necessary to remove the compounds that prevent successful crystallization of artemisinin from the raw extracts. In the case of ethanol extracts, the raw extract was deposited onto a solid support and then packed into a column and solvent elution was used to recover artemisinin and similar compounds. In the case of the hexane extract, a simpler procedure using solid adsorbents was used. The latter paper also confirmed that methylated flavonoids are responsible for the decreased crystallization yield.

A specific grade of silica gel was proposed as a suitable adsorbent for pretreatment of raw hexane extracts [51]. Notably, waxes and undefined polar compounds were specified as the compounds that complicate crystallization of artemisinin, contrary to findings of Suberu *et al.* that the impact of waxes was statistically insignificant [19]. In addition, it was reported that the efficiency of purification of the extract was highly sensitive to the nature of the silica gel. Other adsorbents screened, including activated carbon and celite, were not as effective or they retained a significant amount of artemisinin, reducing the overall yield.

The use of silica gel flash chromatography was discussed within the concept of a hybrid chromatography–crystallization process for efficient recovery of artemisinin [52]. The approach adopted in this hybrid process is similar to adsorption–crystallization methods described above, although more complex antisolvent crystallization was used. This allowed achieving a reasonable compromise between crystal yield and purity. The problem of this specific methodology is the use of toxic and environmentally unfriendly solvents. As an example, the use of dichloromethane at an industrial scale should be avoided as a principle. Similarly, the use of acetonitrile for any process apart from its analytical applications should not be considered as a viable option. This statement requires a clarification: a hazardous or environmentally unfriendly solvent may be used if it can be completely recovered and run within a closed-loop tightly controlled system. This is generally impossible for mixed solvents in large-scale extraction processes and, therefore, should be avoided.

Adsorbents described thus far are nonspecific, for example, silica, activated carbon, and its combination with celite. However, it is possible to design adsorbents with a higher degree of specificity to enable much higher purity of the desired molecule to be obtained from the step of adsorption separation. Thus, molecularly imprinted polymers were designed to recover artemisinin from raw extracts or from pretreated extracts after nonselective adsorbents [51]. The best affinity towards artemisinin and the best elution properties were obtained with the polymer based on *N,N*-methylenebisacrylamide. This polymer also had shown the best stability over 10 loading-elution-purification cycles, attaining 96% recovery of artemisinin with respect of the composition of the raw extracts. The authors suggested a sequential purification protocol using nonselective silica adsorbent as a primary purification step, followed by selective imprinted polymer adsorption–elution. This process allowed them to attain a 75% purity of artemisinin crystals from the first crystallization step.

Another imprinted polymer, based on a copolymer of 3-aminopropyltriethoxysilane and calix[4]arene grafted onto porous silica particles, was developed for a similar separation process [53]. However, the authors have only shown equilibrium and kinetic adsorption of pure artemisinin solution, which limits the usefulness of the study, as the real interest is in purification of extracts and in purification of solutions before recrystallization and real system tests should have been shown. The use of $scCO_2$ in combination with imprinted polymers was suggested as a means of improving both the capacity and the kinetics of artemisinin uptake [53]. A significant increase in the equilibrium uptake of artemisinin was observed, but this was only done for pure artemisinin and must be validated for the real systems.

10.3.2
Column and HPLC Chromatography

Introduction of an intermediate adsorption step as described for several systems above allows improvement in the yield and purity of artemisinin obtained in the first crystallization step. This is generally a preferred method of separation as it reduces the necessity for chromatographic column separation at the end of the purification sequence. Taking chromatographic separation approach to its most efficient realization leads to preparative HPLC. A recent application of prep-HPLC for isolation of artemisinin and its co-metabolites artemisinic (**2**) and dihydroartemisinic acids (**3**), which are believed to be the biosynthetic precursors of artemisinin, has been reported [17]. Since a number of HPLC protocols for identification of multiple co-metabolites in raw extracts have been published (Section 10.4), it is clear that HPLC separation can be scaled for preparative use. High cost of materials and equipment, however, may still preclude the technology from commercial use.

The use of cheap loading materials for column chromatography may be one alternative solution to adsorption–crystallization sequence. Thus, the use of preparative charcoal chromatography was described for the case of the ethanol extracts of artemisinin [27]. The efficiency of artemisinin recovery from extracts was reported at 87% with the overall recovery of 72%. These are high values and need to be statistically validated as well as confirmed for other solvents more frequently used for extraction. The authors used chlorinated solvents for elution, which is completely unacceptable on a large scale. Therefore, this process needs further validation and development, as well as economic assessment compared to the more established adsorption–crystallization protocol.

10.3.3
Countercurrent Chromatography

As a complete alternative to solid adsorbents, the use of liquid chromatography for purification of artemisinin has been described as early as in 1986 [50]. The principle of countercurrent liquid–liquid chromatography is the generation of efficient mixing and settling zones between two fluids using a variable gravity field (or a constant gravity field in another variant of the process) [54, 55]. This enables a relatively compact engineering solution to the problem of establishing efficient liquid–liquid equilibrium and allows to access large design freedom of liquid–liquid partitioning. The downsides are the complexity and cost of the instrumentation and the large use of mixed solvents. However, in recent years, pilot-scale and even production-scale machines with industrially relevant throughputs have become available. A system of solvent was designed for separation of artemisinin, artemisitene, and arteannuin B by countercurrent chromatography [56]. However, despite good separation efficiency and many successful industrial trials of the technology specifically for artemisinin separation, its use on an industrial scale is at present economically prohibitive.

10.4
Analysis of Artemisinin and Co-metabolites

Although this issue is being considered at the very end of the chapter, analysis of artemisinin in raw extracts represents a considerable challenge and a number of recent studies have appeared with new methodologies. There are several critical questions that require precise quantification of artemisinin. The first one is the ability to quantify the content of artemisinin in biomass. This is important for the extractors and the growers. A highly desirable method of quantification of artemisinin in fresh leaves is a cheap handheld device. However, thus far no such method has been developed. Instead, several methods for rapid analysis requiring small amounts of fresh or dried leaf were published, using either thin-film chromatography, which can be relatively portable, or standard analytical instrumentation, which requires rapid access to a laboratory. The easiest method to use in a handheld device would be a spectroscopic technique, such as UV-*vis*, IR, NIR, or Raman. However, artemisinin has a very low molar absorption coefficient in UV: between 163 and 183 $l\,mol^{-1}\,cm^{-1}$ in the region between 205 and 215 nm, with a maximum at 210 nm [57]. Since almost everything absorbs at 210 nm, it is impossible to use UV detection of artemisinin without chromatographic separation first. As for other spectroscopic methods, the electronic spectra of artemisinin is rather complex [58], which means that low-resolution portable instruments would not be able to quantify a specific spectral band of artemisinin. This is further complicated by the presence of very similar co-metabolites, whose spectral fingerprints overlap with that of artemisinin, again requiring chromatographic separation first.

Thin layer chromatography (TLC) is a sufficiently simple and low-cost method, which can be portable. TLC methods for detection of artemisinin have been continuously developed for a number of years (see, e.g., Refs. [59, 60]). The high-performance version of TLC enables quantification of several related co-metabolites with a good degree of accuracy and reliability, thus making this method potentially applicable as an in-field analytical method.

All other reported methods require rapid delivery of a plant sample to a laboratory to use one of the more sensitive methods. Thus, a number of GC-MS-based methods were developed for quantification of artemisinin and its co-metabolites [61–63]. A more interesting method is based on an indirect analysis of artemisinin content. A head-space GC-MS method was developed, which reveals the profile of volatile components of leaves, the key link being between the content of artemisia ketone (**4**) and artemisinin: the lack of artemisia ketone seem to correlate with high concentration of artemisinin [64]. An advantage of the indirect method is that it does not require the use of solvents or high-temperature plant sample/extract treatment, which may affect unstable compounds.

In addition to GC-MS-based methods, a number of HPLC and LC-MS-based methods have been published. Since the most common method of detection in LC instruments is UV and artemisinin has a low absorption in UV, the earlier methods used pre-analysis derivatization [65–67]. However, proper selection of columns and mobile phases enables better separation of peaks and thus allows to use UV detection directly [57, 68, 69]. The more recently published protocols are geared toward quantification of artemisinin, its co-metabolites, and minor impurities such as deoxyartemisinin (**5**) directly in raw extracts.

Other detectors have also been used after LC separation, most notably the light scattering detector (ELSD) [57, 70, 71]. ELSD has a nonlinear response and thus requires more careful calibration than a UV detector. However, response of artemisinin is very much better than in the case of the UV detection, enabling better accuracy of detection.

The considerable advance in mass spectroscopy seen over the last decade has been reflected in the proliferation of new instruments and the reduction in their cost. Owing to the much lower limits of detection and the ability to couple either with liquid or with gas chromatography, mass spectroscopy has become an essential tool for natural products analysis, especially in identification of unknowns or trace impurities profile. Not surprisingly, a number of mass-spectroscopy-based methods for analysis of artemisinin have been developed [68, 72–77]. For example, for the analysis of crude extracts and simultaneous quantification of several metabolites, a multiple reaction monitoring technique has been applied to LC-MS/MS measurement using a TQD instrument [72]. This allows quantification of six most important co-metabolites within 6 min, thus providing analytical methodology for high-throughput experimentation. In this method, the epomer of artemisinin, 9-epi-artemisinin (**6**), has been quantified in raw extracts for the first time.

Least important today are the NMR-based methods. Various NMR methods are routinely used for identification of new molecules, especially structurally complex biomolecules. However, this requires rather expensive high-field instruments. Analytical methods for quantification of artemisinin based on proton NMR have been published (see, e.g., Ref. [78]). However, until low-cost benchtop NMRs are ubiquitous in research and process development labs, the use of NMR for routine analysis of extracts is not a viable option.

10.5 Conclusions and Outlook

Extraction of artemisinin from *A. annua* is a fascinating multifaceted problem that has been tackled by many different experts and entrepreneurs. Successful commercial extraction and purification of artemisinin was established decades ago on the basis of empirical observations and heuristic knowledge of practitioners, whereas fundamental research into the topic has been gradually catching up. The chemical complexity of this extraction process has been discussed in the open literature only in the last few years, when the interactions of co-metabolites with solvents and their impact on both extraction and purification began to be understood. This now enables better analytical methods for quantification of artemisinin and important co-metabolites, such as deoxyartemisinin, artemisitene, arteannuin B, and 9-epi-artemisinin. New insights into the interactions of co-metabolites in different solvent systems and their interactions with adsorbents or column packings enable targeted design of better purification protocols that have significantly higher overall yields of artemisinin from dry leaf.

Many cleaner approaches to extraction and purification of artemisinin compared to PE extraction have been suggested: microwave-assisted extraction, ultrasonic-assisted extraction, $scCO_2$ extraction, and extraction with tetrafluoroethane or ionic liquids. Of these, only $scCO_2$ and R134a extraction have seen limited commercial uptake, despite the fact that economics are in favor of R134a extraction, various safety and environmental aspects are also much better for the $scCO_2$ and ionic-liquids-based extraction as was shown in the earlier comparative study [6]. The stronghold of conventional PE-based extraction as the main commercial method for artemisinin production is due to sunk capital, lack of investment in new technologies, and the structure of the market, which is dominated by small companies with limited technical and financial capacity. The world of artemisinin is dominated by public subsidies and charitable projects, which also makes sustained effort into development and implementation of new technologies in this field rather haphazard. However, this topic attracts altruistic industrialists, academics and students, and as it is a very good model for in-depth study of biopharmaceutical processing in general, the work is continuing in many laboratories around the world to develop cleaner extraction of artemisinin. Recent introduction of semisynthetic artemisinin will have a significant impact on the artemisinin extraction market once manufacturing volumes and costs reach their target values. Future introduction of a vaccine is, of course, the most desirable solution to the problem of malaria and will significantly reduce the need for plant-based artemisinin. But, these are yet to come and there are likely to be new biomedical applications of artemisinin, which remains an exciting molecule to work on.

References

1. Hsu, E. (2006) The history of qing hao in the Chinese materia medica. *Trans. R. Soc. Trop. Med. Hyg.*, **100**, 505–508.
2. Wright, C.W., Linley, P.A., Burn, R., Wittlin, S., and Hsu, E. (2010) Ancient Chinese methods are remarkably effective for the preparation of artemisinin-rich extracts of Qing Hao with potent antimalarial activity. *Molecules*, **15**, 804–812.
3. Haynes, R.K. (2006) From artemisinin to new artemisinin antimalarials: biosynthesis, extraction, old and new derivatives, stereochemistry and medicinal chemistry requirements. *Curr. Top. Med. Chem.*, **6**, 509–537.
4. Wright, C.W. (2002) *Artemisia*, Taylor & Francis Group, New York, London.
5. Weina, P. (2008) Artemisinins from folklore to modern medicine-transforming an herbal extract to life-saving drugs. *Parassitologia*, **50**, 25.
6. Lapkin, A., Plucinski, P.K., and Cutler, M. (2006) Comparative assessment of technologies for extraction of artemisinin. *J. Nat. Prod.*, **69**, 1653–1664.
7. Withers, S.T. and Keasling, J.D. (2007) Biosynthesis and engineering of isoprenoid small molecules. *Appl. Microbiol. Biotechnol.*, **73**, 980–990.
8. Kopetzki, D., Levesque, F., and Seeberger, P.H. (2013) A continuous-flow process for the synthesis of artemisinin. *Chem. Eur. J.*, **19**, 5450–5456.
9. Stringham, R.W. and Teager, D.S. (2012) Streamlined process for the conversion of artemisinin to artemether. *Org. Process Res. Dev.*, **16**, 764–768.
10. Fan, X., Sans, V., Yaseneva, P., Plaza, D., Williams, J.M.J., and Lapkin, A. (2012) Facile stoichiometric reductions in flow: an example of artemisinin. *Org. Process Res. Dev.*, **16**, 1039–1042.
11. Yaseneva, P., Plaza, D., Fan, X., Loponov, K., and Lapkin, A. (2015) Synthesis of antimalarial APIs (artemether) in flow. *Catal. Today*, **239**, 90–96.
12. Duke, M.V., Paul, R.N., Elsohly, H.N., Sturtz, G., and Duke, S.O. (1994) Localization of artemisinin and artemisitene in foliar tissues of glanded and glandless biotypes of *Artemisia annua L. Int. J. Plant Sci.*, **155**, 365–372.
13. Duke, S.O. and Paul, R.N. (1993) Development and fine structure of the glandular trichomes of *Artemisia annua L. Int. J. Plant Sci.*, **154**, 107–118.
14. Lapkin, A.A., Peters, M., Greiner, L., Chemat, S., Leonhard, K., Liauw, M.A., and Leitner, W. (2010) Screening of new solvents for artemisinin extraction process using *ab-initio* methodology. *Green Chem.*, **12**, 241–251.
15. Klamt, A. and Eckert, F. (2000) COSMO-RS: a novel and efficient method for the a priori prediction of thermophysical data of liquids. *Fluid Phase Equilib.*, **172**, 43–72.
16. Klamt, A., Eckert, F., Hornig, M., Beck, M.E., and Bürger, T. (2002) Prediction of aqueous solubility of drugs and pesticides with COSMO-RS. *J. Comput. Chem.*, **23**, 275–281.
17. Tian, N., Li, J., Liu, S., Huang, J., Li, X., and Liu, Z. (2012) Simultaneous isolation of artemisinin and its precursors from *Artemisia annua L.* by preparative RP-HPLC. *Biomed. Chromatogr.*, **26**, 708–713.
18. Lapkin, A., Adou, E., Mlambo, B., Chemat, S., Suberu, J., Collis, A.E.C.,

Clark, A., and Barker, G. (2014) Integrating medicinal plants extraction into a high-value biorefinery: an example of *Artemisia annua L. C.R. Chim*, (in print).
19. Suberu, J.O., Yamin, P., Leonhard, K., Song, L., Chemat, S., Sullivan, N., Barker, G., and Lapkin, A. (2014) The effect of O-methylated flavonoids and other co-metabolites on the crystallisation and purification of artemisinin. *J. Biotechnol.*, **171**, 25–33.
20. Christen, P. and Veuthey, J.-L. (2001) New trends in extraction, identification and quantification of artemisinin and its derivatives. *Curr. Med. Chem.*, **8**, 1827–1839.
21. Ferreira, J.F.S., Zheljazkov, V.D., and Gonzalez, J.M. (2013) Artemisinin concentration and antioxidant capacity of *Artemisia annua* distillation byproduct. *Ind. Crops Prod.*, **41**, 294–298.
22. Asfaw, N., Licence, P., Novitskii, A.A., and Poliakoff, M. (2005) Green chemistry in Ethiopia: the cleaner extraction of essential oils from *Artemisia afra*: a comparison of clean technology with conventional methodology. *Green Chem.*, 7, 352–356.
23. Briars, R. and Paniwnyk, L. (2013) Effect of ultrasound on the extraction of artemisinin from *Artemisia annua. Ind. Crops Prod.*, **42**, 595–600.
24. Hao, J.-Y., Han, W., Huang, S.-D., Xue, B.-Y., and Deng, X. (2002) Microwave-assisted extraction of artemisinin from *Artemisia annua L. Sep. Purif. Technol.*, **28**, 191–196.
25. Fleming, A. and Freyhold, M. (2007) Assessing the Technical and Economic Viability of the Ethanolic Extraction of *Artemisia annua*. A report available at, *http://www.mmv.org/IMG/pdf/3_ethanolic-extraction-december-2007.pdf* (accessed 9 August 2014).
26. Liu, Q.N., Schuehly, W., Freyhold, M., von Kooy, F., and Van der Kooy, F. (2011) A novel purification method of artemisinin from *Artemisia annua. Ind. Crop Prod.*, **34**, 1084–1088.
27. Xu, J., Luo, J., and Kong, L. (2011) Single-step preparative extraction of artemisinin from *Artemisia annua* by charcoal column chromatography. *Chromatographia*, **74**, 471–475.
28. Van der Kooy, F. and Verpoorte, R. (2011) The content of artemisinin in the *Artemisia annua* tea infusion. *Planta Med.*, **77**, 1754–1756.
29. Suberu, J.O., Gorka, A.P., Jacobs, L., Roepe, P.D., Sullivan, N., Barker, G., and Lapkin, A. (2013) Anti-plasmodial polyvalent interactions in *Artemisia annua L*. Aqueous extract - possible synergistic and resistance mechanisms. *PLoS One*, **8**, e80790.
30. Bioniqs (2006) Extraction of Artemisinin Using Ionic Liquids. (Report by Bioniqs Ltd (UK) commissioned by the Medicines for Malaria Ventures (MMV), *www.mmv.org* (accessed 9 August 2014).
31. Bioniqs (2008) Extraction of Artemisinin Using Ionic Liquids. Report available at *http://www.mmv.org/newsroom/publications/extraction-artemisinin-using-ionic-liquids* (accessed 9 August 2014).
32. Sanders, M.W., Wright, L., Tate, L., Fairless, G., Crowhurst, L., Bruce, N.C., Walker, A.J., Hrembury, G.A., and Shimizu, S. (2009) Unexpected preferential dehydration of artemisinin in ionic liquids. *J. Phys. Chem. A*, **113**, 10143–10145.
33. Mendes, R.L., Nobre, B.P., Cardoso, M.T., Pereira, A.P., and Palavra, A.F. (2003) Supercritical carbon dioxide extraction of compounds with pharmaceutical importance from microalgae. *Inorg. Chim. Acta*, **356**, 328–334.
34. Leitner, W. (2002) Supercritical carbon dioxide as a green reaction medium for catalysis. *Acc. Chem. Res.*, **35**, 746–756.
35. Coimbra, P., Blanco, M.R., Costa Silva, H.S.R., Gil, M.H., and de Sousa, H.C. (2006) Experimental determination and correlation of artemisinin's solubility in supercritical carbon dioxide. *J. Chem. Eng. Data*, **51**, 1097–1104.
36. Quispe-Condori, S., Sanchez, D., Foglio, M.A., Rosa, P.T.V., Zetzl, C., Brunner, G., and Meireles, M.A.A. (2005) Global yield isotherms and kinetic of artemisinin extraction from *Artemisia annua L* leaves using supercritical carbon dioxide. *J. Supercrit. Fluids*, **36**, 40–48.

37. Kohler, M., Haerdi, W., Christen, P., and Veuthey, J.-L. (1997) Extraction of artemisinin and artemisinic acid from *Artemisia annua L.* using supercritical carbon dioxide. *J. Chromatogr. A*, **785**, 353–360.
38. Gong, X.-Y. and Cao, X.-J. (2009) Measurement and correlation of solubility of artemisinin in supercritical carbon dioxide. *Fluid Phase Equilib.*, **284**, 26–30.
39. Kohler, M., Haerdi, W., Christen, P., and Veuthey, J.-L. (1997) Supercritical fluid extraction and chromatography of artemisinin and artemisinic acid. An improved method for the analysis of *Artemisia annua* samples. *Phytochem. Anal.*, **8**, 223–227.
40. Tzeng, T., Lin, Y., Jong, T., and Chang, C. (2007) Ethanol modified supercritical fluids extraction of scopoletin and artemisinin from *Artemisia annua L. Sep. Purif. Technol.*, **56**, 18–24.
41. Xing, H., Su, B., Ren, Q., and Yang, Y. (2009) Adsorption equilibria of artemisinin from supercritical carbon dioxide on silica gel. *J. Supercrit. Fluids*, **49**, 189–195.
42. Wilde, P.F., Skinner, R.E., and Ablett, R.F. (2002) ITFM extraction of oil seeds. WO Patent 03/090520 A2.
43. Wilde, P.F. (1996) Fragrance extraction. US Patent 5,512,285.
44. Corr, S. (2002) 1,1,1,2-Tetrafuoroethane; from refrigerant and propellant to solvent. *J. Fluorine Chem.*, **118**, 55–67.
45. Mustapa, A.N., Manan, Z.A., Mohd Azizi, C.Y., Setianto, W.B., and Mohd Omar, A.K. (2011) Extraction of β-carotenes from palm oil mesocarp using sub-critical R134a. *Food Chem.*, **125**, 262–267.
46. Han, Y., Ma, Q., Lu, J., Xue, Y., and Xue, C. (2012) Optimisation for subcritical fluid extraction of 17-methyltestosterone with 1,1,1,2-tetrafluoroethane for HPLC analysis. *Food Chem.*, **135**, 2988–2993.
47. Han, Y., Ma, Q., Lu, J., Xue, Y., Xu, J., and Xue, C. (2012) Development and validation of a subcritical fluid extraction and high performance liquid chromatography assay for medroxyprogesterone in aquatic products. *J. Chromatogr. B Anal. Technol. Biomed. Life Sci.*, **897**, 90–93.
48. Saul, S., Corr, S., and Micklefield, J. (2004) Biotransformations in low-boiling hydrofluorocarbon solvents. *Angew. Chem. Int. Ed.*, **43**, 5519–5523.
49. Yu, G., Xue, Y., Xu, W., Zhang, J., and Xue, C.H. (2007) Stability and activity of lipase in subcritical 1,1,1,2-tetrafluoroethane (R134a). *J. Ind. Microbiol. Biotechnol.*, **34**, 793–798.
50. Acton, N., Klayman, D.L., and Rollman, I.J. (1986) Isolation of artemisinin (qinghaosu) and its separation from artemisitene using the Ito multilayer coil separator-extractor and isolation of arteannuin B. *J. Chromatrogr.*, **355**, 448–450.
51. Piletska, E.V., Karim, K., Cutler, M., and Piletsky, S.A. (2013) Development of the protocol for purification of artemisinin based on combination of commercial and computationally designed adsorbents. *J. Sep. Sci.*, **36**, 400–406.
52. Qu, H., Bisgaard, K., Frette, C.X.C., Tian, F., Rantanen, J., and Christensen, L.P. (2010) Chromatography-crystallization hybrid process for artemisinin purification from *Artemisia annua*. *Chem. Eng. Technol.*, **33**, 791–796.
53. Yang, B. and Cao, X. (2011) Synthesis of the artemisinin-imprinting polymers on silica surface and its adsorption behavior in supercritical CO_2 fluid. *AIChE J.*, **57**, 3514–3521.
54. Foucault, A.P. and Chevolot, L. (1998) Counter-current chromatography: instrumentation, solvent selection and some recent applications to natural product purification. *J. Chromatogr. A*, **808**, 3–22.
55. Ito, Y., Weinstein, M., Aoki, I., Harada, R., Kimura, E., and Nunogaki, K. (1966) The coil planet centrifuge. *Nature*, **212** (5066), 985–987.
56. Weiss, E., Ziffer, H., and Ito, Y. (2000) Use of countercurrent chromatography (CCC) to separate mixtures of artemisinin, artemisitene, and arteannuin B. *J. Liq. Chromatogr. Related Technol.*, **23**, 909–913.
57. Lapkin, A.A., Walker, A., Sullivan, N., Khambay, B., Mlambo, B., and Chemat, S. (2009) Development of HPLC analytical protocols for quantification of

artemisinin in plants and extract. *J. Pharm. Biomed. Anal.*, **49**, 908–915.
58. Galasso, V., Kova, B., and Modelli, A. (2007) A theoretical and experimental study on the molecular and electronic structures of artemisinin and related drug molecules. *Chem. Phys.*, **335**, 141–154.
59. Widmer, V., Handloser, D., and Reich, E. (2007) Quantitative HPTLC analysis of artemisinin in dried *Artemisia annua L.*: a practical approach. *J. Liq. Chromatogr. Related Technol.*, **30**, 2209–2219.
60. Quennoz, M., Bastian, C., Simonnet, X., and Grogg, A.F. (2010) Quantification of the total amount of artemisinin in leaf samples by thin layer chromatography. *Chim. Int. J. Chem.*, **64**, 755–757.
61. Woerdenbag, H.J., Pras, N., Bos, R., Visser, J.F., Hendriks, H., and Malingre, T.M. (1991) Analysis of artemisinin and related sesquiterpenoids from *Artemisia annua* by combined gas-chromatography mass-spectrometry. *Phytochem. Anal.*, **2**, 215–219.
62. Liu, S., Tian, N., Li, J., Huang, J., and Liu, Z. (2009) Simple and rapid micro-scale quantification of artemisinin in living *Artemisia annua L.* by improved gas chromatography with electron-capture detection. *Biomed. Chromatogr.*, **23**, 1101–1107.
63. Ma, C., Wang, H., Lu, X., Xu, G., and Liu, B. (2008) Metabolic fingerprinting investigation of *Artemisia annua L.* in different stages of development by gas chromatography and gas chromatography–mass spectrometry. *J. Chromatogr. A*, **1186**, 412–419.
64. Reale, S., Fasciani, P., Pace, L., De Angelis, F., and Marcozzi, G. (2011) Volatile fingerprints of artemisinin-rich *Artemisia annua* cultivars by headspace solid-phase microextraction gas chromatography/mass spectrometry. *Rapid Commun. Mass Spectrom.*, **25**, 2511–2516.
65. Zhao, S. (1987) High-performance liquid chromatographic determination of artemisinine (Qinghaosu) in human plasma and saliva. *Analyst*, **112**, 661–664.
66. Qian, G.-P., Yang, Y.-W., and Ren, Q.-L. (2005) Determination of artemisinin in *Artemisia annua L.* by reversed phase HPLC. *J. Liq. Chromatogr. Related Technol.*, **28**, 705–712.
67. Elsohly, H.N., Croom, E.M., and Elsohly, M.A. (1987) Analysis of the anti-malarail sesquiterpene artemisinin in *Artemisia annua* by high-performance liquid-chromatography (HPLC) with postcolumn derivatization and ultraviolet detection. *Pharm. Res.*, **4**, 258–260.
68. Stringham, R.W., Lynam, K.G., Mrozinski, P., Kilby, G., Pelczer, I., and Kraml, C. (2009) High performance liquid chromatographic evaluation of artemisinin, raw material in the synthesis of artesunate and artemether. *J. Chromatogr. A*, **1216**, 8918–8925.
69. Pilkington, J.L., Preston, C., and Gomes, R.L. (2012) The impact of impurities in various crude *A. annua* extracts on the analysis of artemisinin by liquid chromatographic methods. *J. Pharm. Biomed. Anal.*, **70**, 136–142.
70. Avery, B.A., Venkatesh, K.K., and Avery, M.A. (1999) Rapid determination of artemisinin and related analogues using high-performance liquid chromatography and an evaporative light scattering detector. *J. Chromatogr. B*, **730**, 71–80.
71. Liu, C.-Z., Zhou, H.-Y., and Zhao, Y. (2007) An effective method for fast determination of artemisinin in *Artemisia annua L.* by high performance liquid chromatography with evaporative light scattering detection. *Anal. Chim. Acta*, **581**, 298–302.
72. Suberu, J., Song, L., Slade, S., Sullivan, N., Barker, G., and Lapkin, A. (2013) A rapid method for the determination of artemisinin and its biosynthetic precursors in *Artemisia annua L* crude extracts. *J. Pharm. Biomed. Anal.*, **84**, 269–277.
73. Saraji, M., Khayamian, T., Hashemian, Z., Aslipashaki, S.N., and Talebi, M. (2013) Determination of artemisinin in *Artemisia* species by hollow fiber-based liquid-phase microextraction and electrospray ionization-ion mobility spectrometry. *Anal. Methods*, **5**, 4190.
74. Wang, M., Park, C., Wu, Q., and Simon, J.E. (2005) Analysis of artemisinin in *Artemisia annua L.* by LC-MS with

selected ion monitoring. *J. Agric. Food Chem.*, **53**, 7010–7013.

75. Bilia, A.R., Malgalhaes, P.M., de Bergonzi, M.C., and Vincieri, F.F. (2006) Simultaneous analysis of artemisinin and flavonoids of several extracts of *Artemisia annua L.* obtained from a commercial sample and a selected cultivar. *Phytomedicine*, **13**, 487–493.
76. Reale, S., Pace, L., Monti, P., De Angelis, F., and Marcozzi, G. (2008) A rapid method for the quantification of artemisinin in *Artemisia annua L.* plants cultivated for the first time in Burundi. *Nat. Prod. Res.*, **22**, 360–364.
77. Sahai, P., Vishwakarma, R.A., Bharel, S., Gulati, A., Abdin, M.Z., Srivastava, P.S., and Jain, S.K. (1998) HPLC-electrospray ionization mass spectrometric analysis of antimalarial drug artemisinin. *Anal. Chem.*, **70**, 3084–3087.
78. Castilho, P.C., Gouveia, S.C., and Rodrigues, A.I. (2008) Quantification of artemisinin in *Artemisia annua* extracts by 1H-NMR. *Phytochem. Anal.*, **19**, 329–334.

Index

Green Extraction of Natural Products: Theory and Practice, First Edition.
Edited by Farid Chemat and Jochen Strube.